The Four Cosmic Pillars

In terms of

Of Singularity

EXPLAINING THE

"Four Cosmic Pillars"

IN TERMS OF

APPLYING PHYSICS

ISBN-13: 978-1537679945
ISBN-10: 1537680145

All rights are reserved.
No part, parts or the entirety of this book may be reproduced by publishing, electronically copied, duplicated by whatever means that form reproduction or duplication of any description, without the prior written consent of the copy rite owner.

WRITTEN BY PEET SCHUTTE

© KOSMOLOGIESE EN ASTRONOMIESE TEGNIKA

WHOM IT MAY CONCERN,
I do find much pride in my status as being Afrikaner and would like to have my names used by pronouncing it in the manner Afrikaans dictates...therefore I would sincerely appreciate the courtesy when readers will take note that my name and last name are pronounced in Afrikaans, which is originally from Dutch and must be pronounced that way. Peet one would pronounce "here" which is the closest English to the pronouncing of the "ee". The "Sch" in Schutte is pronounced exactly as school is where both actually are pronounced Skutte or "skool". By pronouncing my name in Afrikaans you do me the utmost courtesy any one can. Being an Afrikaner is what I am most proud of. Another point I wish to highlight is that I feel compiled to produce this work in a comic-like format. I have found that the more intellectual and the more educated Academics are, the less they understand the most primitive or classical mistakes in science as well as physics.

As I said my mother tongue is Afrikaans and my second language is English. I have per suiting this theory that I partly present in this book, of which the investigating research was done the past thirty years. Then I compiled my presentation thereof for the past nine years on full time basis whereby I was tying to introduce my findings to many academics without much joy. This past nine years saw me go without any income as I tried to get my theorem recognised. Going without a steady income left me almost destitute and in order to find a manner to get my theory across to the attention of influential readers, I decided to publish these books electronically as to try and get around the stranglehold of Newtonian bias controlling science at present worldwide. I decided to publish these articles through LULU.com which I saw as way the only manner whereby I could generate funding by which I would be able to have the twenty seven books I already wrote linguistically edited and then to have the books published on a Print-On-Demand basis. With my first language not being English and the books not linguistically checked by an expert there are bound to be language errors that readers will notice. In the past I tried to check my work myself but after checking say one hundred and fifty pages for language corrections, instead of having corrected work I ended instead having four hundred pages of new written information which is still not language corrected but holds a lot more information. This is because my priorities lie elsewhere. I aim to spend money on correcting the work as far as language goes, as I receive money and in the hope that I will receive money. I will have all my work including the one you are reading edited professionally and corrected as I find money to do so . . .

In the book that deals with gravity there are just too many and numerously wide ranging facts that form the complete picture as a whole, which leaves me unable to include a full introduction in a space as small as that which page will allow. The explaining include for instance those phenomena, which I call the four cosmic pillars, but wise as you are, you would not believe me at this point that I have cracked the coconut because I guess in your vast experience you have seen too many idle explanations in the past proving to be senseless and little impressive, therefore my mentioning my success would not matter much either way. The proof I bring is true about gravity being formed as a result of these phenomena, 1) the Lagrangian system 2) the Roche limit 3) the Titius Bode law 4) the Coanda affect, which I explain by delivering mathematical proof as to how they fit into the overall picture of gravity and which I mention just below. I prove the fact that every individual one of those phenomena is forming a unit that is in total being what we think of as gravity. The phenomena altogether constitutes a unit that forms the process working as gravity. Nevertheless my mentioning these facts will be just completely unbelievable to you without you reading the book, because I guess you have heard some attempt to explain the phenomena before but when I say you have not heard it in the context I put it, you might still be most sceptical because you have never heard it in the correct manner that I explain it and that poses the difference. Still you may not be convinced about my claims and although my explaining the phenomena is correct, does not change the fact that you don't believe me. The phenomena form an intergraded unit that results in gravity forming where each forms a part of gravity. You may still be you would be sceptical ...but convince yourself that I did manage to:
 1) Find the location, position of singularity as a factor forming space-time
 2) Finding space-time by dissecting Kepler's formula in relation to valuing singularity
 3) Finding and proving space-time and aligning space-time with gravity
 4) Find the working principals behind gravity as a cosmic occurrence.
 5) Find the reason for the Roche limit and explaining the resulting of gravity from that.
 6) Find out why the Lagrangian system, becomes the building form of the Universe.
 7) Find why the Titius Bode law mathematically provides the foundation of gravity
 By proving that the Coanda affect is gravity through activating space-time
 By using the above the four cosmic pillars, it enable me to present the proof where I now can explain what conditions bring on the sound barrier. By proving it is gravity that the individual structure generates motion above and beyond the gravity the Earth provide is what is producing individual motion that the independent object earned within the sphere of motion that the Earth's

gravity provides where the <u>independent and individual motion</u> put the relevance that gravity has beyond the conserving means gravity has where **the space** that is serving the **independent object** is independently in motion. **The adding to the independence on top of the normal structural independence is creating more individualism by the independent motion of the individual structure being apart from the motion that the gravity of the <u>Earth provides</u>.** The fact every one misses is that any structure that is not part of the Earth's crust has an independent gravity and the form this gravity applies is stronger than the Earth's gravity which is why the structure maintains its form and this provides the independent individuality the structure has giving the unique structural space. The gravity of the Earth strives to incorporate everything into the Earth's sphere and into the Earth's structure and therefore the fact that the object is not incorporated into the Earth shows defiance and individuality, which gives it, mass.

By applying individual motion on top of the structural individuality that increases by the motion that the Earth provides, the independence of the individual object is becoming further exaggerated by having independent motion, which is further defying the incorporation the Earth strives to achieve. As the motion of the independent object grows more independent by applying more excessive motion to such an extent **where motion creates almost the ultimate independence that may free the individual object with independence from the motion the Earth creates** is what is breaking the restraint gravity has on all objects with independence formed by their structure. The structure show independence at all times by not forming part of the structure of the Earth within the sphere of the Earth's gravity. Moving about shows even more reluctance on the part of the top when spinning allows the top to eventually become part of the Earth. **Breaking the sound barrier is the motion** in space duplicating space by crossing over gravity borders, which is the limit to what constraint the Earth may produce in accordance with what full independence would allow.

These are the definitions underwriting cosmology and while my work is that much ignored; let's see how far I stray from these definitions in comparison of how much Mainstream science underwrites these definitions by bringing indisputable proof in presenting unwavering hardcore facts.
Quoted directly from the Oxford dictionary of Astronomy the following:

The definition of space-time is as follows:
Space-time is a four dimensional position of the Universe where the position of an object is specified by three coordinates in space and one position in time. According to the theory of special relativity there is no absolute time, which can be measured independently of the observer, so events that are simultaneous as seen from one observer occur at different times when seen from a different place. Time must therefore be measured in a relative manner as are positions in three-dimensional Euclidean space, and this is achieved through the concept of space-time. The trajectory of an object in space-time is called world line. General relativity relates to curvature of space-time to the positions and motions of particles of matter.

The definition of singularity is as follows:
Singularity: a mathematical point at which certain physical quantities reach infinite values for example, according to the general relativity the curvature of space-time becomes infinite in a black hole. In the big bang theory the Universe was born from singularity in which the density and temperature of matter were infinite.

The Oxford dictionary of Astronomy defines gravitation as follows
Gravitation is the force of attraction that operates between all bodies. The size of the attraction depends on the masses of the bodies and the distance between them; gravitational force diminishes by the square of the distance apart according to the inverse square law. Gravitation is the weakest of the four fundamental forces in nature. I. Newton formulated the laws of gravitational attraction and showed that a body behaves as though all its mass were concentrated at its centre of gravity. Hence the gravitational force acts along a joining of the centres of gravity of the two masses. In the general theory of relativity gravitation is interpreted as the distortion of space. Gravitational forces are significant between large masses such as stars planets and satellites, and it is this force, which is responsible for holding together the major components of the Universe. However on the atomic scale the gravitational force is about 10^{40} times weaker than the force of electromagnetic attraction

I have to give potential readers this fair warning that *The Cosmic Code as the Absolute Relevancy of Singularity* **requires a somewhat higher level of understanding and needs a greater degree of insight that the other books in this series does namely**

1 Explaining Physics in terms of the Absolute Relevancy of Singularity,

2 Explaining the Sound Barrier in terms of The Absolute Relevancy of Singularity,

3 Explaining the Four Cosmic Phenomena in terms The Absolute Relevancy of Singularity and

4 Explaining the Cosmic Code in terms The Absolute Relevancy of Singularity
Which all are also available from Lulu.com.

I have no chance that what I state as my theory on **The Absolute Relevancy of Singularity** will be read, or much less that it will be seriously considered and I have not a snowballs hope in hell that it will be accepted by those with the authority to change physics principles. The theory I introduce here and now would never be accepted in my lifetime because science in the Newtonian way is bent on believing in the marvellous, the facts bordering the supernatural, the outrageously inconceivable and the magic of what can never be explained, although they claim to use facts. It is **the marvellous** to think that mass can create gravity. It is **bordering the supernatural** to think that with nothing between stars, yet by the magic of mass, mass has an unexplainable ability to attract another star many astronomical units away. It is the **outrageously inconceivable** to argue that life started on Mars, then overcame the quite impossible to escape the gravity that Mars holds on all things held captured in its gravity, and after overcoming the unthinkable, then made a dive for the Earth just to come and evolve over here. Science think they my have the ability to create a Black Hole in a Manmade atom-accelerator because science thinks of the Black Hole as **the magic of what can never be explained** and therefore that proves that science has no idea of what a Black Hole is while I can prove what a Black Hole is singularity reaching limits That fact that I can explain what a Black hole is, that ability the Wizards of Oz will never allow because the explaining I present is clashing with Newton head on and it is to be done in as simple manner as I am about to explain. However, when I prove what a Black Hole is I am going to destroy the fantasy world everyone makes believe as physics. To science a Black Hole is a world of magic where gravity has the ability to go mad and a Black Hole is something that man could manufacture by creating an atomic accelerator tunnel, or so science thinks. In other words the best science at present can do to explain the gravity in a Black hole is to give gravity a level of superior intellect and then take it away (by allowing gravity to go mad as it seemingly does in Super Novas and in Black Holes). Why can I prove what a Black Hole is…it is because I can prove what the Roche limit is and believe me that is one thing science this far could never get around in proving. The facts they use is as much fiction as Little Red Riding Hood's talking wolf…when it comes to explaining the integrating details of how gravity comes about. In science, when following my theory, everything can be explained by using physics, but using my explanation will make all present science become fiction, make all present science look like a fairy tale and make all present science seem to be good bedtime stories deprived of truth…and the money spent on Newtonian fiction-science will never allow me to have success because that would be too costly for the industry money-wise. Why would I call science a fairy tale…well this is just one of many, many reasons. Science wishes to promote something as impossible as time travel, which I show, is impossible. Science believes in travelling at speeds unlimited that could exceed the speed of light. I prove all such thoughts are impossible because I show that gravity and time is the very same thing. No one can beat gravity because gravity as time maintains the structural integrity of the Universe. In beating gravity one wishes to beat the cosmos that hold us secured. That is why time can manifest as what is known as the Hubble constant. Time is the redeploying of space by extending the absolute relevancy of singularity and that is only one of several factors that serve as time. Every time I declare Newton was mistaken and therefore science is wrong in presenting the most basics of physics, the workings of gravity, I am barraged by rejection and silent ridicule. Every time I challenge the Members of science to either prove Newton correct or to prove me wrong, I am ignored…my challenge goes unmet, so please forgive me for showing much antagonism…it is a result of Mainstream Science rejecting my efforts unfairly for many years. What I write is undeniably and undisputedly correct, but the instant science admits to my work being correct, that admission demotes most of the work science has accepted in the past as correct to the level of science fiction. It will destroy the groundwork of mainstream science and demote what is accepted to become fairy tales, which is what most Newtonian based theories are. Let Newtonian science explain what the cosmic purpose or the function is of a star…of a galactica…of an atom…of gravity…they have no idea. By the time you have finished this book you would have found answers to all the above questions in detail.

Mainstream science has so little idea of what a Black Hole is or what could cause a Black Hole that they devised a "Mini Black Hole" to suit there marvellous misinterpretations of gravity. That is a form of fantasy that fairy tale writers can't compete with. Science is so misguided in understanding life that they put life in all places throughout the Universe without ever finding one shred of evidence of the presence of life. Yet they say they work only with proven facts alone. They hold the opinion that life could have come from Mars but fail miserably in explaining how it will be possible for life to escape the gravity of Mars and then fly all the way, ever so precisely guided; directly to the Earth. How would it be possible for life to escape the gravity of Mars without them when explaining such a possibility by employing realistic physics, going into so much fantasy it leaves the story of the three pigs and the blowing wolf seem real. Science has the explaining of the exploding Super Nova down to the last detail where they explain that a Super Nova is gravity that has gone mad without ever proving how gravity can go mad because the truth of the matter is that gravity has no intellect to "go mad" in any way. Mainstream science always places new object found where their findings prove that the newly found object is on "the edge of the Universe", meaning where the Universe ends by forming an edge. This fantasy they dish up to anyone willing to believe him or her without ever telling what is beyond that edge. All they can see is an end of the Universe but in reality where there is an end there has to be a beginning of something else…this is physics. The Universe I show can't have an edge because I show where the point is that could never start and I show where the point is that could never end. I show that which can go no smaller and I show that which can go no bigger. I am about to introduce a Universe that mathematically can never start and the same Universe can mathematically never end.

I have been on a self-teaching mission that lasted thirty years and now that I have the answers and from which I have drawn the conclusions, I now find so much resistance from mainstream science in getting the findings my research uncovers out in the open. I offer tot academics many books in which I use diagrams, sketches, mathematical explanations and cosmic photos including other tools I employ to promote the required understanding needed to bring the ideas across that I wish to promote. However, publishing in this manner is very costly and money is one thing I do not have and therefore sending it to academics with no reply is an expense I cannot endure. Any academic feeling confronted by my accusation, please show how you prove $F = G \frac{M_1 M_2}{r^2}$ is applicable and is true. Show how the use of the formula could be applied meaningfully to present an answer worth of anything. Use the Newton's formula to show when the Moon is going to hit the Earth as the mass of the Earth pulls on the mass of the Moon. Better still, prove that mass does contract to create gravity and then explain how this is done…and please leave out the graviton because that is a joke! The idea that mass draws mass closer $F = G \frac{M_1 M_2}{r^2}$ is mathematically proven as an untruth, which means it is not true. What is the truth? …When you have completed this introduction you will have had a peeping view, a tiny glimpse of the truth…but as little as you would gain from reading this introduction alone, when put in comparison to what any person can gain from reading all of my work in total, you will gain endlessly more than what science is to explain about the truth, because what you then have gained by reading this document is much more than what science know about the truth. What I try to convey is that there is a good reason why academics block any and all publishing of my work, and when finishing this book, in comparison to what I offer, you have not even opened a first page of what I offer as new information when judging what my other work uncovers. Still, your effort in reading this document allows you to discover so much more of true science than what previously was known If you think I am boasting I challenge you to show where any of my explaining gravity requires superior intellect to understand... however in my simplistic approach to gravity I prove everything I say by applying the simplest mathematics there is.

The effort that this book represents the informing about an entire new way of cosmic appreciation meant to show that there are grounds for concern in the way science thinks and this book does not even bring all such arguments indicating concern in full. That one can only find when reading the first ten letters forming books named as with a title beginning with **Open Letters…**and those titles are included as books which I mention on my website, having the same name as this book namely www.gravitysveracity.com.

I am about to prove that gravity is **the Coanda effect** and gravity comes about from four cosmic phenomena never yet understood since it was never yet explained. Science doesn't believe there is something such as **the Titius Bode law** but science does believe that mass would generate gravity. Science has no clue about **the Roche limit** but science believes in spite of the Roche limit that big craters on Earth are reminders of massive asteroids that hit the Earth in giant collisions. With the Roche limit in place these crates are the result of something else because it can't be from asteroids colliding with the Earth. We all know how the bicycle rides and we all think we understand how the bicycle rides but having the bicycle ride on two wheels have little to do with balance and everything to do with the Coanda effect.
The bicycle rides forward when peddled but also the bicycle rides downwards when peddled and the two are both linked to gravity. I am going to prove that the Coanda effect forms gravity. I am going to prove that the **Coanda effect** comes as a result of the **Titius Bode law**, **the Roche limit** and the **Lagrangian positioning system** but most of all how these are related to singularity. That means I am going to prove that mass has no effect on gravity but mass comes as a result of gravity. I am going to prove what singularity is and that there are two types of singularity that in the end is only one type of singularity.

Teaching ever since time began forms a pillar on which memory and remembering what you are taught is the most prevalent part of tutoring. One is expected to remember what those coming before and which are tutoring you, wish you to remember. The Tutor lays a foundation by ensuring that everything known and accepted coming from the past are well and truly founded in the mind of the student. In that there is no problem. The problem arises where the information studied is flawed and no one ever realised that. Fortunately this does not occur regularly, but if and when it does, notwithstanding the exceptional part it forms, it then becomes a major problem to deal with. Therefore what comes form the past are carried on into the future as unblemished truth and no person meddles with the thoughts called information given as study material. However, as unlikely as it could be, this did happen and it is part of the basis of physics. When the student is taught, the student is expected to accept without argument. What comes from the past are considered as tested beyond suspicion of inaccuracy and proven to what is absolute unwavering accuracy! It is way beyond doubt. There is this motto that students are mindless and students can only start to think after receiving information that came from the past.

Students are incapable of arguing by reason to introduce new thoughts. This ability to reason only comes after the learning process secured knowledge through the memory process and only when testing shows facts learned by memory is well established and it then forms a solid base for everything the student knows, then the students may form an ability to reason and to argue. This mostly takes about all the time that living one lifetime presents. Well, what happens when that everything that everybody believed in the present, inherited by all from the past, was totally flawed? This has happened to physics and no one in physics so far yet realised it. Not one in physics shows the ability to realise the flaw coming from the past as part of the legacy. Then the mistakes will carry on from forming facts the past, carried over as flaws into the future for as many generations as it takes to realise the mistake that is dragged along and this carrying on of a flaw could continue indefinitely, if there is no clear minds working to recognise the mistake and correct what needs to be corrected. I ask of you not to judge me according to what you have already achieved for in that sense I fall short of receiving your recognition in status. Judge what I present to you, for then you will realise with all my shortcomings, I present you with a truth that exposes short fallings in the basics of physics.

Here and now and before the beginning of what this document may be to any potential reader, all parties reading take note that I state it emphatically that all members forming the community of science in physics judges me being not sufficiently educated and certainly not to the level where I am able to form any opinion on matters concerning Sir Isaac Newton or his physics. Any and all of my self-tutoring goes begging in their eyes notwithstanding and regardless of the fact that I did my private and individual studies by which I furthered my insight. That allowed me to show with clarity what destructive force Sir Isaac Newton released in order to corrupt the laws of mathematics, contaminating science along the way and mostly raping the work of a great man, Johannes Kepler and what Sir Isaac Newton did to derail the truth and disguise scientific correctness where such violation can only be expressed as being blatant criminal fraud. What his deeds amount to, is to corrupt the laws of mathematics, to render the laws of cosmology useless and to rubbish all of science. By your reading, you will learn what it is that those academics that are guarding science never wanted published and read by the public at large. What I say is don't run and hide from my attack and coward away from my confrontation as so many of the most intellectuals amongst the Physics Paternity did when I confronted their thinking. On every occasion where I confronted members of the Academic Paternity in the past, those I confronted acted in precisely such a manner, such as cowardly ending all reading by throwing the book down, and then pretending to show the utmost disgust in what I say.

I researched the work of Kepler and found science doesn't even recognise his work, while it is his formula that forms the basis of all physics. Everyone thinks that Kepler found planets rotating, with Newton being able to explain Kepler, which makes everyone more concerned about how Newton saw Kepler's work. The formula used in physics as a principle is $F=mV^2$ which should be $F^3=mV^2$. $F^3=mV^2$ is replicating Kepler's formula in detail as $a^3=T^2k$. By using Kepler's formula we have $F^3=mV^2$ that is a precise replica of $a^3=T^2k$. The duplication is so obvious that we have (F^3 becoming a^3) while (m is k) and (V^2 is T^2). Einstein also only duplicated Kepler's formula by putting $E=mC^2$, which also should read $E^3=mC^2$. Again that is precisely Kepler's formula $a^3=T^2k$. (E^3 is a^3), (m is k) and (C^2 is T^2). In $E^3=mC^2$ Einstein mimicked $a^3=T^2k$, Kepler's formula. (E^3 is F^3 is a^3), (m is k) and (C^2 is V^2 is T^2). So what is so brilliant about Einstein's formula if Kepler had it centuries before? $E^3=mC^2$ is $F^3=mV^2$ which is $a^3=T^2k$. Newton corrupted the formula when he added $4\Pi^2$ to the formula and removed k that Kepler introduced while $a^3=T^2k$ Newton ignored. Newton changed $a^3=T^2k$ by using the symbols G (m + m$_p$) to replace k and then declared $a^3 = T^2$. I still wish to see the proof confirming Newton's changes as being correct notwithstanding that everyone thinks physics is entirely based on this conception. Whether the formula used is $F^3=mV^2$ or is $E^3=mC^2$, it still remains duplicating what Kepler introduced as $a^3=T^2k$. So I changed it back to Kepler's version of $a^3=T^2k$ as to better the understanding of the foundation of astrophysics and mainstream physics. The entirety of physics is not based on Newton. Physics precisely duplicates Kepler's findings while science doesn't even recognise Kepler's formula. By giving Kepler the credit due, the entire Universe becomes completely understandable…but then for my audacity to show mistakes in physics I am ignored flat! All I ever ask is prove the truthfulness of G(Mxm)÷r² because it is F³=mV² that forms the basis of physics and that accuracy comes from Kepler's view of **a³=T²k** that became Einstein's E³=mC².

Whilst recognising the work of Johannes Kepler, Mainstream science bluntly ignores the impact of his work, and in that they miss the full vastness of the wide influence of his work. Newton shrouded Kepler's work under a blanket of alterations which I show was most unwanted since Kepler's work needs no alterations or corrections and every one since then kept Kepler's work hostage under Newton's changes. It is therefore almost absolutely realistic to say that all information what you are about to read in this letter and article sent to you for your attention was never yet printed in the near or the distant past although Kepler's work has been with us for about four hundred years, during which time it went unnoticed. It seems to me that any research predating Newton never came into use or into practise. My investigation of Kepler's work brought about a conclusion that no one yet arrived at concerning them with the findings of Kepler because no one scrutinised Kepler's formula before. Everyone is satisfied with Newton's version notwithstanding the incorrectness of it. The world seems satisfied with the idea that Kepler found planets rotating around a centre formed by the Sun and because of that Newton saw a circle. Where Newton saw a mathematical circle and was unable to understand **a³ = T²k**, Newton added what he thought is mathematically required to indicate such a circle. Newton added a mathematical $4\Pi^2$ to the formula of Kepler and removed the distance symbolising measure that Kepler introduced using **k**. On the other side Newton changed the symbol of **k** by using the symbols G (m + m$_p$). This is just a longer and probably a more detailed manner of indicating **k** and better defining of **k** but it symbolises precisely to the point what **k** stands for nonetheless. I wish to draw your attention to the matter of Johannes Kepler's findings that Mainstream science considers as resolved and closed for many a century while it is not. My investigating Kepler helped me too resolve other unresolved matters but it was only possible by using Kepler's work. This brought about the idea that the Universe is in a state of contracting towards a centre of sorts where mass will form this contracting. This was prevailing until a man by the name of E. P. Hubble came to the forefront.

E. P. Hubble (1889-1953) confirmed an expansion through out the Universe, which contradicted all that science thought was known about our Universe. According to the accepted Newtonian cosmology everyone is of thought that the Universe is in a normal state of contraction because that is what $F = G \frac{M_1 M_2}{r^2}$ implies. Every person is very aware of the idea that the universal expansion would not last for ever, but has to start with some contracting effort at some point. Then all the heavenly bodies will collide and destruct, without any thought about any wavering on the matter and on the matters reliability there is evidently no doubt. When $F = G \frac{M_1 M_2}{r^2}$ apply, there should not be any force, which is able to keep the mass that is producing all the gravity that contains the Universe apart. Known for almost a century, science has failed to give any explanation about this cosmic phenomenon of a Universal expansion except for some silly notion about dark matter being dormant and not forming gravity, as it should. If the dark matter is present as is claimed, then why doesn't the mass form gravity as it should and contract? What does our ability to see or not to see or the luminosity that the dark matter does not have, got to do with the mass bringing about pulling power, that is if mass brings about any pulling power. If the mass is there, visible or not, then the dark mass has to pull because light has no standing in the forming of gravity and if mass does pull, it has to pull to form gravity. However Hubble's law contradicts this idea of a collective contracting Universe totally. This phenomenon about Hubble's constant finding the cosmos expanding should not occur with Newton's perception about gravity envisaging the contraction that must come by the force created by mass in $F = G \frac{M_1 M_2}{r^2}$. If the Universe is on a contracting as Newton said it has to, we have to first find proof about the location to where such contraction is pointing. In order to locate the contracting we have to locate the centre of the Universe, which means we have to locate singularity. With singularity eternal small, holding the place where the Universe started, we first have to differentiate between singularity and zero, should we wish to find singularity. In modern science the phenomenon we know as the Roche lobe comes more and more to the foreground, indicating an undeniable interaction between orbiting structures sharing a common axis.

That axis science at present does not recognise, notwithstanding the reality and undeniable proof there is behind all evidence. As apparent as it is to me, I went about divorcing $F = G \frac{M_1 M_2}{r^2}$ from all ideas forming cosmology and applying the roundness we have in Π to specific positions where one may locate singularity, which we have to locate if we wish to find gravity.

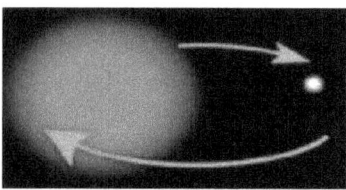
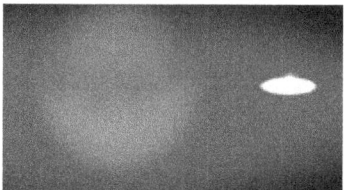

The Roche limit in the practical sense

The formula $F = G \dfrac{M_1 M_2}{r^2}$ cannot explain the comic occurrence shown in the pictures above called the Roche limit, I should find some attention when I say I can explain what is occurring in this instance and this occurrence connects directly to the Roche limit, as explained above. Not only does the Roche limit explain this phenomenon, but also it ties directly to the Titius Bode principle, also being another inexplicable factor in light of the formula $F = G \dfrac{M_1 M_2}{r^2}$.

According to the formula of $F = G \dfrac{M_1 M_2}{r^2}$ all orbiting structures should collide with a bang, but instead they do the tango until one drop, but when dropping it still does not collide with the larger structure, as would the formula $F = G \dfrac{M_1 M_2}{r^2}$ suggest that is used by science. The position where the formula applies is most surprising. Where the formula $F = G \dfrac{M_1 M_2}{r^2}$ applies, one has to find singularity applying because the position of r is pointing to a specific pinpointing of space contracting.

The Coanda effect

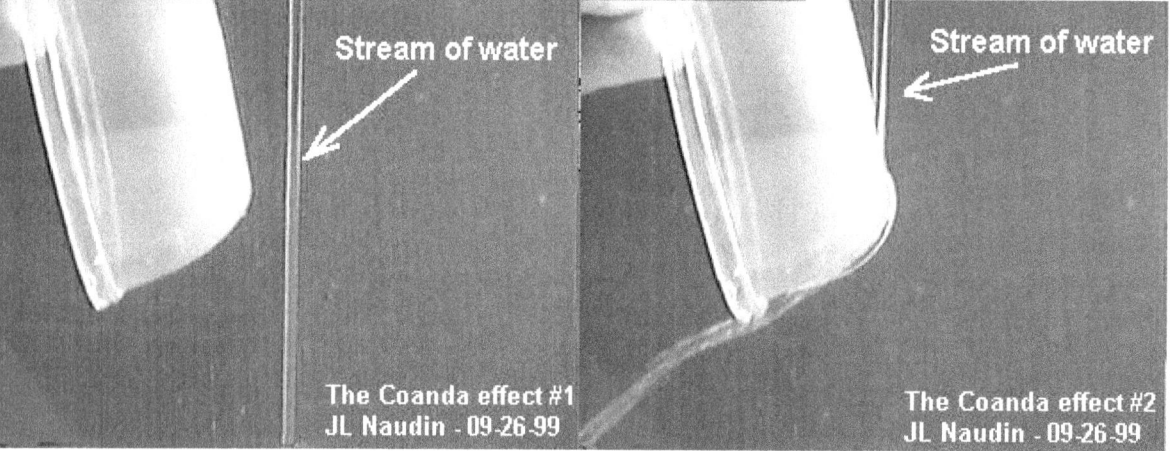

The Coanda effect where a liquid concentrates around the surface of a solid and by movement concentrates the density of the liquid to gather and compact while maintaining a relevance to the centre of such a round solid. I discard the idea that mass could be responsible for forming gravity because in almost four hundred years all evidence is indicating that the truth is to the contrary.

This is not only limited to planets in our solar system. In the Universe, there are giant stars spinning around each other. These stars are binaries, which are also one form of double stars where double stars are another such a form. The difference between the types depends on the distance they remain apart. They keep a certain distance apart and do not collide. In the case of the sun and its planets, it could be a case that the systems might be to small, or they might be to apart. However, this is not the case with binary stars. They are close, they are big, and they spin around a mean axes called the Roche limit.

The Roche limit is:

The Four Cosmic Pillars In terms of Singularity

The region surrounding each star in a binary system, within which any material is gravitationally bound to that particular star. The boundary of the Roche lobes is an equipotential surface, and the lobes touch at the inner Lagrangian point, L_1, through which mass transfer may occur if one of the components expands to fill its lobe. It names after the French mathematician Edouard Albert Roche (1820-83).

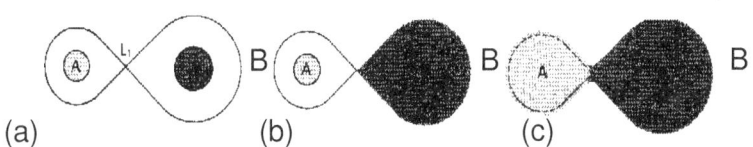

(a)   (b)   (c)

THE ROCHE LOBE: In a binary system, the Roche lobes of components A and B meet at the L_1 Lagrangian point. (a) In a detached system, neither star fills its Roche lobe. (b) In a semidetached system, one massive component, B, fills its Roche lobe. (c) In a contact binary, both components overfill their Roche lobes and share a common envelope.

LAGRANGIAN POINT:
The Lagrangian points are five equilibrium points in the orbit of one body around another, such as a planet around the Sun

From singularity there comes three values each holding 180^0 and this fact science is familiar with. The straight line is always a potential triangle with on side apparent and the other side in infinity.

Planet	Mercury	Venus	Earth	Mars	Ceres	Jupiter	Saturn	Uranus
Bode's Law distance	4	7	10	16	28	52	100	196
Actual distance	3.9	7.2	10	15.2	28	52	95	192

Bode's Law:
A numerical sequence announced by J.E. Bode in 1772, which matches the distances from the Sun of the six planets then known. It is also known as the Titus-Bode law, as it was first pointed out by the German mathematician Johann Daniel Titius (1729-96) in 1766. It is formed from the sequence 0,3,6,12,24,48,96, and 192 by adding 4 to each number. The planets were seen to fit this sequence quite well – as did Uranus, discovered in 1781. However, Neptune and Pluto do not conform to the 'law'. Bode's Law stimulated the search for a planet orbiting between Mars and Jupiter that led to the discovery of the first asteroids. It is often said that the law has no theoretical basis, but it does show how orbital resonance can lead to commensurability. The importance that becomes known is the sequence the Titius – Bode law saw in the number arrangement of 3; 6; 12; 24; 48; 96 etc. The incorrect application of the Titus Bode law lies in subtracting the figure of 3 from 10 leaving 7. The other way of reasoning is to add four each time to the first value of three starting with 3 and so on. The true significance of the Titius-Bode law is that it points directly to a circular growth of 7 stages. The 7 relating to 10 is a precise derogative of the Roche limit or the Roche limit is a precise derogative of the Titius Bode principle because he two systems interlink.

The question immediately springing to mind is how on earth have I manage to come to find gravity being formed with the Titius Bode law applying. It is shockingly simple!

The importance that becomes known is the sequence the Titus – Bode law saw in the number arrangement of 3; 6; 12; 24; 48; 96 etc. The incorrect application of the Titus Bode law lies in subtracting the figure of 3 from 10 leaving 7. Then secondly is cross applying this into the number arangeemnt hels by the Earth in the Tiitius Bode law which is 7 and 10.

When looking at the Titius Bode law the alignment does not make sense because the distance doubles every time a new planet is positioned. Mercury is 3 and Venus is 6 and the Earth is 12 and in that the meaning of this is very much hidden.

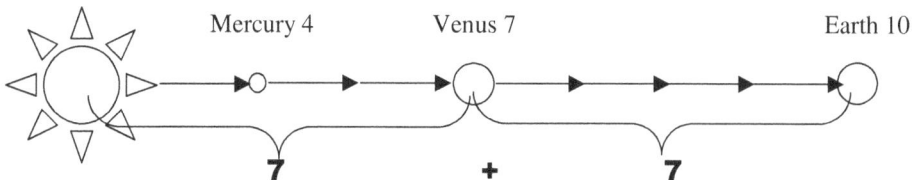

Looking at the Titius Bode principle and not the method we see that Venus, which is the Earth's immediate inner planet, holds a position of 7 in relation to the Sun and when this doubles we will find the Earth also holding a position of 7 from Venus, which the immediate inner planet is doubling from Venus to the Earth. If the distance doubles every time, then the frequency between Venus and the Earth must be the same as the distance or frequency between Venus and the Sun. In this same table the Earth holds a position of 10 in the method of measure applying. This puts the earth at a double 7 and also a factor of 10.

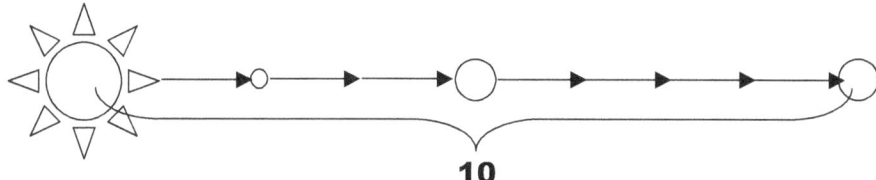

From this comes the value of gravity because as time used gravity, where gravity is time and gravity is used to form space, this forms the pattern whereby the building blocks were laid down by singularity to form space. The space we see is the remembrance of gravity applying that formed space as time formed gravity.

The ratio of 7 to 10 would apply as seen from every planet as the planet circles the Sun. The fact that we see 7 to 10 applying is because we are within the governing singularity of the Earth by forming a part of the controlling singularity of the Earth. The same ratio of 7/10 will apply when standing on another planet.

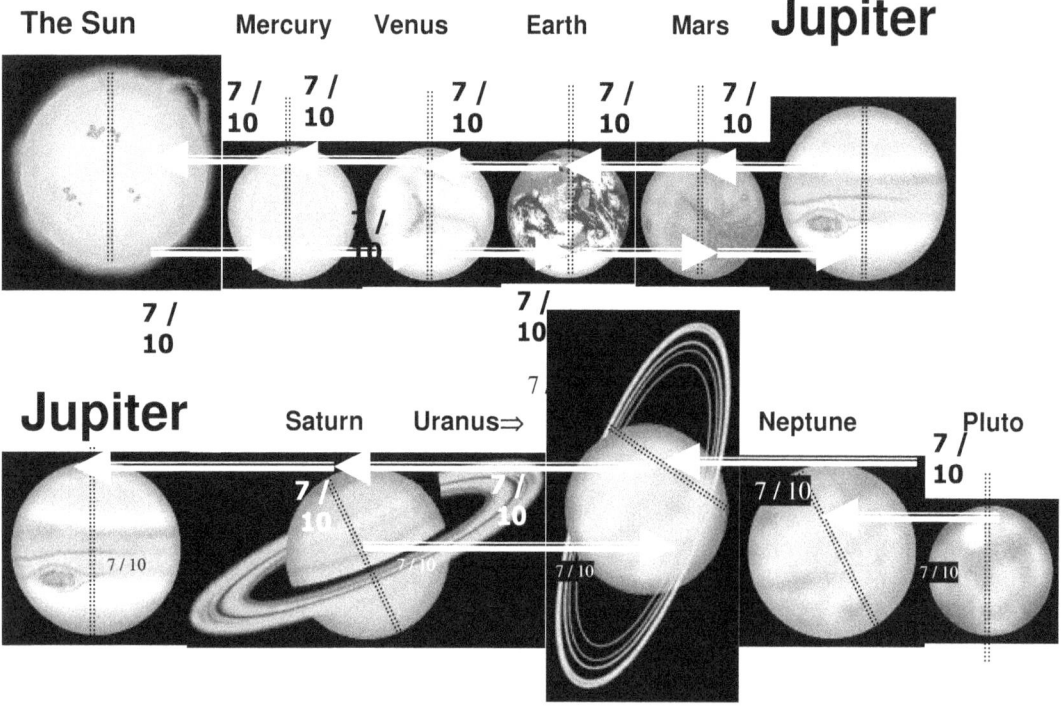

In that way gravity is the Titius Bode law applying where it places a relevancy of circling which is 7 in relation to straight-line moving which is 10 and from that configuration we have Π coming into place.

Inclining by 7°

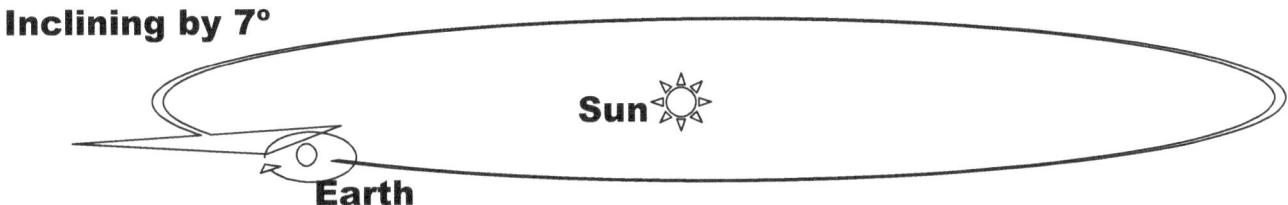

Inclining by 7° as the Earth goes around its axis.

I repeat this very basic law because it is evident that Professors forget the most basic mathematical principles such as that multiplying zero with anything leaves only nothing therefore zero is not a mathematically usable number; still they want to fill the Universe with 0. In mathematics gravity is the **Pythagoras's theorem**, which is a relation in Euclidean geometry among the three sides of a right triangle. I am sure most know this but for those professors that forgot how this works since they forgot so much of the most basics of mathematics, this is how it reads: **The Pythagorean theorem**: The sum of the areas of the two squares on the legs (a and b) equals the area of the square on the hypotenuse (c).

The theorem is as follows: In any right triangle, the area of the square whose side is the hypotenuse (the side opposite the right angle) is equal to the sum of the areas of the squares whose sides are the two legs (the two sides that meet at a right angle). This is usually summarized as follows: **The square of the hypotenuse of a right triangle is equal to the sum of the squares on the other two sides. If we let c be the length of the hypotenuse and a and b be the lengths of the other two sides, the theorem can be expressed as the equation:**

$a^2 + b^2 = c^2$ or, solved for c: $c = \sqrt{a^2 + b^2}$.

This equation provides a simple relation among the three sides of a right triangle so that if the lengths of any two sides are known, the length of the third side can be found. A generalization of this theorem is the law of cosines, which allows the computation of the length of the third side of any triangle, given the lengths of two sides and the size of the angle between them. If the angle between the sides is a right angle it reduces to the Pythagorean theorem. This says we can solve the riddle of the Titius Bode law!

We have the Earth spinning by 7° around singularity, once spinning around its axis ands the spinning around the axis of the Sun. The 7° are movement and therefore by being movement it has to be calculated by the square thereof. Spinning around singularity brings equality to 7 duplicating because 1º = 1º.

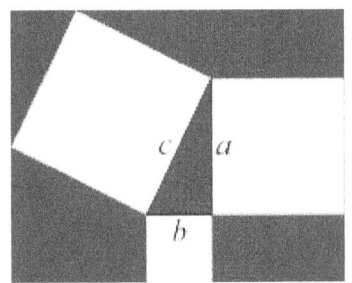

$(1^2 + 7^2) = (1^2 + 49) = 50$

This applies double (around the Sun and around the Earth axis and therefore it is **50+ 50= 100**

Because 1º = 1º we can add 7^2 and that brings about a double 10 that mutliplies.

When we take the square root of 100 it becomes 10 and there the Titius Bode law applies as the law of gravity and as the law of time moving by forming space forever. The Titius Bode law shows the Earth is circling by 7 and is moving forward by 10 and that forms gravity that forms space. From this we can deduct that the Universe moves 7/10 in a three-dimensional sphere ($Π^6$) forms as it starts at 7/10($Π^6$)÷6 = 112, which is a value forming the start of the element table and that I explain in the Cosmic Code. One is 7/10 which is the Titius Bode law which is the interaction of gravity spinning and by spinning is forming a sphere ($Π^6$) within a cube (÷6) and that is how the cosmos forms usingΠ. The dimension of ΠºΠ is flat but by spinning $Π^3=(ΠΠ^2)$ using 6 dimensions, the Universe goes in a sphere ($Π^6$) spinning in a cube 6. In this I prove that for instance amongst so many other things that the Universe is a sphere spinning in a cube. By ticking ΠºΠ time forms space by becoming space as time moves into the future leaving the past behind as space. This is what we see from how the Titius Bode law is employed whereby the Titius Bode law is the way of building the solar system just as it then has to be the building of the cosmos. Time is a substance and is the only renewable substance with the ability to come into the Universe because from the start it came into the Universe to form the Universe as space. Time renews what is by securing what was as space. As time moves on space grows by the margin of singularity Π°Π leaving spots that form dots. The proof of this is in the value of Π being 3.14159 where 3.14159 -3 = 0.14159 x 7 = 0.9911, which is singularity as the spot (0.9911) becoming singularity 1 as the dot. That is gravity and not some idiotic cry about mass performing magic to bring on gravity.

Now every physicist, show your academic worth and your educated dignity and accept the challenge I put to you and to all physics educators: I challenge you all: **PROVE ME INCORRECT IN ANYTHING I SAY!**

Whatever gravity is, gravity has to be Π. If gravity is linked to mass as Newton stated, then mass has to be very closely connected to Π. Looking at every aspect that forms gravity, it is formed by a circle. The Earth as much as the Sun as much as all stars and galactica holding gravity is round and the roundness are Π. The curvature of space-time, the fact that gravity bends light into a curve, this bending comes in the form of a circle that is formed by Π. The Sun for instance spins around and that is formed by Π. The Earth holds the Moon captured while the Moon circles around the Earth and the circle is a result of Π. If it is with gravity that the Moon circles around the Earth, then in all of this we must locate gravity holding Π as a value.

Looking at the Solar system we find that all planets and objects not classified as planets and all things that is just simply forming solar debris has one thing in common…all apply the value of Π in the process where they orbit the Sun, which also uses the formation value of Π to construct the roundness the Sun has. Gravity has much more in common with Π than it will ever have with mass that produces gravity. Wherever singularity forms gravity, it involves Π which then results in gravity manifesting as some or other form holding Π as a major factor.

Being at.   **Π Going too.**

Singularity: a mathematical point at which certain physical quantities reach infinite values for example, according to the general relativity the curvature of space-time becomes infinite in a black hole.

Singularity = $Π^0$

Coming from = Π

Singularity = $Π^0$

With no line starting from zero because there is no zero as a mathematical fact, then all particles hold the point of infinity and not merely the Black Hole.

Where singularity holds position in the centre of any and all rotating objects as a value of Π merely applying movement (in the form of atoms) qualifies all matter to be space-time. It does not only fit the description of space within Black Holes but it fits all stars where singularity becomes part of all the stars from the minute to the largest cluster of matter.

Through rotation encircling the point of singularity and matter is (1) coming from, (2) being at, (3) as it is going too in one movement in relation to the specifics of the centre point being singularity, all matter then qualifies to form space-time.

From that argument one may conclude that all stars will become Black holes depending on the gravity increase they may generate.

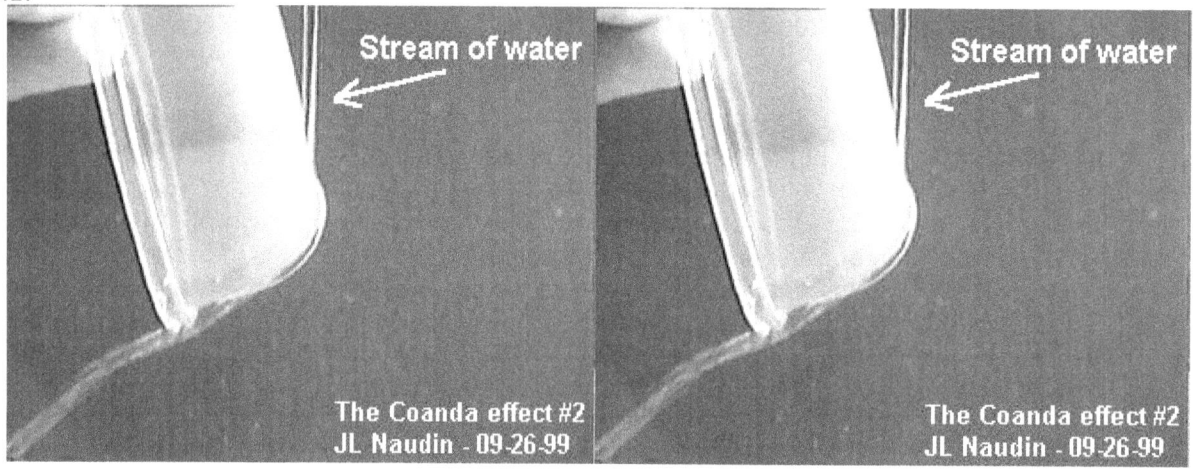

I say this phenomenon called the Coanda effect is gravity. I say mass is a product of gravity whereas Mainstream Science has been saying for centuries that gravity is a product of mass. Science says that gravity is due to mass establishing gravity while not one person could ever explain the least detail as to how it is done. I went on to research Kepler and I discovered gravity through discovering Kepler. I concluded that gravity is the movement of material through space. By following Kepler's guide as Kepler formulated the process in introducing the equation four centuries ago being $a^3 = T^2k$ he gave us an explanation to what gravity is...if only Newton took notice of this important document. This says material holding space moves through space and proves that gravity has nothing to do with mass while mass is the product of space moving.

What is it about gravity that I say which no one wants to know? No one wants to listen to my point because I call Newton a fraud. He defrauded science and took all the other suckers running after him like sheep that are / were unable to think by there own ability. Now no one wants to find out how stupid the entire lot was that came after Newton and followed in his misguided footsteps. Saying this much in the past had every academic rejecting my work at that point. No academic found my work worthwhile to read after reading this much about Newton. No one wants to know that Newton went on lying for almost four hundred years. If you feel annoyed with my remarks concerning Newton, then explain how mass brings about gravity! No one understands the issues of mass and gravity. No one in science clearly distinguishes between gravity and mass and everyone in science tries to confuse the two issues by making them one and the same. They are two distinct different issues never to be confused.

I have discovered that the Universe is not employing a Special Relevance of singularity, but there is a state of ***The Absolute Relevancy of Singularity*** that is not only controlling the Universe but is what the Universe constitutes of...it forms the Universe …it is the Universe. However, notwithstanding the magnitude in significance ***The Absolute Relevancy of Singularity*** poses to science forming a breakthrough, yet past experience taught me I have no chance that my theory on ***The Absolute Relevancy of Singularity*** will be noticed. I came to the conclusion that members forming the body of Mainstream science in physics will not care to take any notice of ***The Absolute Relevancy of Singularity*** and I don't believe that it will be read, will be seriously considered and much less be accepted by those with the authority to change physics principles. I hold the opinion that the theory I introduce here and now would never be accepted in my lifetime because science in the Newtonian way is bent on believing in the marvellous, the outrageous and the magic of what can never be explained, although they claim to use facts as a basis. Science has no idea of what a Black Hole is and I can prove what a Black Hole is. Science has no idea what "the sound barrier" is and I can prove what it is. The explaining of science coming from this that I prove is almost endless.

Yet, I feel I need to warn you whom are reading this letter that this work contained in this letter strays widely from mainstream science and for that there is a very good reason, but I should add that in the least it is thought provoking. I researched the work of a man that is most exceptional and even more prominent in the history of mankind and yet the meaning of his work went unnoticed all this time. His role in the gathering of information furthering knowledge accumulating of the human species' efforts stands second to none while most of everyone is not even aware of the full implication of his work. While recognising his work Mainstream science bluntly ignores his work and in that they miss the full vastness of the wide influencing of his work. It seems to me that any research predating Newton never came into use or in practise. My investigation of Kepler's work brought about a conclusion that no one yet arrived at concerning the findings of Kepler because no one scrutinised Kepler's formula. Kepler found planets rotating around a centre but Newton saw a circle and added what is mathematically required to indicate such a circle. Newton added a mathematical $4\Pi^2$ to the formula of Kepler and removed the distance symbolising measure that Kepler introduced using **k**. On the other side Newton changed the symbol of **k** by using the symbols G $(m + m_p)$. All of this I change and show why it has to change back to Kepler's vision in order to better man's insight into physics, but in that I change the grain and foundation of mainstream physics, I change the total understanding of what forms the basis of cosmology and that part is what mainstream science avoids.

Not withstanding this, still I hope that this writing may spark interest even at such a low academic level and grade in scientific sophistication development because I am about to prove that I discovered:
1) The location, the position and the value of **singularity** as a factor forming space-time
2) Finding **space-time** by dissecting Kepler's formula in relation to valuing singularity
3) Finding space-time, **proving space-time** and **aligning space-time** with gravity
4) The **working principals** behind and manifesting **of gravity** as a cosmic occurrence.
5) The **Roche limit** and explaining the resulting of a law coming about from singularity.
6) The **Lagrangian system**, how and why that becomes the building form of the Universe.
7) The **Titius Bode law** and I show mathematically how gravity comes about from that
8) The **Coanda effect** and the producing of gravity through reproducing space-time.

9) The **sound barrier** by proving it **is gravity** generated **by motion** in space becoming independent where motion creates independence. Breaking the sound barrier is the motion in space duplicating space by crossing over gravity borders. It is $a^3 = kT^2$ where ($k \leq T^2$) or ($k > T^2$). Most of all, I prove that gravity is the Coanda effect forming, applying as gravity everywhere in the cosmos.

Kepler said $a^3 = T^2k$ but that could also be $k = a^3/T^2$ and could be $k^{-1} = T^2/a^3$ and that is the Coanda effect.

Newton said a sphere is $a^3 = 4/3 \Pi r^3$, which is mathematically correct, however

Kepler said the cosmos told him a cosmic sphere is $a^3 = k T^2$ where that puts the cosmos in completely different mathematical dynamics altogether. There are the two distinct possibilities of a^3 which Newton saw and which Kepler saw and both are most valid, but altogether unequal. Between the two concepts there is literally one Universal difference and the two can never be mistaken as promoting the same principles.

It is true that Newton's method or formula of calculation $a^3 = 4/3 \Pi r^3$ when measuring the sphere is widely used, but Kepler received his code of calculation from a very high authority, which is none other than the Universe and therefore can not be discarded as Newton did. It is the duty of the cosmologist not to reject Kepler's findings, or as Newton did, try to transform it into something that Newton could understand after such transforming, because it then strays from the original meaning…but dutifully to search for the meaning as Kepler received the formula from the cosmos. We can test any of the following symbolic values in the mathematical expression and also test the principal behind the expression in which Kepler stated them. By such testing we will find that time after time there were never any corrections in the translations of Kepler's formula required since the translation thereof was never incorrectly presented by Kepler in the first place and in that a case therefore asked for no alterations as Newton did as to secure the correct reporting of the cosmic information being translated. By taking the formula on face value it can change as follows: $a^3 = T^2 k$ can become $k = a^3 / T^2$ or become $k^{-1} = T^2/a^3$.

When translating Kepler's mathematical expression into English we can see what Kepler said also read as $k = a^3 / T^2$ where **k** is indicating one point from a centre point that is space a^3 relating to time T^2. From a centre comes space-time. The centre **k** brings space a^3 in ratio to time T^2, which are space / time a^3 / T^2 **k**. Reading this correctly cannot bring any dispute…yet it does…and it's been doing it for centuries on end!

Kepler was the very first person to mathematically introduce **space a^3** aligning a **centre k** and relating the resulting movement to **time T^2**. Not only did he introduce **space-time a^3 / T^2k** but he also placed **space a^3** and **time T^2** in a relevancy **k** long before Einstein did and placed **gravity in space-time a^3 / T^2k** even before Newton named gravity. He showed that space **k** is growing in relevance (**k $=a^3/T^2$ and** also the opposite as **$k^{-1} = T^2/a^3$**). The manner in which the Universe attends to space-time is $a^3/T^2 = k^1$ and $T^2/a^3 = k^{-1}$. Kepler was the person who placed gravity as the ingredient in the Universe that determines **space a^3** and **time T^2k** and this proves the Universe is not expanding which it can't do, but is changing relevance by allowing material to grow as time develops space occupied. Kepler was the first one that saw that gravity comprises of two factors being **k** or linear gravity and **circular gravity or T^2** as gravity keeps space in form while all is staying together.

Kepler said $a^3 = T^2k$ and that correctly translates to a mathematical expression $k^0 = a^3 / T^2k$ which in the verbal statement in English translates that Kepler said that there is a **space a^3** which is **equal =** to the motion in **the time duration T^2** thereof between two specific points which holds a relation onto a centre k^0 where from there forms **a straight line k** that is centred on the spot where space begins from k^0 **that produces k** as well as producing the circle therefore that spot $k^0 = a^3 / T^2k$ has hold k^0 at a value of having the least space. The line **k** is centred onto a spot where space begins specifically at k^0. This point not only produces the line **k** coming from a point k^0 but represents also the space a^3 that forms the eventual circle by the rotation of T^2. Therefore from the centre holding k^0, k^0 leads to **k** that forms the revolving space a^3, which is rotating T^2 at a distance **k** where T^2 forms the outer limit of k^0. Mathematically $a^3 = T^2 k$ will also be $k^0 = a^3 / (T^2k)$ because $k^0 = 1$. But $k^0 = 1$ also present the single dimension where all factors are a product of one. If anyone can locate k^0 then also that person will find singularity. That is where gravity is because gravity is strongest where space is least. Then that suggests that gravity is strongest at k^0 because there space is least. That is gravity because that is what keeps the orbiting objects in orbit but also that is what Newton completely missed when he changed Kepler's work. Newton failed to recognise gravity as the only ingredient in Kepler's formula. He admitted he missed this because he admitted he did not know what gravity is while Kepler explicitly showed what gravity is. Gravity is what keeps the orbiting objects in rotation while orbiting. $k = a^3 / T^2$ is **distance^1 = space^3/ time^2** forming from a pivoting centre k^0. That is a cycle and moreover it is a cycle formed **by space/time**. What Kepler said is that space is a^3 **being in motion T^2 k**.

That says **space³ (a³)** relates directly to **time²** that uses the symbols **T² k.** This is also what I refer to when I say one has to read what Kepler did not say when one wishes to see what Kepler meant to say. Kepler introduced space³ –time²⁺¹ long before Einstein's date of birth appeared on any calendar although Einstein is credited with the formulating of the concept of space-time and giving it a name. Going even further Kepler stated that the space a^3 is on the move T^2 around in a circle at a distance **k.** That is what that comet is doing. The space³ (Comet) is circling T^2 the Sun using a radius **k** to establish the cyclic time² as a period of continuous motion and continuous motion is gravity. Remember in this statement I am separating cosmic principles applying from the way that gravitational principles apply on Earth.

As Kepler said **a³ = k T² and therefore k⁰ = a³ / k T² and therefore we have to find k⁰.** As a result of examining this proposition, I located two principle positions both holding singularity. ***What is in the Universe is spinning*. The entirety of everything forming the Universe is spinning inside the Universe and** such spinning are always in the centre of one specific point, wherever such a point might be. In the **precise middle** of all **objects in rotation** is a precise centre where this pre-designated centre is dividing the object in rotation into sectors that will **start the spinning initiation** from that centre point. By spinning the one side is coming towards while the opposing side at that time is going away. Thus, the spinning object **will have a middle point**, a very specific **centre point that does not spin** and only holds Π as a specific value because within that centre being that small, no radius can apply. But also within the one value forming, such a line **cannot have is zero** because the line **is there and holds contact** to the rest of the material bringing about that **zero does not start any** line and therefore the **value of the line must be infinite,** just as described in **accordance** and by **the definition of singularity.**

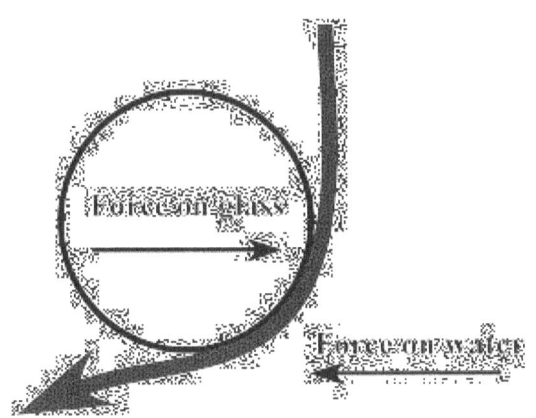

<u>The condition for the presence of this centralised singularity $k^0 = a^3 / (T^2 k)$ is movement $T^2 = a^3 / k$ **of** space $a^3 = k T^2$ in relevancy $k = a^3 / T^2$ going both ways $k^{-1} = T^2 / a^3$ (Newton's 3rd law.)</u>

<u>This explains the Coanda effect and the Coanda effect is gravity and gravity "glues" the water to the glass!</u>
In considering the spinning motion in the fraction of time in the detailed instant every aspect of rotation will turn in every instant of change in time by putting every spot there is in another location in accordance with the centre point that is unable to spin. While spinning the points will change direction every 90° of spinning and will oppose what it was every 180°. Although the points had the same characteristics only one instant before, they oppose the characteristics it had just before and just after the very instant in which they are and to which they relate by similar points also in rotation. The fact of the graph proves my point in quarterly opposing dimensions and values. As every point relocates, therefore every point completely changes its attitude from what it was to what it is in terms of what it will be when it is going there. Going down to the centre, as the rotating direction moves inwards, the rings will become smaller and smaller. In dimensional terms, which I explain later on the value of **2k** relates to **T².** That relation extends to the next value where **T²** relates to **k**, which relates to **T².** The first space in the circle will then be **T²k.** From the centre being in infinity one can realise by applying mental power the single dimension factor not seen but present all the same. Extending that into the 3D comes six **k** and any one of the six will further extend to form a seventh point as **T²** All this is a multiplying of **k⁰ = a³ / (T² k) = 7**

Lines mathematically cannot start at zero because there is no evidence of zero as a factor in mathematics. Should you disagree with my statement the question in need of answering is this: **What will the length of the shortest hypothetical line imaginable be and moreover, what would the total overall length be in that case?**

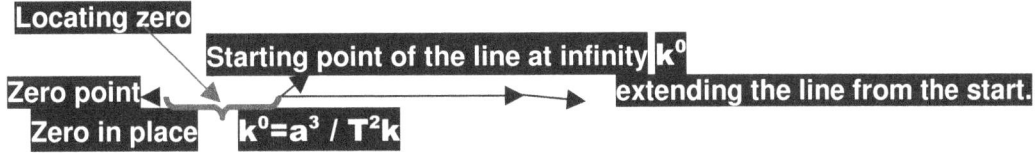

Let's find **k⁰ = a³ / (T² k)** and see where it is hidden. The sphere is a circle in many facets and therefore we will approach the sphere as one multi dimensional circle, however the sphere as such remains one circle to the power of many. When investigating a circle one would draw a line from one edge running through a centre all the way to the other edge. In doing that we would find the measure of the diameter, which is most important when trying to establish the volumetric worth of the sphere. The circle has Π to indicate form and uses r² to establish the worth of such a circle by using the radius symbolised as r in drawing a straight line. In any circle or sphere the size only depend on the fluctuation of r in the square as a component to the circle or sphere but that does not affect the form which comes by indication of Π in any way there may be. The conclusion from this is that no line can start at zero because that will be a mathematical impossibility.

This statement by itself excludes zero and with zero excluded one then begin to appreciate all the rest of the concepts governing corrected cosmology. A line or spot starting at zero would therefore be shorter than the shortest line possible. For obvious reasons can no line, or any line grow or extend from zero because such a line must then quit zero and become something, thus abandon its original value by the adding of the first value. Mathematically said it would be as follows 0+0=0 whereas of it started with something infinitively small it would be $1^0 + 1^0 = 2$ and then from using something infinitively small it will grow into something immense such as the Universe.

In any circle or sphere the size only depend on the fluctuation of r in the square as a component to the circle or sphere but that does not affect the form by indication of Π in any way there may be. The conclusion from this is that no line can start at zero because that will be a mathematical impossibility. If a line started with zero, that would nullify Π (0^2 x Π = 0) and that would leave the form without having any form because Π x 0 = 0. This statement by itself excludes zero and with zero excluded one then begin to appreciate all the rest of the concepts governing corrected cosmology. A line or spot starting at zero would therefore be shorter than the shortest line possible. For obvious reasons can no line, or any line grow or extend from zero because such a line must then quit zero and become something, thus abandon its original value, should such a line wish to progressively become more of what it was before.

If the reader is wondering where this is going, well I am trying to remove zero from lines as I am trying to remove zero from graphs as I am trying to remove zero from the Universe because I am trying to remove zero from filling outer space. When a line starts off with zero while still forming a line that composes of a line, that would mean the start of the line has a different value to the end and a line holds conformity through out…no line can end by applying zero as the concluding worth. When any line is starting from point zero it can never leave zero because of the influence of being zero disqualifies any possibility of growth, or even being present at any point to grow. If the line then had to grow in all directions at the same pace the line must therefore be a circle or being three-dimensional, a sphere. Flowing from this fact is that in the Universe there can be no zero point or unfilled space as every point has to be filled with at least something other than zero. In the case of the growing sphere the value of the circle isΠ, and that has to be the point where creation started. That gave me the clue where to start looking for singularity. One would find singularity in the value Π and the value Π will be in all things rotating in a circle. You might wonder how does that apply to the cosmos and moreover to gravity? Mainstream science promotes the idea that outer space as far as outer space goes, is filled and even overflowing with nothing. The nothing they place in outer space is so much of nothing that the nothing they have filling outer space is overflowing because the nothing is expanding it is growing! That nothing then must be growing because E.P. Hubble proved the cosmos is expanding. If this sounds ridiculous then might I remind everyone that I didn't fill the Universe with nothing because that was the doing of mainstream science? You cannot fit nothing into outer space because it just will not fit; there is just too much space to harbour nothing. If any of the factors in Kepler's formulae represent nothing then that is what you will get, it would be nothing.

The Universe in its total entirety will be nothing. **$a^3 = 0$ $T^2 = 0$ k=0, which then mathematically could only translate to $a^3 \div T^2 k = 0$.** Or in the case of Newton removing **k** then Kepler's formula would read as follows **$a^3 \div T^2 k = 0$** because **$a^3 \div T^2$ x 0 = 0.** If the argument seems ridiculous it is not my mentioning such a fact that is ridiculous but the mere fact of the reasoning that also became an accepting of the valid ness of recognising that it is nothing that fills the cosmos that is the silly part. The basis of such an argument and the fact it could be accepted by science is what is making it ridiculous. It is the fact that one must argue about such a ridiculous matter about an idea that nothing is overflowing while it is filling the Universe, that allows the ridiculous part to enter the conversation because the trend reminds of arguing about fairies and little people existing or not and such argument is nonsense. If space is nothing then explain how it is possible for nothing to have a distance and to have a measurement indicating the number per measured unit in length. Add as many zeros as possible and see how that can form 149 x 10^6 that the Earth is from the Sun. Using zero the distance indicated would be indicating just that value being zero or the capitol O indicating zero while every planet has a precise distance it is located in terms of the Sun as well as in relation to each other.

Try and indicate what is measured and calculated in value in outer space in measured space, while having that going in kilometres or astronomical units and then finding it is nothing in multitude filling that distance. The distance between the Sun and Pluto **is 5900×10^6** kilometres of space, but in that statement we take it that the one as a factor used in determining whatever constitutes to form the measure of a kilometre. By adding one **5900×10^6** times puts 1 present in such a multiplication because adding 0 5900×10^6 times will still amount to 0. The one constitutes the presence of a fact being a statement of a value compiling to present the measure of space as it is in a distance. By saying the distance constitutes of nothing we have to substitute the one factor with a factor of zero. Then the calculation must read **Pluto is $5900 \times 10^6 \times 0 = 0$**. Including nothing as to state the presence of that part contained by the calculation delivers the total of zero. It seems as if science has ignored this issue by simply not thinking about the fact and therefore simply ignoring that what is measured forming the sole value of space has a practical worth, but it is somehow more convenient to put the value of nothing as part of the distance in calculation because that is what is measured and then see how one can multiply by using zero in mathematics and reach a distance holding a value other than zero when multiplying by zero.

I agree that what is filling outer space is invisible, but also it is there, it is present and being present and there while being invisible disqualifies whatever is there from being zero because being zero will mean it is not there and we cannot deny whatever is there of being there. Then what is there will be there, while being invisibly small, but it will still be possible to form a line because every aspect of the Universe forms lines while also it will have the potential to fill space and can still form a measurable unit. That then must be 1 because while $1 \times 1 = 1$, $1 + 1 = 2$ and that qualifies that invisible thing to be present ($1 + 1 = 2$) but at the same time be completely invisible ($1^3 = 1$). When realising this I knew what has to be true about that which I was looking for and that it had to be singularity because singularity can only have one value and that is 1.
To find the invisible I had to locate singularity. I realised that my effort to locate the point holding singularity enabled me to backtrack the exploding Universe to its origins. The Universe is a sphere because it is filled with spheres filling the void spaces (not the nothings) and in that I first had to investigate the visible.
Newton's mathematics says a sphere is $a^3 = 4/3\Pi r^3$ while Kepler said a sphere is **$a^3 = T^2 k$**, and both are equally correct because the cosmos gave numbers to support its statement.

With Kepler **$a^3 = T^2 k$** and with mathematics the volumetric size of space must either be according to the measure of normal mathematics if it is a cube then three sides form $a^3 = L \times B \times H$ and in the case of a sphere the measure will be $a^3 = 4/3\Pi r^3$. This was like comparing a triangle in relation to the half circle and the line.

It predates mathematics where the numerical use of determining a value was not yet established and only form was in use. It is equal to a time when we find in the half circle standing 180° related to the triangle (180°) and both still are equal to the 180° of the straight line notwithstanding the obvious differences used in form. However the starting point of these forms has to be equal and also has to be not zero to have the end be equal and result in all being equal in value in the end.

Kepler said a sphere is **$a^3 = T^2 k$.**

In honesty we have to realise that we cannot dismiss the whole formula that Kepler produced just because it doesn't match the scenario set to determine volumetric size as does the Newtonian version does. Kepler's version holds a foundation based on movement and it is in the movement we find the measure and not in the size as Newton's mathematical formula does.

In Kepler's formula the entire formula is formulating a circle being motion. However with the correct interpretation we find so much more than just motion. The formula is **$a^3 = k / T^2$**: That is what Kepler brought into civilization for all time to come. He saw space **a^3** being in isolation due to the time it uses to move **T^2** claiming such space forming independence according to the lines **k** indicate. Let us look at the factors in more detail before we proceed with the rest.
a^3 symbolises in a mathematical interpretation of implicating the three-dimensional space.

T^2 is representing the period or time that Kepler suggested we should use to calculate time that holds the orbiting planet in direct contact with the space in relation to a very specific centre moving from point **T_1** to **T_2** in relation to a precisely placed centre **k^0**.

k is the space taken from the centre to the end of the line **k** from which the planets must have grown if one accepts the Big Bang growth of particles and the affect of the Hubble constant on all cosmos material. The specific value about the centre is most important because from the specific centre gravity always apply the strongest influence.

The turning T^2 of any circle holding space a^3 is valid only if in reference k to a centre k^0.

Space a^3 will always be circling around as T^2 is in a position referring k to the centre k. That is what Kepler said when he said $a^3 = T^2 k$. Kepler indicated space a^3 will forever fight for independence and show separate individuality in remaining apart as identifiable cosmic components by means of motion. Every space will cling to independence indicated by k through fighting off the integrating of another coverall unifying unit by applying the motion of T^2! The problem we have to solve is what will the cosmos use to secure such independence between all particles? What sets space apart from the rest of space? First we have to admit that Kepler was the one that introduced the following.
Kepler gave us the answer to the following but no one ever took notice!
Kepler was the one that discovered **space / time** as **space** a^3 = **time** $T^2 k$
Kepler was the one that discovered **singularity** as $k^0 = a^3/T^2 k$
Kepler was the one that discovered **gravity** is holding **space-time** relative by the measure of distancing k as $k = a^3/T^2$ and $k^{-1} = T^2/a^3$

Kepler said gravity in space is about the area a^3 that would always keep equilibrium with the time T^2 it takes to travel the distance of the full circle position placed by the indicator k, therefore adjusting k as the need arrives. With k shifting in length a^3 will have to readjust and therefore T^2 will find a new relating value each time. This was the finding of Kepler and came after his intense study of orbiting planets.

Translating Kepler's mathematical expression $a^3 = T^2 k$ correctly to the verbal statement in English Kepler said that there is a **space** a^3 which is **equal** = to the motion in the **time duration** T^2 thereof between two specific points which is a straight line k that holds a relation from a centre k^0 to an end k where the two ends run from the beginning of k^0 to connect at the end of k. I might not be the smartest boy on the block but I'm not that stupid either. I know how to translate mathematics into English… and I translate as follows:

a^3 must have a volumetric interpretation because the third dimension is sure evidence of multiple conjunctions of dimensions put together in three sides opposing three sides having the third dimension in place. The fact that any symbol uses a value to the **third power** a^3 indicates **space** or a volumetric established and separate unit. Using a cube by three dimensions symbolises a cube, a room, a space to be filled, a unit able to hold other ingredients on the inside when empty or partly filled. It is space because it is volume using the third dimension.

T^2 is an indication of something having a cubic nature other than the square forming motion that is provided by the motion the square indicates, which is where the moving object is representing a third dimensional object that is moving from point to point and it is this point to point that multiplies into the square. The space is moving as a unit from one point to another point and the moving between the points are represented by a flat square or following a flat distance between two points. The cubic space was in one instant in one place and then the second instant in the other and because time can never stand still or become single dimensional (this I am about to prove) insisting that time must always support the motion it consist of or space as well as time in time cannot be. It is motion that is taking time, which is motion in the second dimension moving the space in the cube.

k^1 is the symbol used to indicate a straight line between two points with a definite beginning and a specific end position. It is the location where the form in question is holding space and where the space was and where the space are going to be in very next split instant that follows. This indicates points of representing k in different time positions to which the points will then be multiplying indicating form as a result of the square. The movement indicating not a square surface but movement by the square indicates the time the journey took to move the space from one point where k is indicating the location of the space to where the next indication of k. T^2 will shift k where k indicates the space a^3 that formed as a result of the movement T^2 of being the space a^3 indicated the point end of k. However, since time represents the square T^2 and with k being the distance that prove that the k represents the distance the space a^3 representing the form it is obvious that T^2 represent the time that represents the space a^3 in the square T^2 through the motion. It is the distance moving space a^3 in the cube to complete time in duration in the square of motion T^2; therefore k is permitted to be in the single dimension.

In the circle we may locate a straight line by reducing r that is symbolising the radius of any circle where such reducing will be indefinitely to the tune of halving r each time, then r would become infinitely small, even beyond human calculating means and become not a line, but a dot. However as mentioned in the case of the smallest dot holding one spot, r would become insignificant beyond human comprehension even, but never reaching zero and still Π would remain intact and dictating form. I believe one can begin too

The Four Cosmic Pillars In terms of Singularity

see where my suspicions are heading in my quest to locate singularity because the flaw comes about in the manner mathematics are practised for thousands of years and even today in our so called "modern times". The radius represents the initial line, the first that ever was. The very first instant in time was when Π appeared from infinity holding Π^0. With Π resulting from Π^0 the line will eventually become r^0 at a point ext to Π where the line began as a spot that grew into one dot and the dots added eventually to become a line. Finding this line made up of dots is most important when trying to decipher Kepler's formula.

Let us find the smallest possible line first. We already have reached the conclusion that by reducing the line, the reduced line will eventually leave all sides on the same spot on the condition that the circle spins. Such a spot must be round in form since it still holds Π as a factor next to r^0. We now are entering the domain of singularity where the visible is no longer traceable and only intellect can bring understanding of the scenario. With the line being the smallest line, such a line will start off as a dot Π that moved away from a spot Π^0. With all possible sides being in precisely the same spot we have all possible sides onto one spot. I chose to differentiate the dot and the spot by giving the spot a value of Π^0 while the dot holds Π next to r^0. Mathematically the spot is placing even form being Π in the single dimension Π^0 where the space is one (1) and holding exponentially zero (1^0). There the space moved over to form the spot Π^0 and by introducing form changed to the dot Πr^0 forming a circle as a dot.

Again I must draw the attention to the fact that we now are reaching into areas only the human mind can venture by understanding and seeing nothing more than with the eye of intelligence. The understanding of this concept demands our reaching the point where the mind of the animal cannot reach. If it starts with a line it then is there where that line only represents two sides being one and as such that is representing rather a flat Universe. At the dot Π we have roundness because we have Πr^0 while at the spot there is not yet any round form because of Π^0 and only when Π being round it then is requiring a shape or form and this lies beyond or before space at a time when any form of shape comes into the cosmos scenario. This part of the Universe came in a place at a time in a period where shape and form was a part of the distant future hidden in and beyond the developing eternity. The spot is located at a point that when entering the domain where the spot is located, it is also at the same time crossing the spot and landing on the other side where the radius becomes the diameter. Nothing can enter the allocated position the spot holds because entering the spot is crossing over to the other side of the spot. It serves us well to realise that the entire Universe was that small at a point where everything started forming because the spot that developed into the dot is still with every spinning circle...and the Universe is a multitude of spinning circles. It is also very wise to remember that once anything becomes a part of the Universe, it can never leave the Universe since it then has no place to go or no gate to pass through in order to leave the Universe. With the spot becoming a dot, there must have been a time when everything in the entire Universe was that big as the spot is, and that then moved on to form the dot and in that it went on growing in relevance.

There was a Universe forming in a time that everything present at this moment was so small that only form was in place and this was when the triangle, the half circle as well as the straight line was equal $180°$, with no numerical values in place yet. At that point the line must have been so small it had reached a point not yet mathematically dividable in any way. The dot that formed was so small during that time that if at present any further dividing that took place, such dividing would have brought growth because there then would form space between the sides going in the opposite direction. However it is important to realise that anywhere we might locate Π^0 we also locate 1^0 because Π^0 is 1^0. The dividing brings all there is having all sides moving literally on the precise same spot, and I have located singularity in just such a spot.

I came to the conclusion that the spot I found had to be singularity purely on the grounds that that spot holds only one side to serve as a start to the starting point of all directions possible. In that side is only one spot where there is only one side applicable and one dimension present. In that spot space ended. That point is serving as a position for all possible points and cannot allow further dividing as it is in the smallest line or spot there may ever be. In the very centre of any and all circles spinning we find this point holding no space and therefore forming Π^0 that is 1, which is singularity.

Again I indicate the precise location of such a point. What is in the Universe, is spinning and therefore what I am referring to, applies to everything holding a place in the Universe and therefore this which I mention directly links everything holding any space whatsoever in the entire Universe. In the **precise middle** of all **objects in rotation** is a precise centre dividing the object in sectors that will **start the spinning initiation** from that centre point. Thus, the spinning object **will have a**

The Four Cosmic Pillars In terms of Singularity

middle point, a very specific **centre point that does not spin** and only holds Π as a specific value because no radius can apply. But also the one value such a line **cannot have is zero** because the line **is there and holds contact** to the rest of the material bringing about that **zero does not start any** line and therefore the **value of the line must be infinite**, just as described in **accordance** and by **the definition of singularity.** As I am introducing a very new idea, I wish to explain in better detail what I try to convey. While the toy top is spinning one will find singularity by moving the rotating line or radius progressively to the middle by reducing the length the line has from the edge to the middle. At one point all further reducing must end but the ending cannot include zero or nothing because the rest of the line still attach the rest of the top. As the rotating direction moves inwards, the rings will become smaller and smaller.

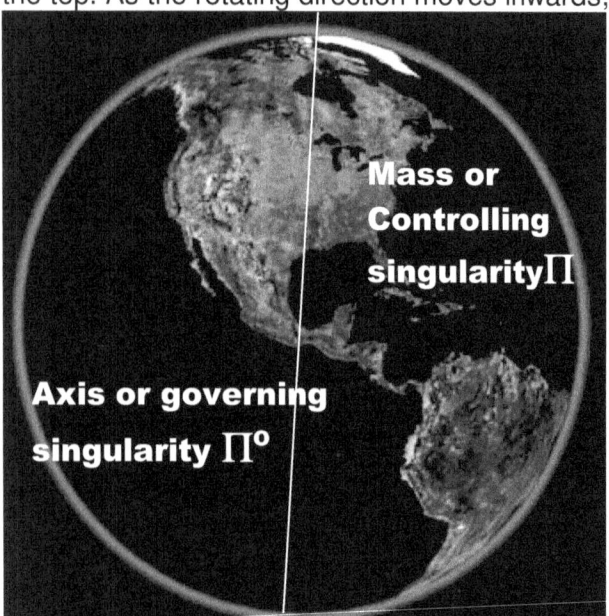

Mass or Controlling singularity Π

Axis or governing singularity Π°

When an object stands on the Earth the object is in mass because the Earth holds control over the atoms in the object. All the atoms spin around their governing singularity and all the atoms in the star connect to by bonding with the Earth's governing singularity. The Earth provides the centre around which the movement spins while the object with mass does the spinning as it then is part of the controlling singularity the Earth provides. The governing singularity of the Earth establishes the worth of the controlling singularity of the Earth by providing gravity or time in relation to the space moving. When an object falls towards the Earth the governing singularity of the Earth captures the atoms of the falling object while trying to establish control over the object falling. In that manner the atoms within the Earth forms the earth by becoming the controlling singularity in relation to the Earth's governing singularity.

The Roche limit is the occupying of a dominant governing singularity that takes the control of the relevance of the controlling singularity that falls within the spherical dominance of a better developed star or structure. The governing singularity places the relevancy applying to the major star on the controlling singularity of the minor star, which is the governing singularity of the atoms forming the minor star. As the relevancy can't form a constructive minor mutual singularity, the minor star structure conforms to the relevancy applying to the major star and with not enough material to form a bonding the minor star expands to the relevancy applying to the major star. We see the minor star expand to the size of the major star.

The Four Cosmic Pillars In terms of Singularity

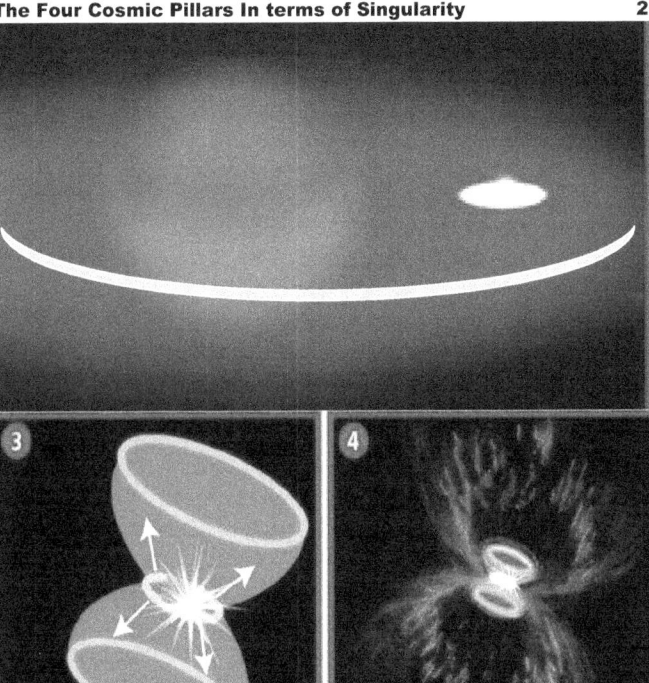

With the minor star spinning (sharing space-time) within the influencing range of the major star's governing singularity the governing singularity extends its control it has in the range of $\Pi^2 \div 4$ and takes charge of the governing singularity of the minor star. This capability puts the controlling singularity of the minor star also in control o the governing singularity of the major star making the entirety that formed the minor star part of the controlling singularity of the major star. With the governing singularity applying the terms of the controlling singularity, everything within the minor star becomes part of the major star since it is under the direct control of the governing singularity of the major star. The density applying in the minor star can't be confirmed or maintained with the spin that the atoms bring about in the minor star and the relevancy applying becomes subject to the terms and density conditions within the major star. Then by applying the Coanda effect the major star accumulates all that once formed the mutual singularity to become the mutual singularity that is part of the governing singularity control of the major star.

The governing singularity is controlling the atoms by means of placing the atoms in a relation where the atoms for the controlling singularity. In this expanding and the governing singularity losing control over the spin and therefore over the controlling singularity we have the Super Nova expanding the star's relevancy.

To understand what I just said we have to investigate singularity and where singularity hides. Singularity is Π producing Π^0 where this establishes Π and by movement Π^2 specifies space ending at Π^3. In order to locate Π we have to locate Π^0 and this we can achieve by producing movement Π^2 that would allocate Π^3.

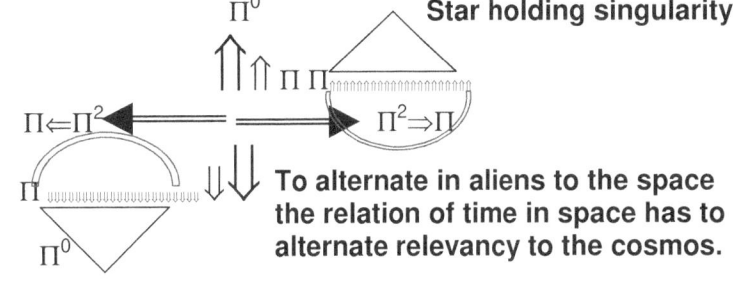

All stars hold a centre point in singularity where that centre point has the value of being the equivalent of all the atoms where each atom holds a centre has an atom's worth that combines to form an equal to the value equal that the star's

Star holding singularity

To alternate in aliens to the space the relation of time in space has to alternate relevancy to the cosmos.

The point is in everything that rotates and everything in the Universe combines to form a connection that connects everything to all other things.

That point albeit hypothetical, is also as much a reality none the less and is placed where that point **must be standing still** because every line **running from that point** in **opposing directions** is also **in opposing directional spin the other or opposing side.** In considering the spinning motion in the fraction of time in the detailed instant every aspect of rotation will turn in every instant of change in time. Although the points had the same characteristics only one instant before, they oppose the characteristics it had just before and just after the very instant in which they are and to which they relate by similar points also in rotation. The fact of the graph proves my point in quarterly opposing dimensions and values.

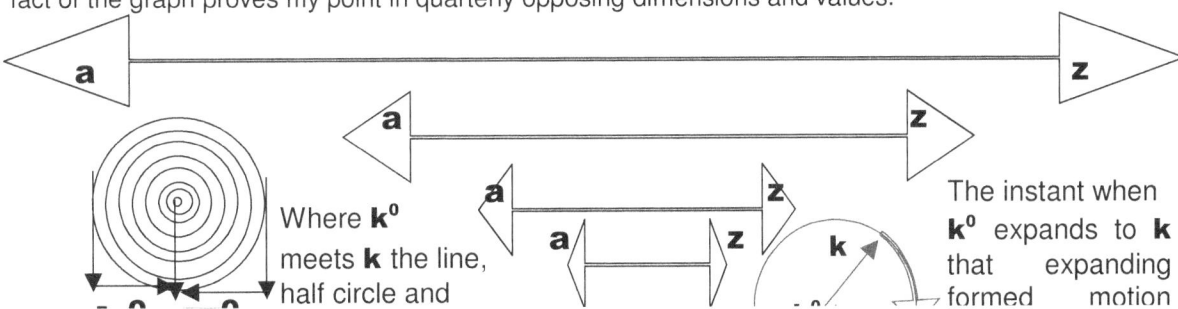

Where k^0 meets k the line, half circle and

The instant when k^0 expands to k that expanding formed motion

> The instant **k** comes about from **k⁰ k** apply further motion as **T²** and with **k** producing motion by expansion and **T²** by contraction…

> …gravity comes into space forming where space expanding **a³** and gravity **T² k** is the same result of singularity **k⁰** setting motion

In the sketch I made it aims to show below each of the lines that with the continuous reducing there is a space left open between the two ends of the line that is symbolising the end of the line in reducing. In the very end where numerical value becomes infinite the two ends will share one location even by having one single point holding each one. There is no chance that I can present any sketch reducing the line to a point where the points are sharing one location literally in the single dimension. Yet although sharing one spot the points are there and with the points being present they may not be dismissed as nothing. From there no reducing in a natural manner can lead to nothing without changing the rules of mathematics in such reducing. But the two ends has reached a position where any further effort of reducing must bring about the start of extending the line once more because every point possible share space with every other possible point at the point of singularity where all points share one common space.

By moving any of the points, such moving by further decreasing at that point must then bring about an increase of space once more since the space at that point in the centre spot of the circle spinning has gone infinite. This also applies to the sphere that is a multitude of circles because the circle uses a line to indicate size running from a centre to an edge. By reducing the line and by reducing the circle the reducing will end up having the ends in the same position in the very centre of the circle. It is this fact of the moving in any direction of any point from that spot holding singularity that such motion will introduce space as the space exceeds the previous limits of singularity. What I am trying to say is by moving from the spot Π^0 to the dot Πr^0, such movement evokes Π^0 a spot to Πr^0 forming the unseen line and without the movement of the spot Π^0 to the dot Πr^0 to form the line Πr^0 the allocating or positioning of singularity Π^0 will not take place.

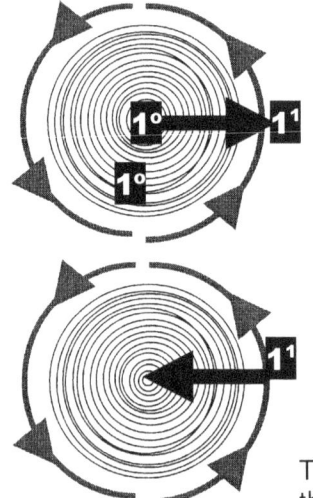

In the Universe there are one substance forming the entirety and that is singularity. Since singularity holds no space it is not a reality but is only in place functioning in terms of all other points holding singularity that spins. Therefore the crucial connection in terms of singularity is found in the movement applying. There is one type of singularity, which is one, but there are two forms of singularity where the one type spins in control of and the other spins with no seeming control applying except time serving as the control by producing the applying movement also known as the Hubble constant. In that I named the one being condoled by spin 1° and the other being controlled by linear expanding through time 1^1. The one is material performing in relation to what never is able to spin but holds control over everything by that which can never stop spinning and in that that which can never stop spinning serves as the controlling singularity to that which can never spin that becomes the governing singularity.

The spinning established a value between the object's movements in relation to the movement of the electrons spinning around the atoms. The evidence that proves the statement is the fact that the minor star relinquishes density when the major star takes control of the minor star's governing singularity and thus takes control of the movement making the control of the atoms dependent on the major star's governing singularity.

The movement of the atom is interlinked with the movement the star establishes which we call gravity. This movement depends on the spin but also depends on the movement of the structure in its rotation of the secondary controlling object as the Sun is in the case of the Earth. This is in relation to time applying to the atoms versus the space that is either dismissed or displaced. The movement is as all movement is, space – time incorporated.

In the spinning top we find that singularity Π^0 can be generated by motion. But singularity Π^0 has no motion within the dimensions we find allocated to the Universe

The Four Cosmic Pillars In terms of Singularity 23

in which we live. Since the singularity found in the centre of the spinning top is in truth just a mathematical point, which means in mathematical terms the point with no sides cannot even be calculated as a factor since the measure thereof goes beyond what mathematics ever can calculate. Mathematics has a use within the 3D Universe but singularity that keeps the spinning top attached to singularity governing the gravity of the Earth, that singularity is truly single dimensional and beyond mathematical measure. It is singularity Π^0

If we put this in terms of singularity (Π^0) we find the Earth (Π^3) is in relation as viewed from Alfa Centauri (Π) four point six years (Π^2) while moving in that space that is time that has gone by. That secures the three dimensional status the Earth has (Π^3) in terms of a present (Π^0) that depends on a location (Π) secured by a future (Π^2) that will come by movement where the future ($\Pi = \Pi^3 \div \Pi^2$) moving forward that also doubles as a past ($\Pi^{-1} = \Pi^2 \div \Pi^3$) by the light coming from and thereby confirming the past. That is space formed three dimensionally by keeping time in infinity apart from time in eternity. The relevance (Π) that forms in relation to the present (Π^0) will relate to movement (Π^2) and the movement is circular which ensures that the relevancy forming is circular (Π) by securing that the movement is circular (Π^2) in terms of one specific point (Π^0) in infinity which then secures a roundness (Π^3) that forms an everlasting eternity ($\Pi\Pi^2$) which validates a never ending circle Π^3. In this time in infinity (Π^0) that secures that there is an everlasting eternity ($\Pi\Pi^2$) in space (Π^3), it is not the space that is everlasting but the movement of time by the line ($\Pi\Pi^2$) that is everlasting. The **governing singularity** (Π^0) holds a **positional validity** (Π^3) of three dimensions Π^3 =($\Pi\Pi^2$) in terms of any **relevance** (Π) formed by the **controlling singularity** ($\Pi\Pi^2$) thus mathematically it equates to $\Pi^0 = \Pi^3 \div (\Pi\Pi^2)$. If a **relevance** ($\Pi$) did not validate a **positional validity** (Π^3) securing a **governing singularity** (Π^0) in terms of movement formed by **the gravity** (Π^2) that produces the **controlling singularity** ($\Pi\Pi^2$) in space, with a three dimensional status Π^3, then space (Π^3) would not be obtained and thereby the Universe would not be secured. That is why space-time is $\Pi^0 = \Pi^3 \div (\Pi\Pi^2)$. However this must be seen where it applies. It applies where singularity as time meets space, which means it applies at a point in the Universe where time still grows and that is at the position that predates the Big Bang. It is where material forms before material forms. It is where the visual will never come. It is where singularity Π^0 forms space Π^3 by singularity (Π) moving (Π^2). This means there is a time space delay validating a connection where any connection is (to us) absent. Stars in an axis dispute are not yanking each other around because mass is pulling one another because of some magical medieval territorial dispute. There is control over atomic control coming as a result of singularity charged by and charging movement. By virtue of a controlling singularity being effectively in place and this is in turn due to a governing singularity controlling the spin, this is done in much the same way as a spinning top stands erect while spinning it tries to surge into the air when spinning too fast or try to fight of the control the Earth takes on the top just before the top is grounded by movement deprivation coming from the Earth.

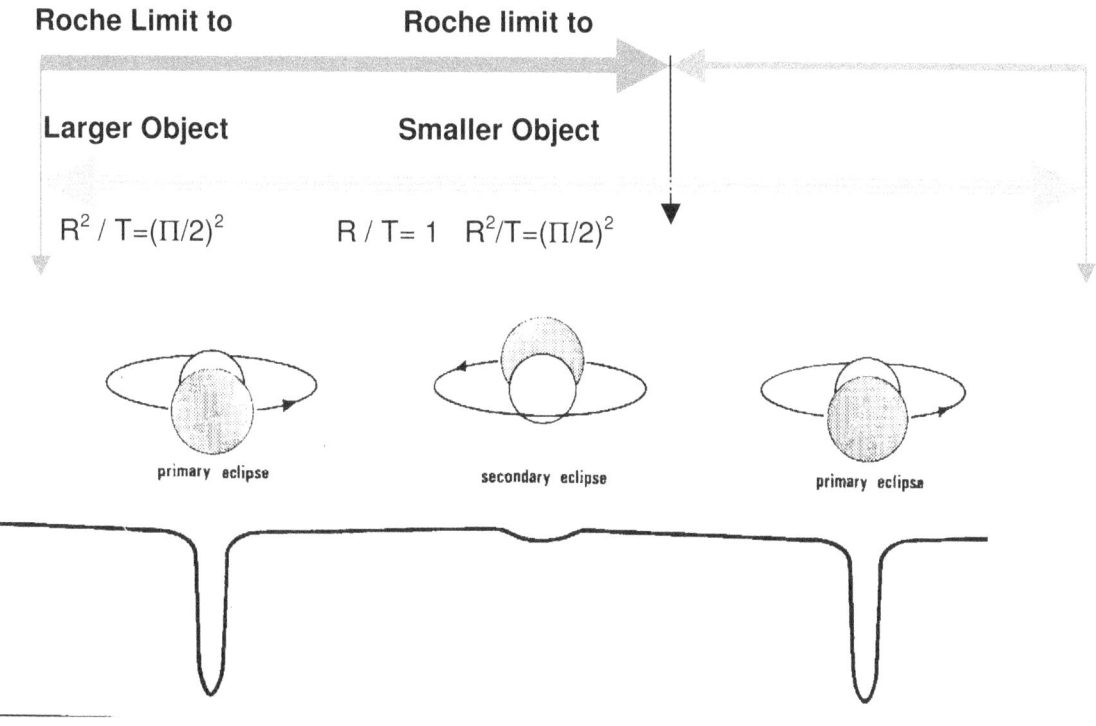

The Four Cosmic Pillars In terms of Singularity

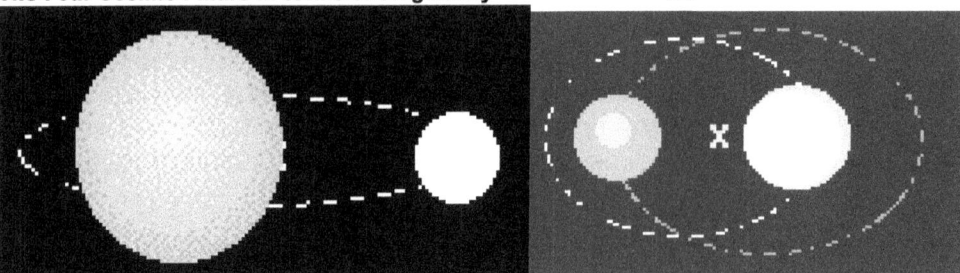

This is the reason why stars would form massive binaries, where they share a common combined circular displacement or a controlling singularity in terms of a governing singularity, separated only by each stars Roche limit with no linear value. As the Titius – Bode Principle comes into effect the linear displacement would once again grow, or the common spin value will be to grate for either one, or both, and their structural composition will collapse, forming smaller structures with less space to occupy the time in which they are. This proves that atomic movement shows loyalty to the control of a governing singularity that forms a president in movement and a governing singularity takes charge of the entire atomic movement. The governing singularity of the one star takes charge of the other stars movement and thereby the controlling singularity which is also taking charge of the mutual singularity and in that it takes charge of the individual atomic singularity as the star deprives the other star's atoms of any independence adhering to the atomic governing singularity in the star. It is this evidence by which we can gauge that singularity cross refers and establishes space-time worth in not only the star forming the mutual singularity but also the star of which it took charge of the controlling singularity.

Where the one governing singularity can't takes control of the other governing singularity a fight proceeds where movement is detrimental in which star takes control of the next star. This has nothing to do with Newtonian mass. The one star has atoms spinning around a governing singularity and all of the atoms by that forms a spin that vests a value into the star's centre singularity. Because the atoms are 1 and the centre is 1 the atoms is not only equal but the atoms' centre is the star's centre and

Important to note is in the case of Binary stars the two stars "lock out" space-time. By finding where it all began is equal to finding where the line began we have to trace the line in order to trace the development of the entire Universe. As seen the first development went beyond where mathematics may take us. The Universe did not become more but only focussed better on detail. What is was present because nothing can be new to the Universe that started out to be what it presently is.

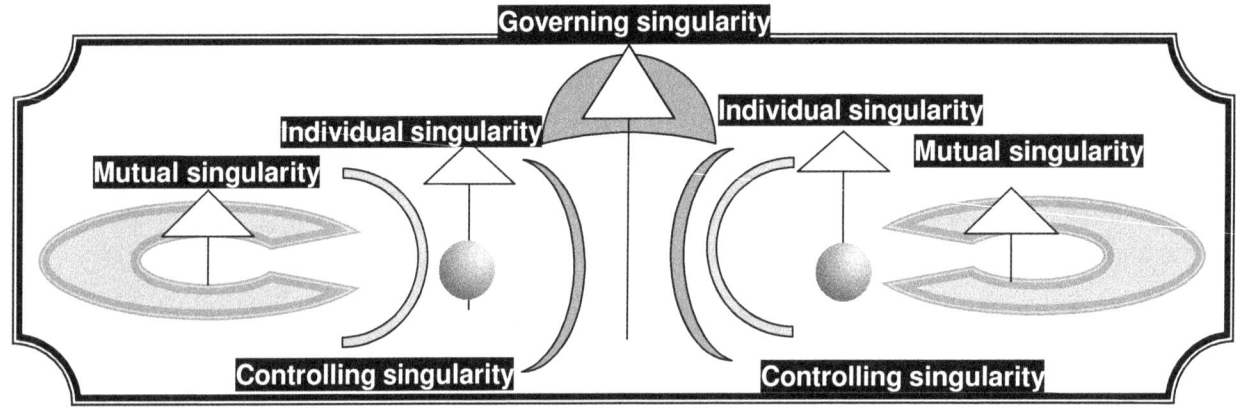

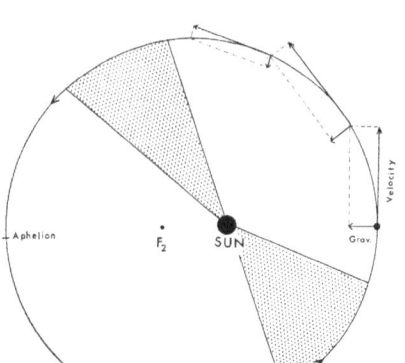

An object can rotate in outer space as long as it can maintain a speed that will keep the object rotating in that orbit. No line can go straight because part of all lines in the cosmos is the curve. There is always some relation between the factor of how much k influences or how much T^2 influences and the combined unit determines a^3.

That reason too has its footing in the Titius Bode law because 7/10 is half of 10/7 and by going seven the factor also has to go 10. The compliment of this is gravity at Π^2

One of the four most important values in the Universe is the Roche limit. On this and the Titius –Bode connecting to the Coanda effect by means of the Lagrangian points and on the inter connecting of the four principles rests the growth of the universe, as space relate to time. The sound barrier is the four cosmic principles applying to form the relevancy between the governing singularity and the controlling singularity. The Roche limit comes into effect when the linear displacement factor reaches a value of one and part of the circular displacement value. In this is the value $\Pi^3/\Pi^2 = \Pi$. When two stars are at the Roche limit, the linear displacement reaches a value of the lesser one, and excites in an electrical sense the singularity within the lesser one charging its atmosphere to extend the atmospheric level to that of the major one.

$$\Pi^3/\Pi^2 = \Pi$$

$(\Pi/2)^2$

$\Pi^3/\Pi^2 = \Pi$ The circular displacement reaches its full complement of half Π^2 which is the Roche limit. The Roche limit forms a centre by the value of the atoms spinning inside the structure and the atoms all foaming a governing singularity secures independence while the structure holds a unity by measure of the governing singularity that the star controls the atoms by.

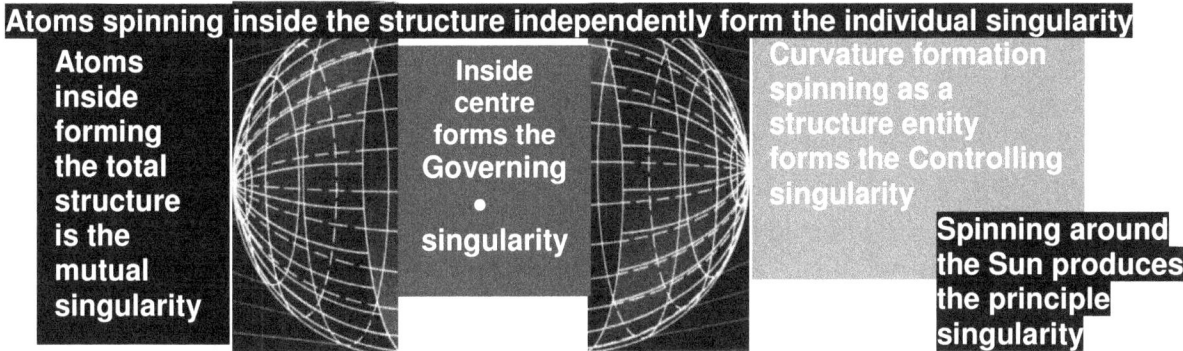

This indicates four factors forming singularity that absolutely dictates the cosmos in terms of movement. Holding that in mind, I therefore had to name the four positions that equally form singularity by dictating gravity. To argue this concept of singularity guiding movement, let's take the Sun that provides a centre k^0 for the Earth a^3 forming a centre where k points a line that forms the orbital circle T^2 wherefrom the edge of the line k is pointing at the position of whichever planet a^3 forms a circle T^2 in relation to a line coming from a centre of the Sun k^0. The line k indicates the distance from the Sun's centre to the planet that orbits and this forms the circle as the planet a^3 orbits T^2 around the Sun. The line k will provide a line from the Sun's centre k^0 and the line k will provide a spot where T^2 produces a circle holding space a^3 in a located position by running around the centre of the Sun k^0. In this view the space a^3 of the Earth rotates and in that forms the **controlling singularity** that holds the value as Π indicated by k forming between k and k^0 being singularity Π°. The Sun holds singularity in the centre, which is forming the **governing singularity** Π° and from that point the circle T^2 comes that forms the orbit Π^2. That means every single point that k indicates there are positions forming space a^3 implicating sides of a double dimension. In the same manner is k not limited to distance or is T^2 lesser by size. If Kepler said $a^3 = T^2 k$ then $k = a^3/T^2$ is also what Kepler said. There are three dimensions a^3 between any two points T^2 flowing as time from the centre of the Sun, which is indicated by the line k. However in the next scenario the Earth holds the **governing singularity** Π° running from the centre k^0 to k forming the edge while the circling rotation T^2 then forms the **controlling singularity** Π indicating the point in rotation. There are also two other points holding **the mutual singularity** and **the primary singularity**, both which I do not explain in this presentation but without which the four phenomena would not form gravity.

The value of k is not to be put in place as a measured value, but is there to bring a reference to the location of singularity $k^0 = a^3/(T^2 k)$ applying as to place a specific singularity in as the **governing singularity** and acknowledge the position of another singularity in place as the **controlling singularity** because there always has to be a **controlling singularity** determining the orbit while there has to be a **governing singularity** determining the spin of the body in relevance performing as the space a^3 in question in the formula $a^3 = T^2 k$ where in that formula k determines the relevance of k^0 as in $k^0 = a^3/(T^2 k)$. However, this burdens k forever with the responsibility of forming a line and a line is what places the Universe in place while the circle T^2 is forming the Universe a^3 at the same time. Every space a^3 in question puts singularity k^0 in position by the motion T^2 in relation k to the position allocated to k in the Universe a^3. Nothing in the Universe can move without moving straight k that is also going in a circle T^2 to form space a^3 in relation to a centre k^0 while in orbit around another centre k^0. In this point k^0 time forms space and space develops as the history of time running from k^0.

a^3 symbolises in a mathematical interpretation of implicating the three-dimensional space holding a specific centre in relation to another specific centre indicated by k that could apply to either centre points in question. This is always a straight-line **k** representing the position of the **controlling singularity** moving in a circle **T²**. The space forming **a³** is a **positional validity** of the space indicated by $k^0 = a^3 / (T^2 k)$.

T² is representing the circle that goes around the **governing singularity** k^0 or Π° that forms in relation to the line **k** pointing to the controlling singularity or Π in reference to the centre k^0. The space that forms holds the orbiting planet **a³** in direct circular contact with the space in relation to a very specific centre k^0 moving from point T_1 to T_2 that then forms Π² in relation to a precisely placed centre k^0. The circle coming about from **T²** is the **controlling singularity** Π, which is always a circle Π relating to the centre Π° that is positioned by the line **k** in relation to the centre k^0 and by forming a circle Π it holds reference to the **governing singularity** Π°. Where **the governing singularity** is the centre of a spinning object such as the Earth, the centre of every atom holds **mutual singularity** Π³ that collectively puts a mutual value of all the atoms' singularity as a combined equal to the **governing singularity** Π°. The solar system will provide a **primary singularity** Π³ = ΠΠ². The one would represent **T²** the other forms **k** that then produces the third singularity forming space **a³**.

k indicates **controlling singularity** from the centre k^0 ending at the line **k**. This line shows the location around which a planet circles. The specific value about the centre is most important because from the specific centre gravity indicates a positional worth. The line forming **k** is pointing the circle or the **governing singularity** formed from a line that ends at a circle **T²** running from the centre k^0 to where the space **a³** is indicated.

The turning **T²** of any circle holding space **a³** is valid only if forming a reference **k** to a centre k^0. $k^0 = a^3 / (T^2 k)$. This depicts a position the domineering singularity k^0 fills in relation to another point serving subordinate singularity **k**. There are always a dominant and a serving singularity interacting. If **k** indicates the centre of the Earth then **T²** rotates initiating the **governing singularity** k^0 where then the centre of the Sun **k** will form the **controlling singularity.** When the Sun rotates, the Sun's centre k^0 forms the **governing singularity** giving the Earth in orbit **k** holds the **controlling singularity**. The measure of **k** is not a specific value but serves only as an indicator to which space rotates or applies by the space rotating in a circle. This role of singularity being **controlling** or **governing** is playing part in movement of gravity forming and is very important when trying to understand the role that the four phenomena play in forming gravity. It is important to understand what happens in the event of an object going through the "sound barrier" or when escaping from the Earth's atmosphere. Where the object is standing still holding a position that allows the object to have mass, the object is part of the Earth while the Earth has the **governing singularity** and the Sun has the **controlling singularity**. As soon as any object moves on Earth, the movement switches singularity by allowing the object to obtain the **governing singularity** while the Earth then for fills the directional circular control in forming the **controlling singularity.** All four phenomena interacts in a manner forming this role where for instance in the solar system the Sun holds the **controlling singularity** and Milky Way forms the **governing singularity.** To find validity in my argument one must draw this statement of motion back to the point where singularity is getting sides or said mathematically Π° is going Π. Π is the **controlling singularity** and Π forming Π² is in relation to the **governing singularity** Π°. When there is singularity there can be no sides. The one forming singularity Π° by measure fills no space while form Π develops Π² into space. The space that even the dot fills being Πr° does not really exist in the manner we humans see space to exist. It is a spot that is there without being there. It does not visually exist because it is not filling any substance and it cannot be recognised since it is not three-dimensional. The spot and the dot have no dimensional worth of any measure but holds relevance. This Universe to which I am referring has never been uncovered. When I am addressing it in the sense of being single dimensional it has never been unveiled in that way by any one since this is the flat Universe one can observe by not observing it. . Einstein tried but failed miserably and I show how much Einstein's wavy Universe full of blocks, which was or is a failure. This Universe holds a line in time made up of dots and spots forming no space but holds a Universe relevant. The Universe has everything moving by a circle in a circle that moves straight. This applies to every atom and that is why the greater the gravity is applying in a star, the smaller that space is the star occupies and the more compact the atoms are in the star.

Let's pretend this is an atom moving in a circle as well as directional

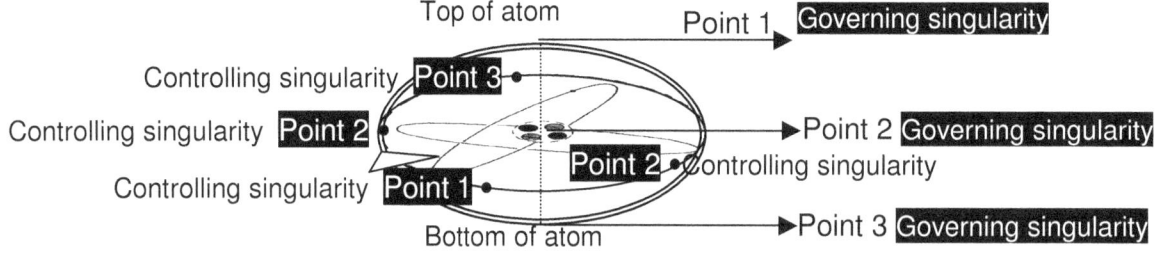

The governing singularity forms a line that never spins but when it moves it repositions the entire atom and therefore technically the governing singularity moves by replacing the atom in a lateral position. The three points placing the governing singularity movement determines the controlling singularity movement by altering the four time reference points, The governing singularity forms a line and time moves the line by repositioning the structure in a new allocation. When these lines move is then when Einstein said the Universe goes flat. When we look at the atom only in regard to the four points forming as movement by circling around the three points we find how movement influences space and then we can see how the Universe is "growing bigger" as these points change their reference to the line having three points while relocating.

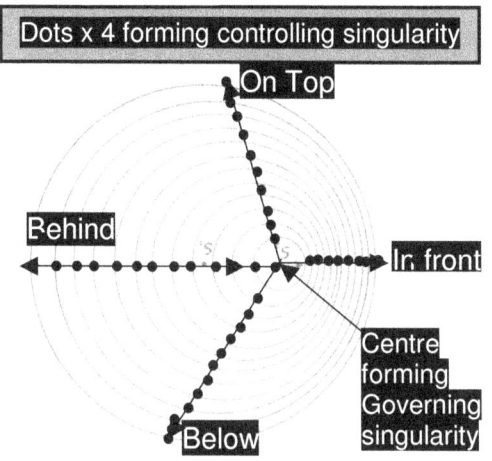

As gravity moves faster the reference points forming the circle $4\Pi^2$, which

The three points forming the line puts the atom or the object in a new reference to every other line forming an atom or object in the Universe. The circle constitutes of four points cross-referencing to one another and those four points form a circle. As the line of three forming the governing singularity repositions, it then puts a new reference to the four points of the controlling singularity forming the circle forming that has the responsibility to be forming the space factor of the atom.

The repositioning of the line is detrimental to the size that the atom can have in the speed that gravity uses to contract the object. The dots forming the controlling singularity by four has to reduce the points it holds and is subject to the faster the atom repositions in a new position in time. During the Big Bang the cosmos moved at the speed of light making space an electron and that made the material be the neutron. Time back then stood still in relation to how time now flies, but time is standing still now in relation to how time will fly say five trillion years from now. Time is giving reference to the space applying at the time of movement applying

holds the value of the controlling singularity will reduce in order to accommodate the repositioning of the governing singularity formed by the line with the three points. Proving this is the fact that the more intense gravity gets, the smaller the stars get and the more compact and dense does the star get by the increase in gravity or the slowing of time as the star moves back to what the Universe was at the point of the Big Bang.

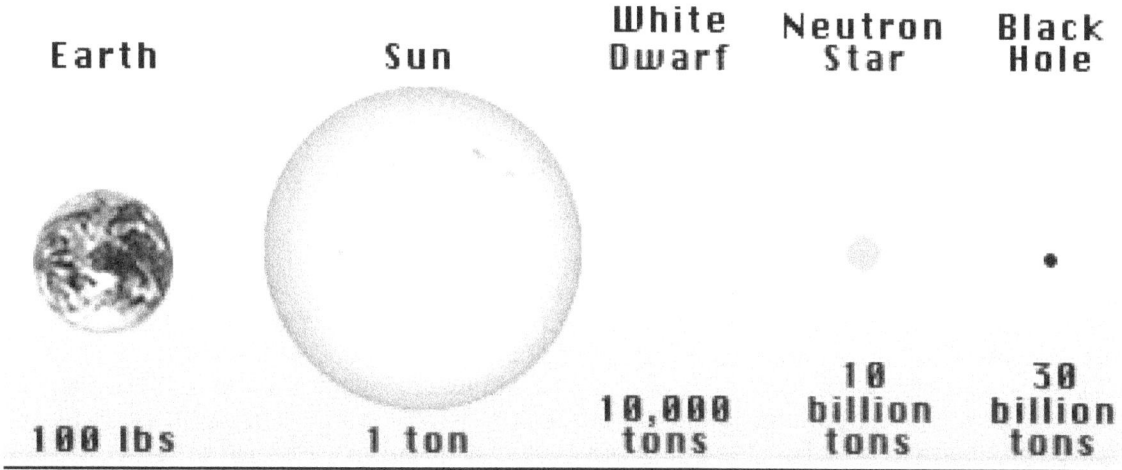

In every case scenario as gravity intensifies the atomic space becomes smaller as the density factor rises in every case and the mutual singularity becomes more compact because of a firmer controlling singularity being the product of a more intense governing singularity.

In determining this behaviour as part of a cosmic process where matter interact with matter in an laid down set of rules, once more we should be asking questions and this time it is whether the top will show the same behaviour in outer space as it does on Earth. With the reply of no it would not come as an admitting that the process involves the interacting of singularity of the Earth with the singularity of the top where the spinning created independent singularity, is as valid as that of the Earth because the Earth has a role in sustaining it or destroying it at the border ends.

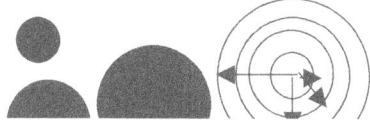

Looking at the affect of gravity it shows the precise quality of no distinctive point, as gravity never seems to end at a point but flows all over affecting all that holds a position in its sphere of influence. The gravity coming from China meets the gravity coming from America at no

Using the concept that gravity applies Π as the circle factor Π as well as Π^2 replacing r^2 the replacing by Π brings two values as Π and Π^2. That I found is the case with gravity and will be apparent when explaining the sound barrier as well as the Four Cosmic Pillars. In order to create a distinction I remained using r as the indicator of the cube or non-circle that has vacant space and by vacant space I refer to non-solid structures. In the solid structure I use Π as a value for reasons that will become apparent in due time.

This spot is the result of a most basic process of reduction as the Hubble constant is a most basic process of expanding during a matter of time. By reducing the line constantly the only value that will eventually remain without dispute from any party arguing about the facts is one followed by an exponential zero (1^0). By only having exponential zero instead of a numerical zero and a radius as one in the square (the radius effectively becomes one holding any and all sides on one point) such a point might become any value of any significant measure implicating anything but zero as the radius. By expanding the line, it will be an evenly spaced structure growing into the most perfect round dot ever possible anywhere at the point when it starts to grow.

The reducing of the line is one dimension in six and although such reducing is representative of two indicators all the other indicators must still be accounted for two. In mathematics there is a line being one quantity and the circle indicator Π being the next circle indicator. Reducing the line will erode the value of Π by ratio. That will eventually lead to having a circle ratio of Πr^2 and eventually lead to Πr^0 but that is not the point where the circle ends. That is where the ratio applying factor ends but it cannot exclude the circle. The circle as a concept can still reduce when it abolishes form to the single dimension. It is not the radius that is responsible for the circle but the figure value of pi and by abandoning π only then does all the aspects fall back into the single dimension.

The circle can reduce one step more when the circle eliminated r completely by returning r to a point of singularity r^0, but the elimination of r as the factor reduced the major factor to the single dimension in Π^0. That will not reduce the cosmos to zero, but it will only eliminate all potential lines r^0 to potential circles $\Pi^0 r^0$ and from there the circle Πr^0 will come about by manifesting as a line but that manifesting can firstly only establish a circle Πr^2.

The only value that singularity can have although the single dimension may host the entire Universe is Π^0. Pick a number and elevate it to the power of zero and in the process one may have established another point holding all points in singularity because that is the value of singularity. Only Π^0 or any other value holding one accompanied by zero as an exponential value can ever be the accurate value to singularity while singularity will then host the rest of all the possibilities in the Universe.

This means that the entire Universe composes of and is made up of singularity... this much I am going to prove. Every point occupied or otherwise constitutes of singularity either under control by movement in a form we call atoms or being passive in a location we call outer space. I wish to repeat the position holding singularity because if I introduce anything new, then this centre singularity is the pivot of everything that I introduce to science, and also I refer to the top because Newton used the top as an example by which Isaac Newton missed everything that Kepler clarifies about cosmology.

In the sketch to the right above the circle to the right would come about from a straight line r growing influencing the appreciation of Π, but to influence Π would lead to a breakdown in r as Π and r are different entities. The circles to the left (black dots growing in size) shows a continuous growth by extending Π every time and since Π is the same part as the previous Π, only extending that billionth of a millimetre each time, the circle will be truly continuous without any signs of a break.

$\Pi = r$

in constant directional change as time flows through rotation

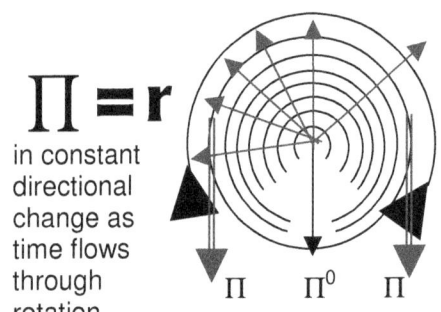

The new direction pointing to a new location in relation to the previous point will oppose the previous point it had in relation to direction considering the centre point.

Let's go back once more and reduce the line by half every time. Then repeat the process until it can repeat no more. The reducing of the line by half every time will get to a point where all the ends land on the same position without any possibility if halving the two ends further. The points share one position and moving the points in any direction will lead too an increase of the line once more.

Locating and finding Singularity

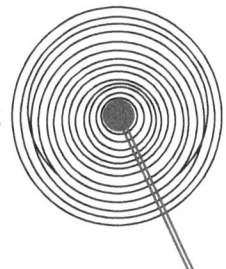

The entire Universe consists of lines running all over.

In the **precise middle $k^0 = a^3 \div T^2k$** of all **objects in rotation** is a precise centre dividing the object in sectors that will **start the spinning initiation** from that centre point. The object has to rotate $T^2 = a^3 \div k$ in order to instate the space $a^3 = T^2k$. Thus, the spinning object **will have a middle point k^0**, a very specific **centre point $k^0 = a^3 \div T^2k$ that does not spin** and only holds Π as a specific value. One value such a line **cannot have is zero** because **zero does not start any** line and therefore the **value of the line must be infinite**, just as described in **accordance** and by **the definition of singularity**

That point albeit hypothetical, is also as much a reality none the less and is placed where that point **must be standing still** because every line **running from that point** in **opposing directions** are also **in opposing directional spin the other or opposing side.**

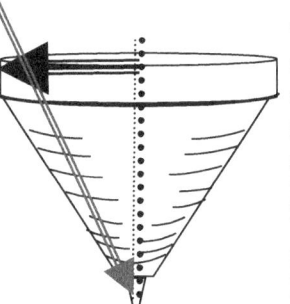

In considering the spinning motion in the fraction of time in the detailed instant every aspect of rotation will turn in every instant of change in time. Although the points had the same characteristics only seconds before, they oppose the characteristics it had just before and just after the very second in which they are and to which they relate by similar points also in rotation. The fact of the graph proves my point in quarterly opposing dimensions and values,

From this centre line that is only theoretical definable, but is still there all the same, an centralised line forms holding opposing values apart and parting the opposing values is what proves that this line forming has no status in space and yet controls all that holds space. The one side will turn left and by crossing this line holding no space, all the space will then turn right.

The parting of directional opposing space will always form what becomes real and distinct when rotating, and by rotating around this line that is only theoretical the spinning is putting this line not rotating even in more distinct prominence. Because of the rotating that evokes the presence of the line not being there, the influence this line holds grows so much it covers all the matter from end to end, to a securing the spinning to a point in the centre that holds no spin value.

It is there as well as not being there by not spinning because it has no space to spin. Yet the line is there and therefore it has the most original value anything can have. When not rotating, the line disappears and only a diameter runs across the material with the diameter going as thick as the material will go.

When rotation begins, the line then forms according to a radius. While the line forms where the radius starts, it shrinks back to a hypothetical position claiming zero spin that through that it is not less distinct but more distinct because from that point every rotating becomes a piece of what ever forms part of what is then spinning. Because it spins the end of the line forming will clearly carry the singularity value of Π^0 to end at Π that then is implicating rotation by the value of Π^2.

When looking at the cosmos from whichever angle it indicates the fact that the cosmos is moving. Everything in the cosmos is moving and all things about the cosmos are moving in relation to everything else in the cosmos. Everything is forever spinning in relation to a point that could be any point that is not spinning and everything is all going towards as much where it is coming from.

Everything is on the move and always encircling something of making that centre point to seem to be of greater importance than what everything is that is spinning. A top can spin but the parameters of its spin are limiting the motion it can apply. By not spinning the top is still spinning as the Earth is doing the spinning on its behalf.

The spinning top that Newton dismissed as $\frac{dJ}{dt} = 0$ brings all the evidence any one needs in order to come to a conclusion that will bring any proof that the singularity governing the top connects too everything anyway. Placing singularity is fair and fine, but what will the evidence be in proving its activeness as part of the creation at large? The reason why we can be sure it is active is that when spinning it shows borders implicating restraining of further movements outside the set limits. By going faster (past the upward border) the spin goes oblong where it actively tries to change the position the top holds to the Earth in relation to the surface of the Earth. By going too slow it once again shows identical characteristics. When going too fast it indicates an attempt to rise into the air, therefore relieve its singularity in an effort to part with the Earth's singularity. It shows unmistakable characteristics of trying to become airborne securing an independent position from the Earth, which holds it down.

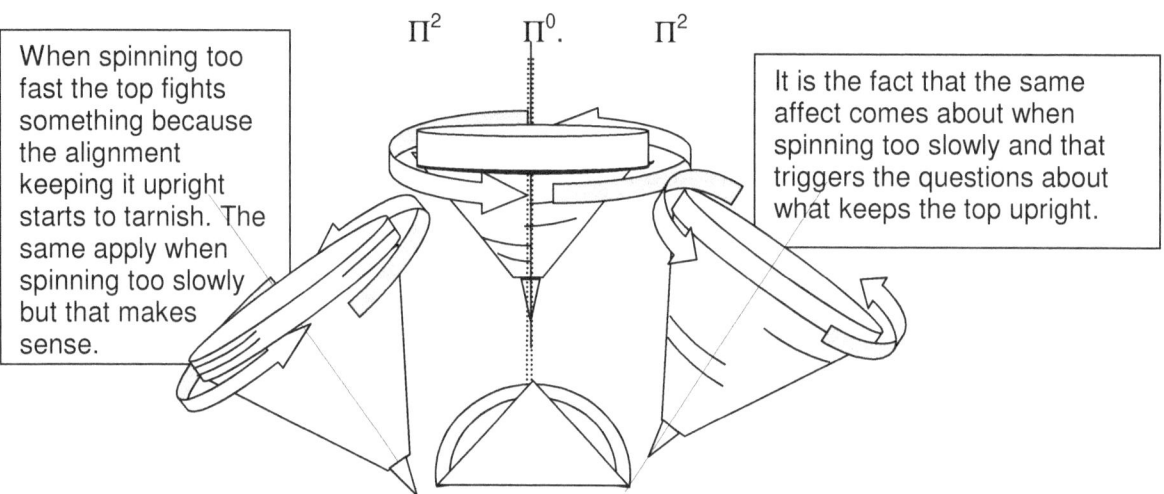

Pinpoint positioning of singularity Π^0 with Π positioning space to either side forming the border set by singularity

$\Pi^2 \quad \Pi^0. \quad \Pi^2$

When spinning too fast the top fights something because the alignment keeping it upright starts to tarnish. The same apply when spinning too slowly but that makes sense.

It is the fact that the same affect comes about when spinning too slowly and that triggers the questions about what keeps the top upright.

The rising above the position the Earth holds the top is clearly indicating the top is trying to generate more gravity as what the mass would be by which the Earth restricts the top and hold it in a position where it will form mass and then become part of the material forming the Earth. When it is at the bottom we surmise correctly that it wishes to topple over and fall down, but something drives the top to put up a fight in order to stay upright. If the top falls down, this action of falling down will kill the centre line holding singularity and it is that centre line the top holds that is keeping the top erect. By destroying the line it will then enforce stopping the independent spinning of the top. Of course the bottoming out shows the same characteristics whereby we gauge that to be the normal process of falling down. If the bottoming is relative to the Earth's singularity and we recognise the process as normal, then the top of the limits should be just as recognisable normal.

When any object is in a state of having mass the object has to be standing still and being secured in a position on the Earth at that point of having mass. The object has to be in a position of absolute rest while it is on the Earth. At a point of standing still in relation to the Earth while excepting only the movement the Earth allows any object to form mass and it is where at that point that the object with mass is resting while all the rotational movement is equal to the movement the Earth delivers where the Earth is rotating. Rotating at the speed the Earth dictates form the factor science call mass. When the object leaves the surface of the Earth such an object will have to move much faster than the Earth moves or have less density than is required to maintain a steady position on the Earth. When any object is standing still being in a state of having mass on the surface of the Earth, an object has micro gravity because the individual gravity left to the object in mass is infinitive small and is left to become an indication of attempting further movement towards the centre of the Earth while the Earth's material blocks the micro gravity to move and hence apply mass in doing so. Mass is not something inherent of the object but is the annexing of the object given mass by the Earth to secure the position of the object to ensure the object becomes part of the Earth structure. Having micro mass (not micro gravity) is where the body in rotational movement extends beyond the limit at the point where the Earth surface would award a mass factor. The movement speed goes beyond the speed required by the Earth at the Earth limit where rotation velocity secures mass as a

The Four Cosmic Pillars In terms of Singularity

factor. By exceeding the rotational velocity at a higher rate, such movement would exceed the movement or gravity of the Earth that is required in order to grant a mass value.

When the top is spinning it is this line that urges the top to excel from the limitation of the gravity of the Earth and extend up into the air and away from the ground. It is this centre line holding singularity that drives the top to lift up from the ground and fight the mass that the Earth inflicts as to retain the top with the limitation of enforcing the top into a state of mass where the mass holds the top onto the ground. The top is fighting the Earth's effort in restraining the top with mass by producing gravity that lifts the top into the air. It is the top's spinning that is producing anti gravity to fight off the gravity of the Earth. I have heard so many scientists refer to man discovering anti gravity as if such a discovery of a force of anti gravity will give humans the power only God can have. It is this mindset that I refer to as science wanting the marvellous, the magical and the unexplained. To the masterminds of science having anti gravity would come to the same as unlocking all the witches' forces and opening the Pandora's Box of forces while anti gravity is simply jumping in the air. If gravity is what is pulling you down, then anti gravity must be something lifting you up.

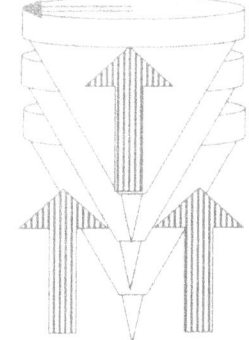

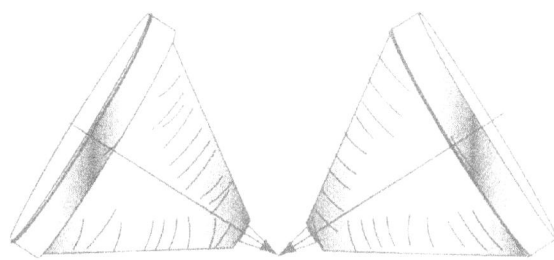

When the spinning has died down so much it will arrive at a point where the gravity of the Earth will reduce the spin of the top to lying still while the Earth secures the top with going into having a state of mass or having no independent movement which is having mass. It is at that point that the centre line within the top that is securing gravity by the spinning of the top will seize to be and the top will once more come to rest in a state of mass. The gravity of the Earth is fighting the gravity of the top which is equal to the singularity of the Earth is fighting to destroy the singularity of the top which is the movement of the Earth is fighting to destroy the movement of the top and all of these relevancies are all the same.

This issue is of cardinal importance and could deliberately be altered to hide the misinterpretation science wishes to connect to mass in order to hide the fact that mass does not bring on gravity but it is gravity that brings on mass. Mass is achieved when the object is resting motionless on the surface of the Earth while it is gravity that is still attempting to obtain movement as to try and move the object down to the centre of the Earth. This movement consists of two parts where one part is following the curve of the Earth while the Earth is rotating and the other part of the same movement is the thrusting of the object educing the object to move to the centre of the Earth. Mass is the result of gravity and not the other way around. Gravity brings on mass and mass depends on gravity to have any value or function.

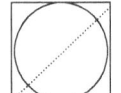 **The value of singularity stems directly from the law of Pythagoras** or **Pythagoras** is the result of **the average of singularity. With the shortest line being a dot, all lines must start from a position implicating Π.**

A circle is a square without corners implementing Π and a half circle is therefore a triangle without corners. The corners are the factor that confused every one in the past. When replacing the value we normally attach to circle being r with Π, the law of Pythagoras becomes quite meaningful and mathematical.

By placing a connecting circle on the sides of the triangle half a circle forms. By implicating Π as a relevancy and not the straight-line r, two values of Π applies to each circle and the straight line is no longer r, but is $Π^2$. This will bring about that each circle holds half the square value implicated to the allocated conditions applying to Π in that specific instance. By adding the two half squares forming the two half circles and then calculating the square root of the total that then forms the average diameter, an average of Π in the connecting line will come about. As both lines are the straight line forming singularity coming from one line being Π, the connecting line then must be the average of the two lines as $Π^2$. That is what **the law of Pythagoras says. Gravity is the result of the Earth spinning around its axis as well as around the axis of the Sun and the dimensional change implicates the law of Pythagoras.**

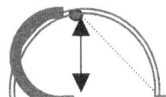 Because every moving line represents one quarter of the sphere in relation to the rest of the sphere and the line also indicate the relevant position between the point indicated and the point in the centre it is a relevancy of singularity in progress. By connecting the

Gravity comes about as a result of the Earth turning in space and with that it pulls objects from space towards and onto the Earth. This is done by the duplication of the law of Pythagoras.

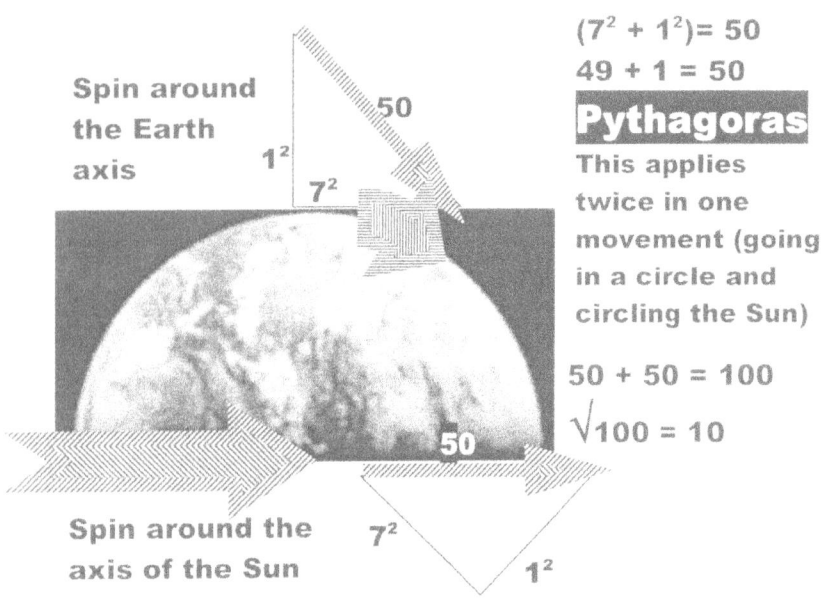

$(7^2 + 1^2) = 50$
$49 + 1 = 50$

Pythagoras

This applies twice in one movement (going in a circle and circling the Sun)

$50 + 50 = 100$
$\sqrt{100} = 10$

When the object moves while being in space or in contact (in relevance) with the spinning Earth, the object wishes to continue moving straight ahead while the Earth also moves straight ahead by turning 7°. Therefore, the Earth by spinning is falling away by turning 7°. That clears space or compresses space by the margin of 7° declining (compressing) of air / space. The Earth spins around its axis by 7° and turns around the Sun by 7°.

The Earth is moving, constantly spinning and in this is contracting space by compression (we call this contracting of space in air the atmosphere) and while the air is getting more compact, it takes whatever is filling with space towards the Earth constantly at a rate of 7°. By the Earth rotating, it is compressing space and with space compressing it is moving objects in the direction of the Earth. That is why objects that is falling, has no mass and only the stupidity of the simple Newtonian mind will force scholars to accept that it is mass that is pulling gravity.

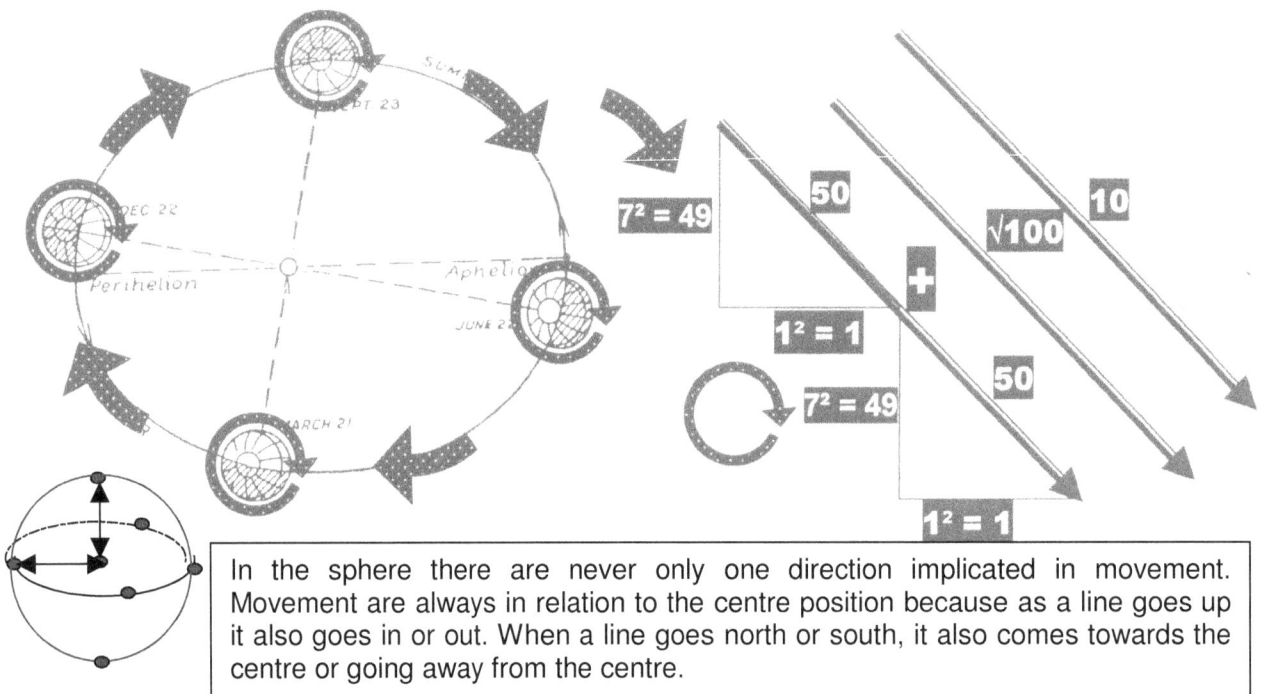

In the sphere there are never only one direction implicated in movement. Movement are always in relation to the centre position because as a line goes up it also goes in or out. When a line goes north or south, it also comes towards the centre or going away from the centre.

There is always relevancy present in movement. As this moving indicates direction it also applies Π^2 for indicating value forming the time factor.

In the sphere there is no radius but only the extending of Π from the centre Π^0 going in six opposing directions relating to one another by the square but remaining Π because of the unity the matter holds in relating to space. Every opposing point unifies by the joint forming of Π. If we wish to bring this reasoning into the cosmos where atomic locations apply it is not possible to draw a precise line that would form a precise ring and not cut some atoms in parts.

The Four Cosmic Pillars In terms of Singularity

Looking for gravity applying there will always be an atom disallowing the precise positioning of the circle, but with gravity the circle continues on a solid basis holding Π as a positional reference and not r. Using gravity one has to apply $Πr^0$ to realise the effect of gravity or Π and not r. In every sphere there then are the seven points confirming Π relating in precise dimensional and positional equality forming equilibrium to the centre $Π^0$ as well as to one another by 90^0 and 180^0 implicating the dimensional positioning. Therefore the sphere holds $_7 // ^Π$ and the cube holds $6 \times r^2$. This argument is very important when studying the Cosmic Code.

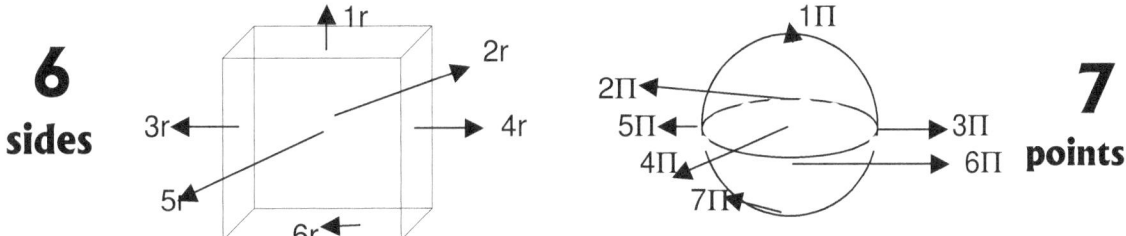

6 sides — **7 points**

Where space comes into contact with the sphere the cube loses one of the six dimensions it has to the more dominating seven dimension points forming a unit by employing Π in the case of the sphere whereby the seven dimensions in equilibrium by Π interacting will dominate the six dimensions or sides of the cube that are loosely connected by r. In this arrangement the domination of the sphere will always remove one point of the cube bringing about that the cube then has 5 sides to the seven of the cube.

This means that in the cube the "bottom falls out" and without a "bottom" to support objects they fall to Earth. Remember that a body "floats" in space, where the cube is not in contact with the Earth as a sphere but at one specific point all objects starts to "fall" to the Earth and it is at that point that the sphere comes into contact with the cube. The fact of gravity is that gravity is in place in relation to form holding reference where mass is not implicated.

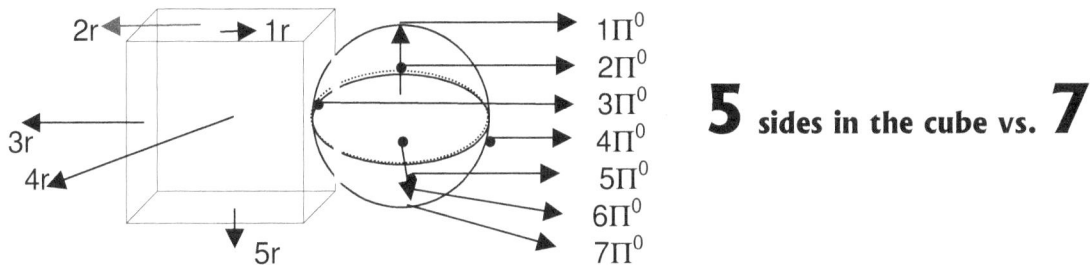

5 sides in the cube vs. 7

That too is the Lagrangian system with five cosmic structures holding relevancy to the centre structure where the centre structure stands in for seven positions diverting from singularity and the orbiting structures standing in for five positions in space.

Everything in the Universe that is spinning and that includes everything anybody can think of because being part of the Universe means spinning in the Universe uses two axis holding singularity around which it spins. The one spin defines T^2 while the other defines **k** and in that the definition is about the space a^3 that forms. Both spins are putting 7^0 in relation with singularity and since singularity is 1^0 it puts T^2 as 7^0 and **k** as 7^0 and that brings about $(7^2 + 1) + (7^2 + 1)$ implements the Pythagorean Theorem. This puts double seven by the square hypotenuse as a right triangle $(49 + 1) + (49 + 1) = 100$. With The Pythagorean Theorem the hypotenuse is 10 ($\sqrt{100} = 10$).

Around axis by 7^0

Around the Sun turning by 7^0

The following is not an in-depth investigation into the four cosmic pillars or the four cosmic phenomena but is a sketchy overview glancing very broadly at what forms the four phenomena. The phenomena use form and are the result of $Π^0Π$ forming space-time where the curvature of space-time results from Π manifesting as space-time. In short, it shows the truthfulness of the four phenomena resulting in gravity forming. The Titius Bode law put double 7 as being found in $(7^2 + 1) + (7^2 + 1)$ in relation to 10 as well as 10 in relation to 7.

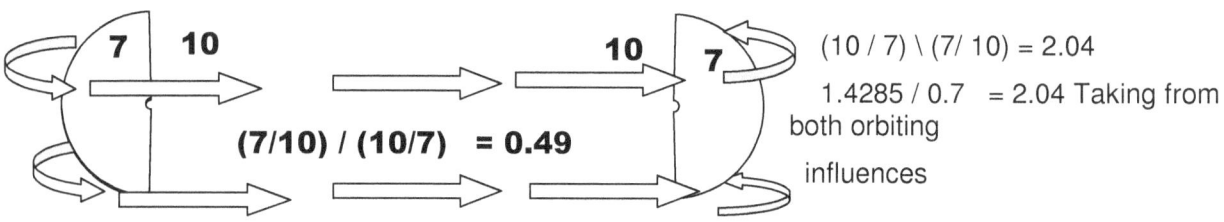

$(10/7) \backslash (7/10) = 2.04$

$1.4285 / 0.7 = 2.04$ Taking from both orbiting influences

SPACE DIVIDED INTO TIME

$(7/10) / (10/7) = 0.49$

$.7 / 1.4285 = 0.49$ Taking from both orbiting influences

THE PROCESS PARTED USING THE ROCHE PRINCIPLE

10 / 7	$(\Pi/2)^2$ The Roche influence on Titius Bode	**Crossing the singularity divide and activating the Roche principal $(\Pi^2/4)$**
7/10	$2.04 \times (\Pi/2)^2 = 5.033$	
$(\Pi/2)^2$	$2.04 \times (\Pi/2)^2 = \underline{5.033}$	
10 / 7	$5.033 + 5.033 = 10.066$ from both objects	

SPACE MULTIPLIED WITH TIME

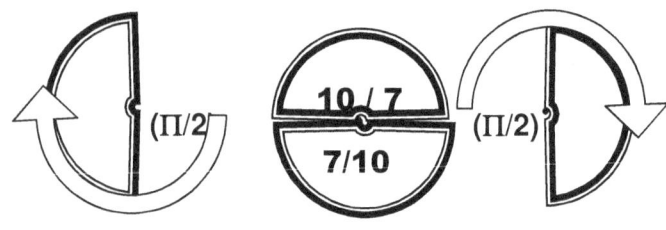

The crossing of the divide will implicate singularity on both sides of the divide bringing about the Roche factor

$10 / 7 (\Pi/2)^2$ The Roche influence on Titius Bode $7/10$ $2.04 \times (\Pi/2)^2$

$7/10 / 7/10 = 1$ and $10 / 7 \times 7/10 = 1$

Those factors all being equal to each other while space holds a value of 10 and material by movement holds a value of seven and therefore equal to one is not influencing change. The space that the motion establishes creates a relevancy of seven factors in space while the direction of motion involves another three dimensions or points, which is incorporating the other singularity in the unit. While the motion is at the same time moving out of ten points in relevancy and only occupying seven points the very opposite comes about through the same action being duplicated. The motion turns ten points to seven by moving from ten and filling seven points through the motion ending before the next cycle starts.

SPACE DIVIDES INTO TIME

On the one side of the Universe

$7/10 / 10 / 7 = 0.49$

on the other side of the Universe

$7/10 / 10 / 7 = 0.49$

And on both sides of the Universe

$.49 + .49 = .98$

$(10 / 7) \backslash (7/ 10) = 2.04$

$.98 \times 10.066 = 9.86 = \Pi^2$

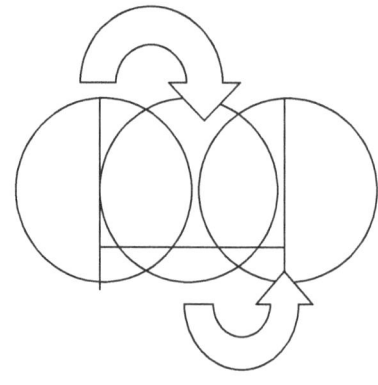

The value science use for gravity is 9.81, which they measured to much detail but the moon has a lot to do with the recorded difference coming about.

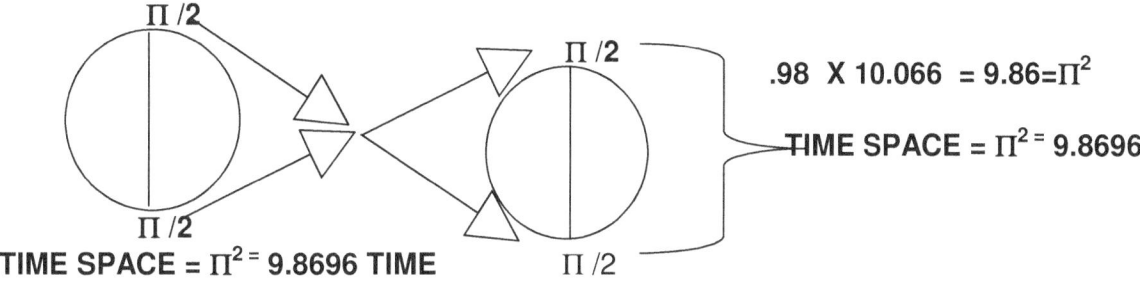

The TITIUS BODE Principle Outside the sphere

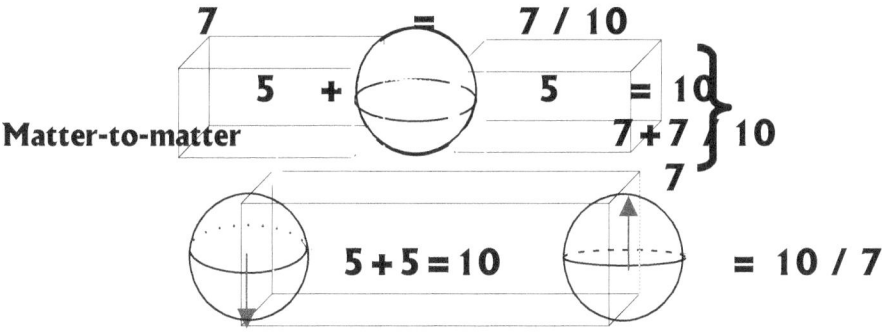

Matter-to-matter

With the dimensional change from space in the cube to space in the sphere a relation of 5 to 7 comes about depicting gravity. The principle of 5 sides in space relating to 7 in the sphere holding matter forms the basis of the Titius Bode and the Lagrangian principles.

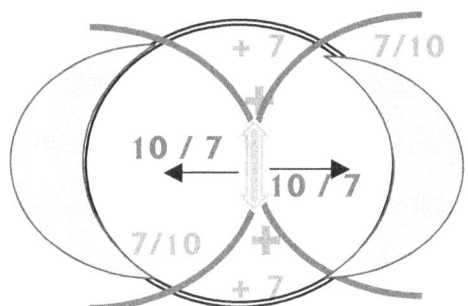

Gravity is motion in space and motion through space. Gravity is a relevancy between space travelling and the time it takes the space to travel. Gravity produces or reduces space during a certain period of synchronized spin of material in motion. Gravity is $a^3 = k\,T^2$ where it then becomes $k = a^3 / T^2$ In the light of this all other explaining fails the test of accuracy. It is no force because mass depends on gravity and gravity does not depend on mass.

The Titius Bode principle is a relation where space is the factor holding ten and material is the factor holding seven. By space being diminished by material, one relation comes about and where material dismisses space, another relation of seven to ten comes about. Gravity is the motion where space conservation is applied by motion control. By applying motion in sequence to space movement Universal harmony is installed. By applying motion at a faster pace than space conserve motion, the motion that is also gravity controls the space by motion duplicating space at an even rate. In that event gravity is applied in the manner we recognise the working of gravity and space.

The Titius Bode law is an extending dynamic deriving from the law of the gravity dimensional factor where the space factor in a square of ten relates to a matter factor in the square by half (half since nothing can be in two places in the Universe simultaneously) of the matter factor of $\pi^{0(7 + 7)}$ or the square of space (10) relate to the matter factor of 7. From such a point every other point will be opposing any other point not pointing in the direction to which the first point is pointing, whereby it extends the direction it holds. No matter what the point is or where the point leads, such a point holding a specific direction will be unique in the direction it is rotating because at that or any other specific point wherever, it will be directing not in the direction it spins but in the direction flowing from the centre point outwards.

The explanation concerning the four cosmic pillars is rooted in the understanding that cosmic growth lies at the heart of relevancy transfer of time or heat from singularity the liquid to the solid or in other words from space to the atom. The Hubble growth is what life would see as healing and aging or "the moon going further away from the Earth" or "the Earth growing bigger around its circumference". It is the atom that is presenting the solid that exceeds its size to the determent of liquid. I explain this process in other books such as **The Veracity of Gravity** and **an Open Letter on Gravity.**

The solar system grows at the heart of time and that is where space becomes three dimensional as it does where Π becomes $7/10\Pi^6/6 = 112.16$ and this code I explain in the **Cosmic Code**. The cosmos grows at Π before it expands by movement to $7/10\Pi^6/6 = 112.16$. The actual process of expanding and the way it involves Π is far to elaborate to explain in a short introduction book such as this.

Gravity is in place when space is in motion producing a duplication of lesser space than time will form as an ongoing sequence. In the other scenario overheating or space expansion or antigravity will come about when gravity cannot sustain space duplication in the preventing of duplicating to bring about reduction. Gravity I the movement of time in which space around material reduces due to the movement of such space filled with material.

With space holding material and remaining at an even duplication without adding heat such space not holding material will become lesser dense and such thinning or loss in density will also increase space as antigravity, which is the same thing as overheating, and space growth comes about. The duplication that motion provides produces cooling in the material sector and this explaining requires far too elaborate detail and that is left to books I have already mentioned.

The duplication that motion secures prevents or supports space in motion by forming harmony in frequency to the motion, which is in truth duplicating space. One side of space is the duplicating of the other side because time is eternal at singularity. It is the double motion applying that performs the gravity between the objects in rotation. But the way Kepler introduced it diverts drastically from the way Newton introduced gravity. The moving away in relation to the coming towards us has when motion applies forms the square that we see as the rotation. The rotation does not bring about an accomplishment of zero as Newton suggested but it brings about a square, which Kepler introduced.

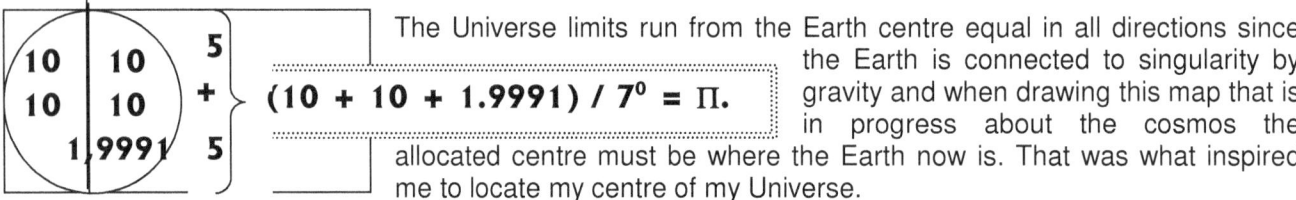

The Universe limits run from the Earth centre equal in all directions since the Earth is connected to singularity by gravity and when drawing this map that is in progress about the cosmos the allocated centre must be where the Earth now is. That was what inspired me to locate my centre of my Universe.

Even admitting to such a notion sounds like madness, but allow me please the opportunity to explain in more detail. I realised that my effort to locate the point holding singularity enabled me to backtrack the exploding Universe to its origins. By applying some basic effort I have located the position from where all movement came and the direction it took moving forward in time…and yes, even time as such. Gravity is the dimensional changing of space holding r as reference in the cube as to the sphere holding Π as the reference. In order to generate spin that is producing time in matter occupying space, therefore creating dimensional change, Π has to be a factor indicating the possibility of spin because by implementing Π the circle sides will follow one another without establishing separation.

As soon as motion takes gravity straight, singularity will reposition the direction changing the direction of motion by 7^0. It is this turning of motion by redirecting the continuing of motion that sets the critical time within the proton connecting to singularity. Instead of r being a line, gravity will inevitably be Π, which is the form value of singularity. That is this 7^0 redirecting in the square of space, which is ten on both sides of singularity and time is that what we find to be the Titius Bode law of $7/10$ and $10/7$ in relation to the Roche limit of $\Pi^2/4$ which is producing the gravity of Π^2.

However the reducing in it is going from ten that is on one side and is crossing over the figure of 1.9991, (which is singularity on both sides of the Universe) and coming into contact with another 10 while turning 7^0 that we find to form Π. In all being the total forming on both sides of the Universe it is $(10 + 10 + 1.9991)/7^0 = \Pi$. The answer must be in finding Π, and thereby locating singularity. If singularity is in affect the original point of the cosmos birth, the reducing path we should follow will indicate the whereabouts such a point must be. That is where cosmology diverts from mathematics.

The seven can never totally separate from the ten, but by singularity being the same but being on the other side it is withdrawing space-time altogether. See it as seven (let us think of that as the cold basis of space) spinning or turning in the ten (which then will represent the hot part in the cold basis) and the ten is part of the seven but the seven is not part of the ten.

The third factor is the axis around which hot as well as cold will turn. Therefore when reading the next page please envisage a cold base turning in a hot and cold space. The purpose of this is not to define whether the argument is correct or not but it is to help **the reader** gain understanding of the process principles involved. But motion also converts space to relate to space by changing relevancies through motion matter is in relation (part of) to the total dimension of space but is not the total dimension of space.

This is the prime element the state where everything started. It started at the time when mathematics was still to be invented by nature and only singularity was in place forming a value of Π and a reference of 180^0 in all directions in relations to other positions singularity established. It is at this point where everything other than **singularity was $\Pi^0\Pi$ becoming Π going on to be Π^2 through motion forming Π^3.**

One must take into account that gravity is different motion of particles forming a relevancy about duplicating space and dismissing space in relation to the effort of the particular and specific element. The motion differences in the motion between two particles bring about relevancies. It is a seven factor standing in relation in motion to a ten factor of which the seven then is included and part of the ten factor. The four time factors that I refer to as the four pillars are applying gravity as time is on the edge of the sphere in relation to the centre line forming singularity in the sphere.

Science (as usual) does no understand the true complexness in the phenomenon applying and then try to substitute ignorance wit fact they too do not understand. In science they mistakenly refer to the Titius Bode effect as the Doppler effect because Doppler found the sound of a train moves in circles which then forma a relation to the sound moving forwards and the sound moving to the back. The Doppler effect is the Titius Bode law and the sound that Doppler discovered works by principle on the very principal that is manifested in the configuration we witness as the Doppler effect. That which Doppler saw the day with the steam train producing sound waves was gravity carrying the sound in wavelengths and this method is a manifestation of the Titius Bode law. It uses the Titius Bode law but the movement of the train distorts the wave pattern of 10 / 7 somewhat.

Looking at the Titius Bode law from the way Newtonian science observes the law they don't even consider to be a law.

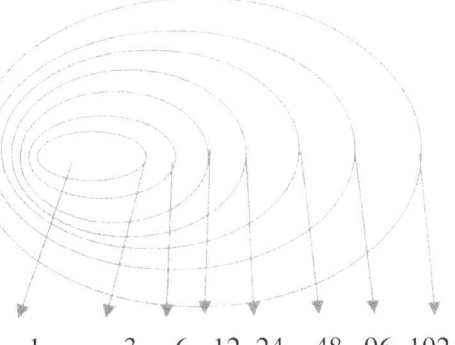

1   3   6   12   24   48   96   192

Planet	Mercury	Venus	Earth	Mars	Ceres	Jupiter	Saturn	Uranus
Bode's Law distance	4	7	10	16	28	52	100	196
Actual distance	3.9	7.2	10	15.2	28	52	95	192

Bode's Law:
A numerical sequence announced by J.E. Bode in 1772, which matches the distances from the Sun of the six planets then known. It is also known as the Titus-Bode law, as it was first pointed out by the German mathematician Johann Daniel Titius (1729-96) in 1766. It is formed from the sequence 0,3,6,12,24,48,96, and 192 by adding 4 to each number. The planets were seen to fit this sequence quite well – as did Uranus, discovered in 1781. However, Neptune and Pluto do not conform to the 'law'. Bode's Law stimulated the search for a planet orbiting between Mars and Jupiter that led to the discovery of the first asteroids. It is often said that the law has no theoretical basis, but it does show how orbital resonance can lead to commensurability. The importance that becomes known is the sequence the Titius – Bode law saw

in the number arrangement of 3; 6; 12; 24; 48; 96 etc. The incorrect application of the Titus Bode law lies in subtracting the figure of 3 from 10 leaving 7. The other way of reasoning is to add four each time to the firs value of three starting with 3 and so on. The true significance of the Titus-Bode law is that it points directly to a circular growth of 7 stages. The 7 relating to 10 is a precise derogative of the Roche limit or the Roche limit is a precise derogative of the Titius Bode principle because he two systems interlink.

Whenever any circle comes about the value of Π becomes an issue not to be ignored. In concerning Π it is most important to look at the factors forming Π and I am not going into that argument very intensely as I try to keep as simple as possible in this book where I introduce the concepts. Looking at gravity is implicating Π and therefore the factors forming Π have to bring about gravity. Notwithstanding any wizards' ideas about mass forming gravity, gravity is about maintaining circles and movement. That is maintainingΠ.

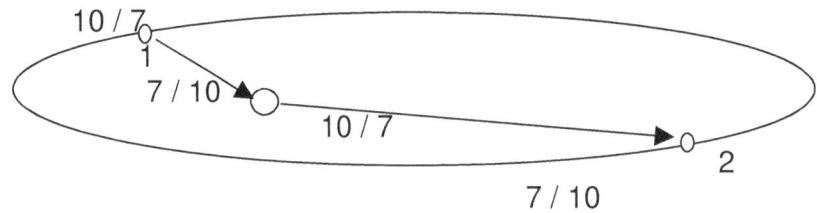

When a circle forms it implements Π as a form and therefore the factors forming Π has to charge the circle forming by the use of Π. In the value of Π we find 7 in relation to 10 and we have 10 relating to seven forming at the same token a value of ((7+7) and 10 / 7), which converts to $Π^2$. Whenever anything orbits another thing gravity comes as such a result and gravity comes about when 7 and 10 interacts. I have already shown that gravity is double 7 interacting with singularity and why that is the case. I have also shown how 10 come as a result from this by implementing the law of Pythagoras.

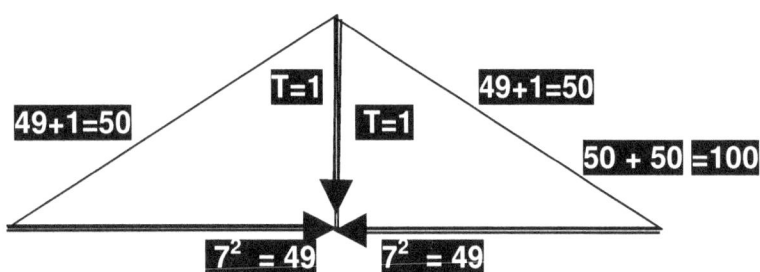

By establishing gravity we find the conduct of singularity interacting and therefore establishing the value of Π. However at the point Π is validated the Universe is still flat. The explaining of the following is a little bit extraordinary but so the understanding of any of the factors forming cosmology. By using singularity the use of singularity not only involves the equal measure of singularity but also in principle being singularity means there can't be two being equal. Being singularity means there is one point holding singularity where it is the same singularity because space is parting singularity is multiplying the very same singularity since singularity is 1^0. Therefore space can be multiplied but not as it involves singularity.

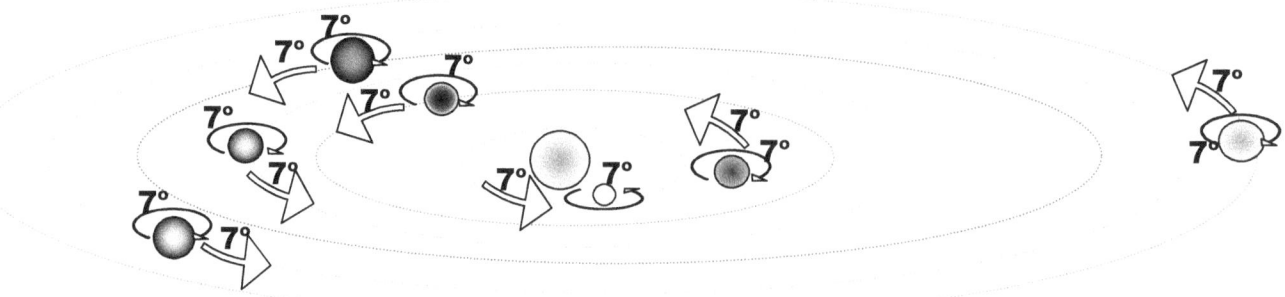

Multiplying one with one result in achieving one 1 x 1 = 1 and therefore one is not just equal to one but is the very same one 1 = 1 notwithstanding whatever space forms between the points representing the same singularity because the space forming is also singularity. By multiplying one with one result in achieving one 1 x 1 = 1 becomes the repeat of one (1 = 1) and not the duplication of one (1 x 1). Therefore by becoming Π the cosmos is still flat. It is when a flat Π interacts with another flat Π in relation to having the same singularity Π^0 that we find gravity Π^2. It is time taking the Universe to a single dimension while space will always be 3 dimensional.

On the side forming space we have 10 being the contribution space makes forming a double as the one planet is responsible for forming 10 and the other planet is also responsible for forming 10 as part of their contribution to space while the third structure stands still (Π^0) as to give time validity to form while time shows growth coming from 0.991 to 1 as the factor forming Π. This is space that establishes Π. This then stands in relation to the 7° of spin that forms the Earth's (or any other planet for that matter) gravity.

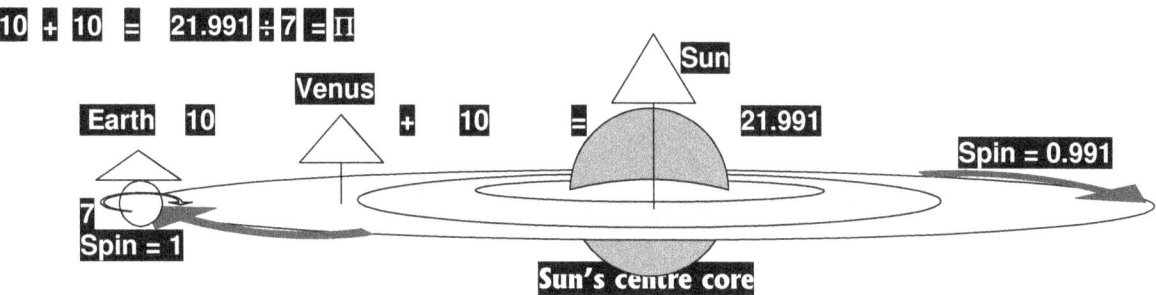

Then on the material side we have three (two planets and the Sun) all spinning which is redirecting movement to the tune of 7° and 7+7+7+0.991= 21.991. This again and once more stands in relation to the 7° of spin that forms the Earth's (or any other planet for that matter) gravity in relation to the atoms within the Earth. By having space forming Π and material spinning forming $\Pi^0\Pi$ the interaction of space-time then will become gravity Π^2. That is the one way that the relevancy of space-time takes place forming 3 dimensions. It must again be said that time $\Pi^0\Pi$ goes flat while space is moving $\Pi\Pi^2$ remains 3-dimensional.

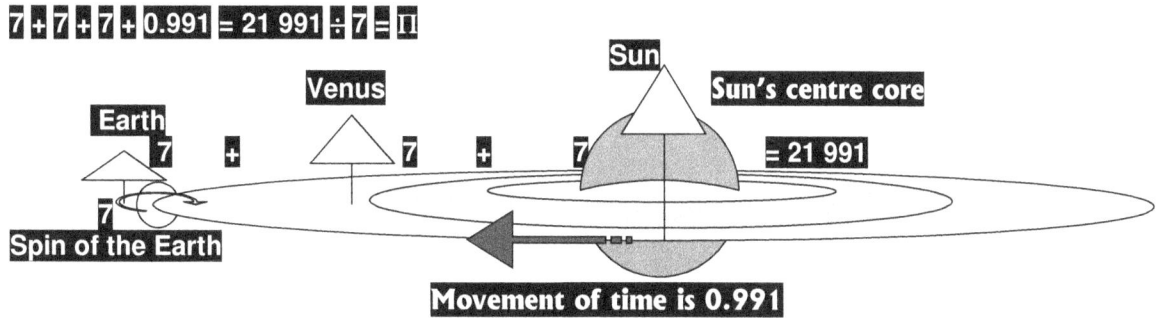

With every planet spinning in synchronised order the three forming the Titius Bode law will each contribute 7° of spin that adds to Π forming. In relation to that we have on the side forming space a double 10 forming in relation to 1 standing still and the movement coming about contributes the growth time contributes that is 0.991. With material forming Π (7+7+7+0.991=21.991) and space contributing (10+10+1+0.991=Π) we fins time forming $\Pi \times \Pi = \Pi^2$. The movement we see forming gravity is not in the one but is in seven duplicating by the square to form 49. The one can never move because the one is singularity and even if singularity goes square (move) it remains 1.

Looking at the overall picture we may find that 7 spins around 1 but since 7 spins around the same one it is spinning the same 7 that then creates a dual in 10 where ten serves as a value that forms space. The mathematical implication in the Titius Bode law works on the principle that we find the number seven doubling as every planet holds a governing singularity that works in tandem with a controlling singularity and the direction changes of the two factors always work in tandem. This is most impotent. One can never see one singularity bring motion about without the other principle also influencing the outcome. From every planet the value of the 7 applying will change and in every case another double seven will be in place since every planet is a Universe apart from the rest of the Universe . In the picture forming space-time by explaining the Titius Bode planet layout we have 7 by the double as well as the planet holding its own position according to the Sun which is 10.

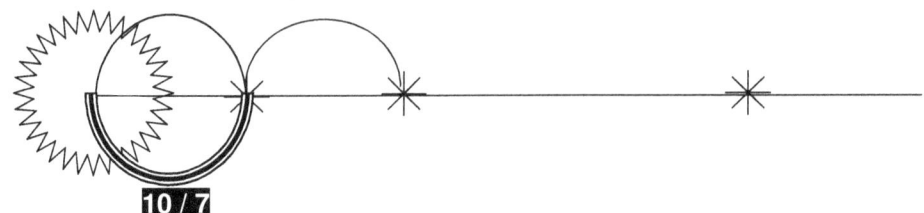

The reason why Mercury has such an "abnormal" orbiting route will be the absence of a double 7 guide.

Any changers occurring in Π will lead to a an unequal triangle providing two different values to r and will alternate the link between r and Π^2 bringing about different form (Π) and time (Π^2). When singularity forming the lines of the triangle is not in equilibrium the triangle will destroy the matching of half circle.

The spherical positioning layout forming the Titius Bode Principle

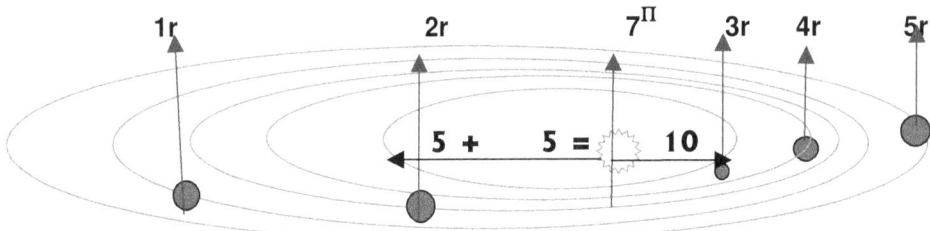

From the matter-to-matter relation in the Titius Bode configuration there are 7 / 10 + 7 / 10 = .7 + .7 = 1.4
From the space-to-matter relation in the Titius Bode configuration there is 10 / 7 =1.42
The 5+5=10 is a position of dimensions as space loses value to singularity. The 7 that matter diverts in points from singularity may seem as coincidental but is valid. Still in accordance to our perception valuing the number in degrees, it seems coincidental but if it is coincidental, it is nevertheless a figure of diverting proven as accountable in all other calculations and plays a most dynamic role.

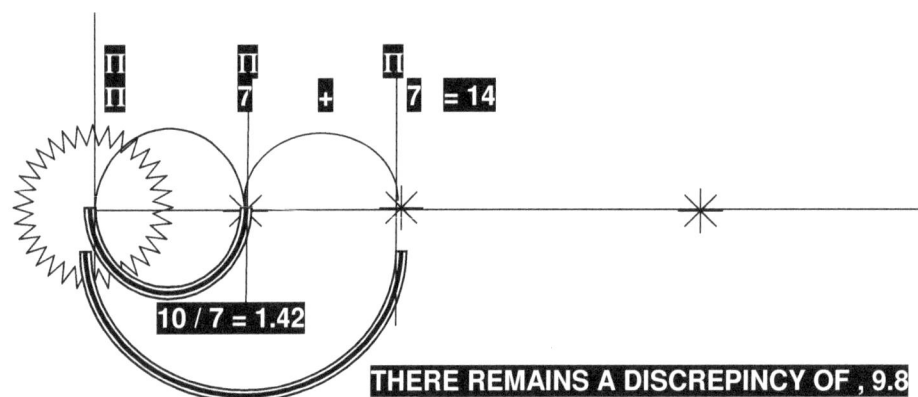

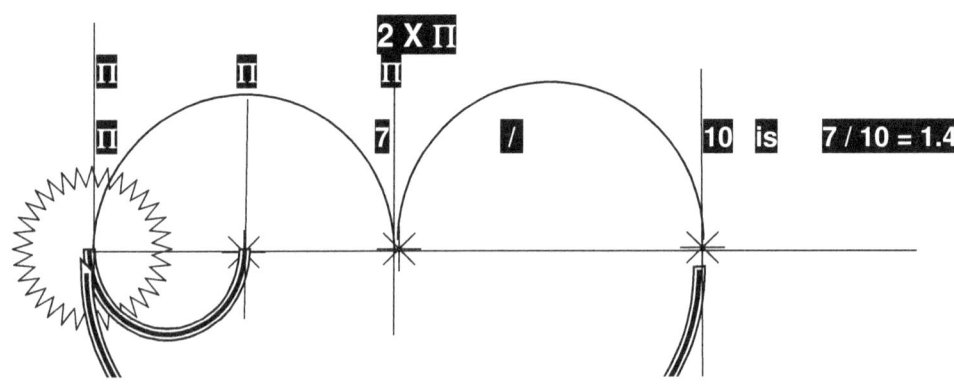

Π^2 is the relation between 7 + 7 = 14 and 10 / 7 = 1.42

The normal application of the Titius Bode Law provides an interaction between the orbiting object going into gravity by spinning in a double seven synchronized manner and into y then forming a compiling value of 10 which is the square of a hundred which is the double fifty which is the result the hypotenuse is the law of cosines which is mathematically $a^2 + b^2 = c^2$ or **(7² + 1²) + (7² + 1²) = 10²**.

It is the manner in which time builds space by spinning those implements Π. It is Π forming space –time by (10 + 10 + singularity 0.991 expanding to 1 being in relation to gravity @ 7) It is Π developing space-time.

In order to try and simplify the explaining there is anther way of expressing the Titius Bode principle and that might be as follows:

Let's say the Earth will be at a larger diameter horizontally (10/7 = 1.42) than at the vertical diameter (7+7)/10 = 1.4. This proposal is that the atmosphere (space-time) at the Arctic is at a value of 7 to 10 and at the equator it is at a value of 10 to 7. The equator holds a much higher relation to heat unoccupied than does the Arctic region and this will allow the atmosphere to contain much less density than the density applying at the equator. This will favour a space value of 10 in the equator regions while the double 7 will represent the governing singularity line running from top to bottom. This puts the line in relation to the circle forming by spin.

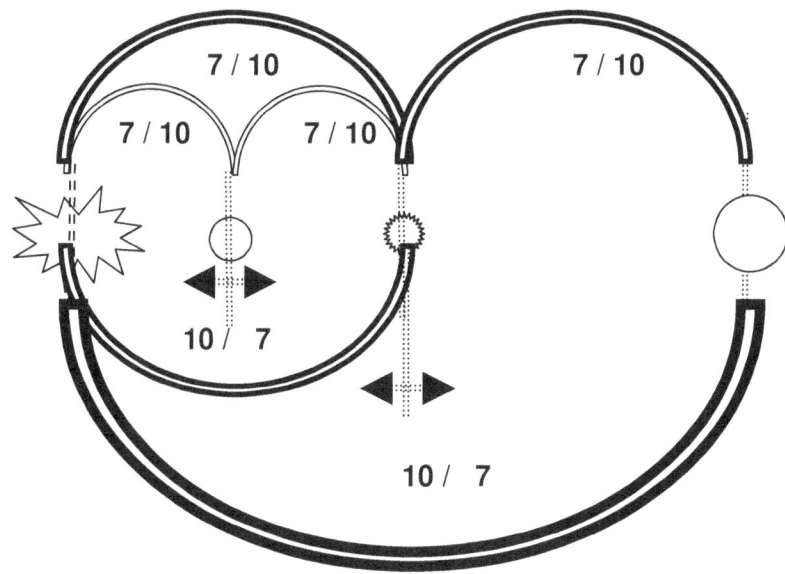

The Titius bode Law is holding 7/10 and 10 / 7 as it extends its influence much wider, but the influence of the extending as far as the Earth is concerned is merely a drop in a bucket to the effect it holds on gravity through the Sun's influence. That is how the Universe's expanding works. Little by little time moves onto become space at the point where singularity introduces time that then becomes space. One must acknowledge that space is the history of time and time leaves space as a reminder of time that has gone by. This is how space-time builds and that is why we detect this as the building form we see in the solar system.

To explain the Titius bode principle in detail one must once again return to singularity and understanding that is as simple as it is totally uncomprehending.

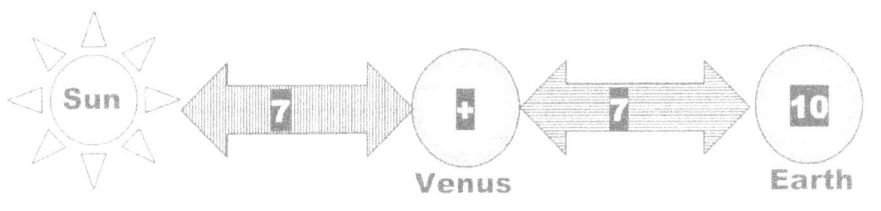

Now we look at the Titius Bode law from the way the cosmos applies the law

Fact 1: The importance that becomes known is the sequence the Titius – Bode law saw in the number arrangement of 3; 6; 12; 24; 48; 96 etc. This goes beyond doubt. That puts Venus at 7 and that puts the distance between the Earth and Venus also at 7 because the distance always doubles according to the Titius Bode law

In this we find the value of space-time compiling. We have 7 as a distance from the Sun to Venus and 7 doubling the distance from the Earth to Venus while the Earth forms the time value in space of 10.

The space between the spheres divide in half, but because of the extending of Π and not applying r as ordinary mathematics will suggest where Π replaces r the singularity extending from Π^0 will be half of Π in the square of Π and that is $= (\Pi/2)^2 = $ **2.4674.**

As I am about to show mathematically how 7 relating to 10 by the same action where 10 relates to 7, this relation forms Π^2 this happens in a double spin. It is an atom that forms the star. The atom spins around its axis. That is one Π^2. Also the atom spins around the star's livers another Π^2.

The lines converging from singularity holds a square to one another and that implicates the oldest mathematical principle that I know of, the law of Pythagoras.

Again we find the presence of a triangle holding a square. This holds space away from matter and therefore we are calculating the square of space depicting singularity and time (always in a square) away from the immediate claim on space by matter.

Matter in relation (part of) with the total dimension of space.

$$\left(\frac{10}{7} \div \frac{7}{10}\right) = 2.04$$

$$\frac{1.4285}{0.7} = 2.04 \quad \text{Taking from both orbiting influences}$$

SPACE DIVIDED INTO TIME

$$\left(\frac{7}{10}\right) \div \left(\frac{10}{7}\right) = 0.49$$

$$\frac{0.7}{1.4285} = 0.49 \quad \text{Taking from both orbiting influences}$$

SPACE MULTIPLIED WITH TIME

$$\frac{7}{10} \div \frac{7}{10} = 1 \quad \text{and} \quad \frac{10}{7} \times \frac{7}{10} = 1 \quad \text{Therefore not influencing change}$$

THE PROCESS PARTED USING THE ROCHE PRINCIPLE

$$\frac{\frac{10}{7}}{\frac{7}{10}}$$

$$\left(\frac{\Pi}{2}\right)^2 \text{ The Roche influence on Titius Bode}$$

$$2.04 \times \left(\frac{\Pi}{2}\right)^2 = 5.033$$

$$\frac{\left(\frac{\Pi}{2}\right)^2}{\frac{10}{7}}$$

$$2.04 \times \left(\frac{\Pi}{2}\right)^2 = 5.033$$

$$5.033 + 5.033 = 10.066 \quad \text{from both objects}$$

SPACE DIVIDED INTO TIME

$$\frac{7}{10}$$

$$\left(\frac{7}{10}\right) \div \left(\frac{10}{7}\right) = 0.49$$

$$\left(\frac{10}{7} \div \frac{7}{10}\right) = .49 \quad \left(\frac{\frac{10}{7}}{\frac{10}{7}} \div \frac{7}{10}\right) = .49$$

$$.49 + .49 = .98$$
$$.98 \times 10.066 = 0.8 = \Pi^2$$
$$\text{TIME SPACE} = \Pi^2 = 9.8696$$

TIME SPACE = Π^2 = 9.8696 = Space and time in a dimensional implication

A STRAIGHT LINE, TRIANGLE AND HALF A CIRCLE WILL ALWAYS HAVE EQUALITY IN DIMENSIONAL CAPACITY PROVIDING EQUILBRIUM BEING 180° BECAUSE EACH ONE SHARES A COMMON DINOMINATOR IN SINGULARITY TO THE VALUE OF Π. As the straight line averts a zero it holds another straight line in place to set about such an averting where the two lines will always carry a relevancy in elation to progress (the triangle) and a common denominator in the start from singularity. This concept we apply as the graph or the vector.

When decreasing the radius to a point where such decreasing will bring about going across the circle centre to the very other side of the centre of the circle and while doing so would bring about that this will lead to the entering a zone with no space where entering this area would consuetude in immediate landing on the opposing side where the spin then will be in an opposing direction. At such a point where there is no further space the decrease of the radius will no longer affect the value of pi. It is only from a straight line r growing that such growth will influence the appreciation of Π, but to influence Π would lead to a increase in r as Π and r are different entities. Therefore when only Π is left and r^0 has gone singular then no space is left because Π only holds form and only form is what then is left. Looking at the affect of gravity it shows the precise quality of no distinctive point, as gravity never seems to end at a point but flows all over affecting all that holds a position in its sphere of influence. The gravity coming from China meets the gravity coming from America at no particular spot but intermingles without distinction.

By tracing the line back to where the circle is no more, the reducing of the straight-line will uncover singularity plus one dimension valuing Π. But while the entire centre forming singularity is still locatable within the Universe we have it is not holding any dimension we may recognise. Reducing the radius r from all angles possible throughout the circle will bring about that all possible direction will eventually land on the very same spot with no more dividing possible. Yet zero cannot be a factor since the sides still hold value, in as much as holding all the value there can rise from such a spot. This is arriving at a point where more reducing will land the one side on the opposite side of the line but it will still not equate to zero.

What this argument further proves is that the circle reducing must then come from all points because the radius might be a line but that line represents a circle through 360° coming from and accounting for all possible directions. Taking that into account, it is important to recognise that notwithstanding the size of a line, which any radius of any size is, there is another line (or dot) eternally bigger as well as eternally smaller than the line in question. While we are in the third dimension being part of the third dimension such being in the third dimension then allows that all parts of the third dimension forever can be divided once more until the line in the third dimension is no longer part of the third dimension. When such a line leaves the third dimension it is still dividable because it might not be part of our dimension any more but it can still reduce further as part of the second dimension. By moving from Π^0 to Π, such movement constitutes to forming the second dimension by forming a square of Π or Π^2. By that time it has left our scope by miles but that does not mean that it ends there because from our perspective that is where it ends. But our perspective does not represent reality. Yet, even then it can still reduce infinitely more from Π to Π^2. By such reduction it forms part of the second dimension and then at last when going Π^0 it forms part of the first dimension. The Universe is Π by never being Π because it will be Π^0 and by moving from Π^0 to Π the movement will bring about Π^2.

However in reducing when the line reaches the first dimension at that point no further dividing of that line is longer possible. We can never grasp what the size of a line that first line is in size that comes about when the first motion breaks the eternal stranglehold of singularity on space. According to our big and small conceptions of what we perceive as large, ultra large, small and microscopic small is just mere words describing thoughts that is totally unrealistic in the context of what the cosmos sprang from as the cosmos

The Four Cosmic Pillars In terms of Singularity

moved out of the spot and formed a dot. Even by the standards of forming the dot, which was eternally bigger that the spot, from that the dot developed and developed all the many dots that came from the spot. The size differentiation that is in place when compared only between those two points exceeds all limits we are able to fathom and divides more than what we wish to create that forms borders that we can appreciate.

When looking at the circle in the conventional manner, we persist with errors brought about in culture and not by applying some significant modern logic. Take a circle and reduce such a circle constantly to where it no longer can reduce. Reduce it to a point where only form remains part of the circle because the radius has gone beyond human measure and becomes so small it is not noticeable with what ever measuring tools man may use, then what remains is pi since pi does not indicate size but indicate form, and form is all that then will remain. In any circle or sphere the size only depend on the fluctuation of r, as a component to the circle or sphere but that does not affect the form by indication of Π in any way there may be. The conclusion I drew from following this process is that from this no line can start at zero because that will be a mathematical impossibility since no line can ever reduce to zero. A line will forever be able to reduce further becoming smaller but it can never reach zero because zero is not part of the scale on which we measure lines. If a line cannot reduce to zero it then cannot start at zero.

A line or spot starting at zero would therefore be shorter than the shortest line possible. For obvious reasons can no line, or any line grow or extend from zero because such a line must then quit zero and become something, thus abandon its original value. That would mean the start of the line has a different value to the end and a line holds conformity through out. When any line is starting from point zero and it uses the factor zero, then it can never leave zero because of the influence of being zero disqualifies any possibility of growth. But when coming from singularity π^0 and the line then had to grow in all directions at the same pace the line must then become a circle π or being three-dimensional, then form a multi circle become 3-dimensional π^3 which is the circle form we named a sphere. Since the Universe is about circles and lines connecting circles I came to conclude that flowing from this fact is that in the Universe there can be no zero improvising as a filling ingredient for the space of a point or be unfilled space. Zero is no valid factor in the Universe. In the case of the growing sphere the value of the circle is Π^0 going Π, and that is where creation must have started. That gave me the clue where to start looking for singularity. One would find singularity in the value Π and the value Π will be in all things rotating in a circle but by measure one dimension smaller. As usual I am again shooting the gun before the hunt started. Lines in mathematics do not start from zero and that is no discovery on my part that was a realisation I came too. The Universe is all about lines and the manner that Kepler pointed to the increasing of the lines by **k= a³/T²** proves growth in the composition of all lines.

To find validity in my argument one must draw this statement of motion back to the point where singularity is getting sides or said mathematically Π^0 going Π. When there is singularity there can be no sides. The one forming singularity Π^0 by measure fills no space while form Π develops Π^2 into space. The space that even the dot fills being Πr^0 does not really exist in the manner we humans see space to exist. It is a spot that is there without being there. It does not visually exist because it is not filling any substance and it cannot be recognised since it is not three-dimensional. The spot and the dot have no dimensional worth of any measure.

It is the point within the Universe I have named as **Infinity** where nothing can go smaller and anything within that point can never reduce. That point is where the entirety called the Universe begins and where everything holding substance begins.

Once one accepts the fact of singularity being present in that location, that accepting of singularity then is contradicting all the things we know and we can measure and we recognise that point being present by merit of the fact that the point referred too is not being formed by any of the things we can recognise. It is made up of everything we don't know and constitutes of everything we are unable to recognise or visualise.

In that spot there is no space. That spot holds **Infinity.**

In that space there can be no motion because there can be no space to have the motion within. It is formed as a line that is so small that our human reality by perception declare that point as not being there and the only reason why we know it is there is because of the results it left as an imprint of its not being there. We cannot detect it but notwithstanding our failure to note it we can recognise the dot on the merits of its absence and while in our Universe it is always absent, reality disallows the dot ever to be absent, because it is never absent. It cannot be absent. It cannot go absent but it can never be there where it should be in a place from where the third dimension forms and it is always present if I wish to locate it. It is infinity that can never go away.

The centre spot to which I refer we cannot see and that what it truly is we cannot detect but we know what there is has no sides to any side and has no space that it fills because what it fills is filled all presenting singularity. What there is, is not valid in our domain and yet, still we know what is there notwithstanding the fact that we are unable to witness what is valid where singularity is for what there is we cannot detect with vision but we can observe what there is with intellect. The only way such a spot can fill space is by doubling whatever it fills to become more than one in the place it has to fill. But the very instant that happens it halves the space it fills because it then cuts the space it has into two parts. On this principle all movement throughout the Universe rests. From this derives motion and nothing in the Universe is without motion because everything moves in terms of all else filling the Universe. That point instigates the Universe to form by the movement of the Universe in terms of that point's inability to move. Any motion from such a point in singularity forms the entire Universe by putting everything forming the Universe in sides we do recognise. Anything within the Universe is in one side or another side or just the other side of the Universe because from that point we have dimensions forming by movement. From the smallest ever possible dot will grow a line in every imaginable direction relating to a prospect of Π because only Π will not favour one specific direction and that puts all directions at equilibrium meaning that any form of what ever might develop from such a spot will have the end and the start being in the same position, which will also have to be a sphere as the flow outward will be equal in all directions. From the smallest spot in singularity comes a sphere. Please think clearly, is that not precisely the commitment we find in gravity, where gravity is flowing from singularity outwards but never favouring any side? We could never explain where the gravity in China meets the Gravity in America. The nature of gravity is to never end and never begin always flowing without favouring any position applying and where it seems to favour, there is a valid explaining concerning singularity. This reasoning prompted me to look for singularity in such a spot because if the prime spot from which all came was a spot holding all, then the spot must hold the shortest line but more prominent it will hold the smallest form including the smallest circle or for that matter the smallest sphere.

That leaves the door wide open for the advancing of any radius in all possible directions. With gravity always being in the centre of a sphere where the space is least available in the entire structure (there is not even space left to fill) one finds a flow of gravity from that centre spot outwards in all possible direction even-handedly. The flow outwards is a flow inwards that concentrates space where no space can ever be. The fact that the original gravity will begin as a circle or will be a circle is the direction it will take when being the first spot created. All progress will be evenly in all direction because no direction will stand out or be in favour above any other direction at first. Moreover, what this information brings home is that through motion and only through movement does space develop in terms of a relevancy dividing singularity. I am about to introduce the second form of singularity.

Kepler said that the space a^3 is equal to the motion T^2 of the space a^3 distant from a specific centre k. That then is $a^3 = T^2 k$. Within the circle $k^0 = a^3 / (T^2 k)$ the centre holding singularity also holds gravity which is centred in the precise middle of the circle. By using mathematics in the way Kepler used it, those rules and laws used correctly in the investigating of the formula that Kepler introduced must form the basis of cosmology. Also such intense investigation then must be without Newton interfering and telling Kepler what he (Kepler) should have found and subsequently Newton's incorrectly correcting Kepler whereas instead Newton should have been looking at what he (Kepler) found because only then he (Newton) could have seen what gravity is. He (Kepler) said that the cosmos said that gravity is $a^3 = T^2 k$. The space is held in check by motion from a centre and that is the way gravity develops. It becomes more than clear that space a^3 is time by dimension T^2 and time is space a^3 without dimension k Gravity is a^3 / k but k is an addition of motion T^2.

Reading this mathematically encrypted coded formula of the cosmos given to Kepler and keeping it removed from Newton it reads as being that the space a^3 is equal to = the motion T^2 of the space a^3 in ratio to a centre k^0, which is relevant to the positioning of k. If we bring in the full equation it will be $k^0 = a^3 \div (T^2 k)$ which means half of space is solid $k = a^3 \div T^2$ and half of space is liquid $k^{-1} = T^2 \div a^3$ where liquid is moving. However, it is also true that everything through movement defines a value in relation to one point holding singularity k^0 and that is what the formula $k^0 = a^3 \div (T^2 k)$ underwrites. What this proves is that gravity is the motion of space provided by time being the liquid. Please allow me to explain. In the formula $a^3 = T^2 k$ the space forms as the space is in motion.

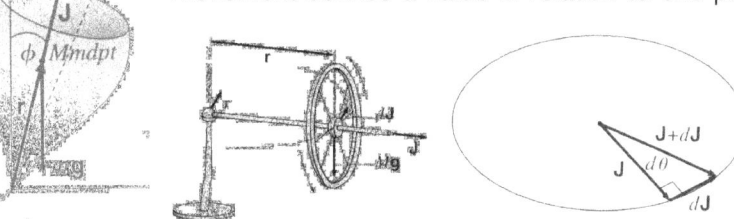

Newton suggested that $\frac{dJ}{dt} = 0$ where he stopped time to have the motion of the circle demolish the work that the circle does. That means he got time standing still or being T^1 and the motion **T= 0**. Let us ponder on that thought for a while, while we remain with the formula Kepler suggested and then it will seem that according to Newton $a^3 = T^2k$ and in that T^2 then becomes **0**. Should that be the case then we have space going flat because $a^3 = T^2k$ where $a^3 = T^2 \times k = 0$ forming a square instead of a cube, and the Universe we have is a three dimensional system in every aspect there is.

I am of a very different opinion about Newton's point of view where he declared that forming a circle moving $\frac{dJ}{dt} = 0$, and by doing such movement removes Kepler's relevancy factor. This places a value of empty space in which a top would spin and Newton missed the difference there is between a top spinning and a top laying on its side on the Earth. There can be no such a thing as empty space. The fact that space is valid removes an empty connection because space can be anything there is in space. The Universe is time contained in space, which makes it space-time. Space has only one value, and this is to contain time and time provides space with a definite value.

I do not disagree for one instant with Newton's calculations and therefore I am not going into repeating the entire calculating process. All of the calculations Newton made are very correct except the eventual and final conclusion Newton came to. Newton never understood the mathematical concept of time playing a part in physics. In the time of Newton singularity and the relevance thereof had no feasibility in any concept regarding physics. Newton had the concept that time could stand still and that is impossible in physics or any other place. Time can never stand still because time is forever moving by establishing space in a three dimensional environment.

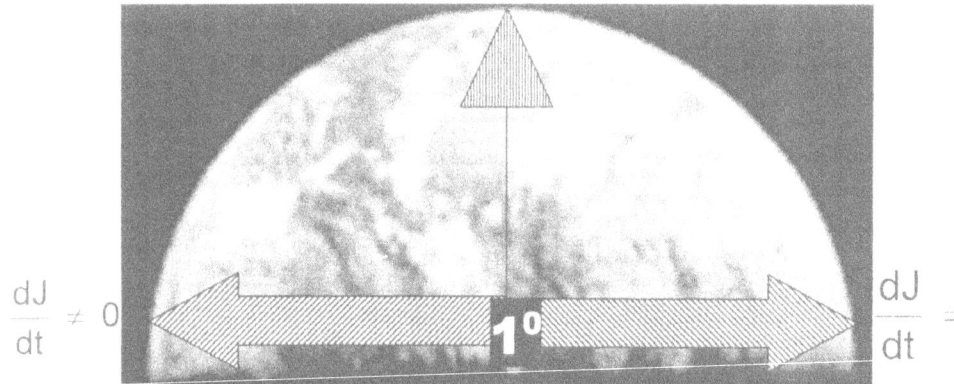

Being the mathematical genius as Newton is so often portrayed as, Newton had very little insight into mathematical possibilities, because when he suggested that $\frac{dJ}{dt} = 0$ he made one huge mathematical blunder. Newton or no other person may place any two objects in a direct relation where the two factors divide and have an outcome that is forming zero. Much surprising is that not one mathematical genius that came after Newton drew the correct conclusion that forming $\frac{dJ}{dt} = 0$ is mathematically not acceptable. Newton saw that dividing something into something else could bring about zero and that is impossible. In concluding that $\frac{dJ}{dt} = 0$ bringing in zero as a legitimate value Newton found a way to replace Kepler's symbolic relevancy value of **k** with using the symbols G $(m + m_p)$. In doing that Newton painted a picture that has no real meaning except where Newton tried and succeeded to put mass into an argument that has no true validity in cosmic principles. This is just a longer and probably a more detailed manner of indicating **k** and better defining of **k** but it symbolises precisely to the point what **k** stands for nonetheless. I wish to draw your attention to the matter of Johannes Kepler's findings that Mainstream science considers as resolved and closed for many a century while it is not. My investigating Kepler helped me to resolve other unresolved matters but it was only possible by using Kepler's work.

I too am well aware that at first glance you will immediately arrive at the opinion that the theme of the book has to be considerably below the standard of an intellectual Master such as the readers must have, due to the position such readers hold, and because of that, the normal research work the readers do. I realise it is dealing with a subject school children learn but in that comes the issue that goes unnoticed. Nevertheless, I hope that this writing may spark interest even at such a low academic level and grade in scientific sophistication and development because I am about to prove that I discovered:

Newton did not think the situation through when he contemplated about gravity. Newton should have thought about factors keeping the gyroscope upright while the gyroscope is spinning. The gyroscope will fall on its side when not spinning and in that position the "Earth's mass" could play a part since the gyroscope fell on its side. However, as soon as the gyroscope started spinning, the balance shifted in favour of a position wherein the gyroscope stands upright. What then comes about has the ability in keeping the gyroscope upright. This is rotational movement and in my other books on the **_Absolute Relevancy of Singularity_** I explain how rotational by the square of the double seven forms Π and Π is forming the curvature of space-time and in that bending of space-time is what we call the atmosphere that keeps the gyroscope square with the Earth and through that the gyroscope stays upright. The gyroscope is acting in accordance with the Coanda effect where the Coanda effect is gravity. By spinning it establishes a solid forming as $k = a^3 \div T^2$ and a liquid forming as $k^{-1} = T^2 \div a^3$. By spinning $T^2 = a^3 \div k$. That is evoking singularity $k^0 = a^3 \div T^2 k$ establishing gravity $a^3 = T^2 k$ in relation to the Earth evoking gravity through also spinning.

Newton found mathematically that the movement of the top by spin removed the value of the radius

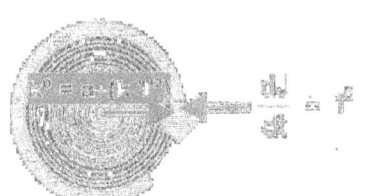

$\frac{dJ}{dt} = 0$ where quite the opposite applies. The spin of the top $T^2 = a^3 \div k$ positions the relevancy that **k** as a factor produces by initiating singularity k^0 on both sides of the relevancy $k^0 = a^3 \div T^2 k$ as well as placing singularity in relation to the spinning top $\frac{dJ}{dt} = 1^0$ because that is the correct mathematical principle coming from the equation. Thee smallest any dividing can be is one and one is the producing of singularity. The spin of the circle does not eliminate the relevance of **k** but institutionalise the measure of **k** by confirming the space a^3 in terms of singularity k^0. However **k** has no confirmed and specifically applying value but puts a relevancy of space a^3 forming in relation **k** to movement T^2 applying. By trying to find a measured value applying to **k** is showing no understanding about what **k** is. The value of **k** is finding the space that **k** indicates in terms of what moves. The indicator **k** identifies the space a^3 that the circle claims in terms of singularity k^0 that the movement T^2 isolates from the rest of singularity $\frac{dJ}{dt} = 1^0$.

The value of **k** is dictated by T^2 as the movement isolate the space a^3. The measure of **k** is the relevance **k** is claiming on behalf of the space a^3, which uses the relevance of **k** to put a limit the space a^3 spinning in accordance with T^2.

Let us have a look at the bicycle. It is said that the bicycle works on a balance and by science mentioning that the rider of the bicycle is applying a balance, in that they think that the entire problem is solved. That is so typical of Newtonian simplicity about a very complex issue. It is the same as putting gravity down to mass pulling by some small particle called the graviton without ever showing any ability to look more intensely to find a solution for a very complex problem. They always go about by creating a graviton or creating dark matter to look for solutions that solves all the unsolved issues and never do true investigative research in what applies to complex issues such as gravity. Saying the riding of a bicycle is due to balance is the same as putting everything in the Universe down to mass taking charge of particles. The simplicity in which a complex issue is solved becomes as laughable as filling outer space with nothing until the nothing runs over and the Universe that can't gain more because it holds all starts to expand. As the wheels spin (T^2) the relevance of the down thrust **k** leaves the bicycle firmly attached to the ground and in doing that it confirms the space in location (a^3) in terms of singularity k^0. Newtonians would call this having mass or whatever. Then when having the bicycle moving forward in terms of individual cycling such forward thrust gives a relevance of (**k**) to the peddling power and the movement (T^2) then is about having momentum in relation to the Earth spinning. That means **k** can push down and **k** can push the bicycle forward.

What Newton suggested while never realising he did suggest is the following, and that is that the rotary movement of objects puts singularity $\frac{dJ}{dt} = 1^0$ in position on the outside of the moving circle. However, by using $\frac{dJ}{dt} = 1^0$ Newton placed emphasis on the turning movement of the circle and saw this as a destroying of the circle while in fact the turning is putting the space that identifies the circle on the cosmic map. That Kepler also found without ever realising what he found. Kepler said $a^3 = T^2k$ which is $k^0 = a^3 \div T^2k$ which is the spin or $T^2 = a^3 \div k$ which is the circular movement T^2 that validates the space a^3 in relation k to a centre k^0 which is exactly and precisely what Newton said when Newton said $\frac{dJ}{dt} = 0$ that actually should read $\frac{dJ}{dt} = 1^0$. The location where Newton placed singularity as being singularity established by the movement of space $\frac{dJ}{dt} = 1^0$ I named **eternity** because there nothing can ever go bigger or become more. Whatever was and is and will ever be is locked in that space I named **eternity**. The "so called expanding" of the Universe $T^2 = a^3 \div k$ is where singularity is shifting relevance k from liquid $k^{-1} = T^2 \div a^3$ to solid formulated as $k = a^3 \div T^2$ and the process whereby this happens is precisely the same as the Coanda effect. Getting back to my first argument about a line and that no line can start at zero but has to use singularity as a starting point, this is all the proof I require. The line k coming from the centre (singularity k^0) forms by forming an initial dot Π^0. However, I went on to say that whatever the line used to start with has to continue in order to repeat the same that began the line. Therefore the line started with Π^0 and it has to continue with Π^0 until such a point, as it must end with Π. Whether the line is Π^0 or is r^0, or uses 1^0 the outcome all refers to singularity being used. By reducing the line we come to the end of the mathematical equation of the circle but the circle does not end there. That is what Newton did not recognise from the figures the cosmos represented to Kepler. The circle only secures the final cosmic figure and the value to singularity where all things have equal value. The movement of the circle splits singularity in two sectors. By forming Π the circle has to form Π^2 due to the movement coming about in securing the space Π^3.

Kepler chose to use different symbols too those being valid, but the concept remains the same. Kepler said that $a^3 = T^2k$ while I show that $\Pi^3 = \Pi^2\Pi$. It still confirms that movement $\Pi^2 =$ is the forming space by three dimensions Π^3 in relation with the movement Π^2 being relevant Π to singularity Π^0.

I shall try and explain what this concept holds in terms of a piston moving while working inside an internal combustion engine. The piston goes up to a point we call top dead centre where the piston stops and according to the crank the piston halts in directional movement. Then the piston starts to accelerate to a point we call bottom dead centre where, again it comes to a dead halt. The piston stops directional movement at T.D.C. and at B.D.C. or that is what we see without seeing anything. This is not the case because if this was the case the engine must vibrate at those two points of stopping. We reason that the piston stops twice and starts moving on the two occasions but if that was the case of stopping at two points without stopping anywhere else, the vibration that the stopping will cause will have the engine disintegrating completely. To us favouring positions the piston stops at two locations but the fact of the matter is that the crankshaft stops every $7°$ of rotation and if the crankshaft stops, then so does the piston stop.

The stopping is a continuous and is an ongoing process that happen every $7°$ of rotating. The crankshaft moves in a straight-ahead position going straight and then it stops and redirects by $7°$ and then it turn by going straight again. It is $a^3 = T^2k$ and then it stops (a^3), it turns (T^2) and then again goes straight again (k) while holding reference with singularity $k^0 = a^3 \div T^2k$ all the time. One cannot part the redirecting and the going straight T^2k because it is the same movement since the space forming a^3 is equal $=$ to the turning T^2 and the going straight k. This is evident when dissecting Kepler's formula $a^3 = T^2k$ that $T^2 = a^3 \div k$ and $k = a^3 \div T^2$ while honouring Newton's 3rd law $k^{-1} = T^2 \div a^3$. Please believe me that this puts movement is a much complicated dimension because this has the material $a^3 = T^2k$ moving $T^2 = a^3 \div k$ in terms of ($k = a^3 \div T^2$ as well as forming $k^{-1} = T^2 \div a^3$) while always referring to singularity $k^0 = a^3 \div T^2k$.

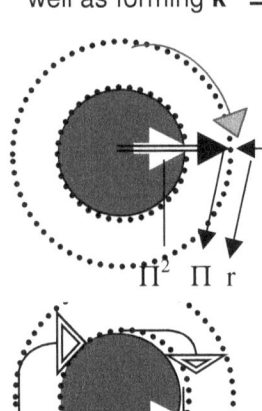

In the circle $\Pi^2\Pi$ the space surrounding the rotating object will also extend by Π as the concentration of the spinning motion draw or drag on past Π^2 extending the influence of Π^2 by the value of Π. This extending of Π^2 to accommodate Π we refer to as the atmosphere, but physics apply to this extending in the normal fashion. From the spinning motion Π does not stop at the end of the solid structure but the influence of Π extend and this then becomes the atmosphere. The influence of Π^2 stops at the end of the solid structure but the influence of Π extending plays a most dominant role in the cosmos, although not yet recognised and that factor is most crucial to a better understanding of the implications of laws governing the cosmos.

On Earth we can measure a distance between two objects to a precise measurement and come back the next day to find the two structures well secured in the same place and distance apart. That how ever we also know will not be the case in space since the objects will always drift apart and away from the position it had. With the circle being $\Pi^2\Pi$ the Π^2 will reflect the circle in the square with Π forming the extending of Π^2. The extending of Π will not end immediately but will

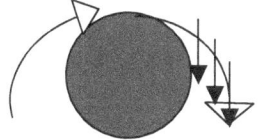

Every quarter is directly opposing the next as well as the previous quarters thereby starting a new set of principles it has to adhere too, but breaks by moving through time anyhow.

In that movement comes about destruction of the self-preserving because any change of what ever small proportions will lead to destruction coming about as if with a snow ball effect. The top can be its personal matter and anti matter just by changing the speed of rotation where the points does not precisely meet the previous points and deformation stars by overheating bringing about an altogether change in relevancies to itself as well as other matter in the same orbiting time.

With all matter having the same start from the same singularity, all matter should therefore be synchronised in growth and in rotation, where the matter is in support of all surrounding matter spinning around the common and original singularity that produced similar growth and rotation speed since time began to the present day.

The Roche limit in the practical sense.

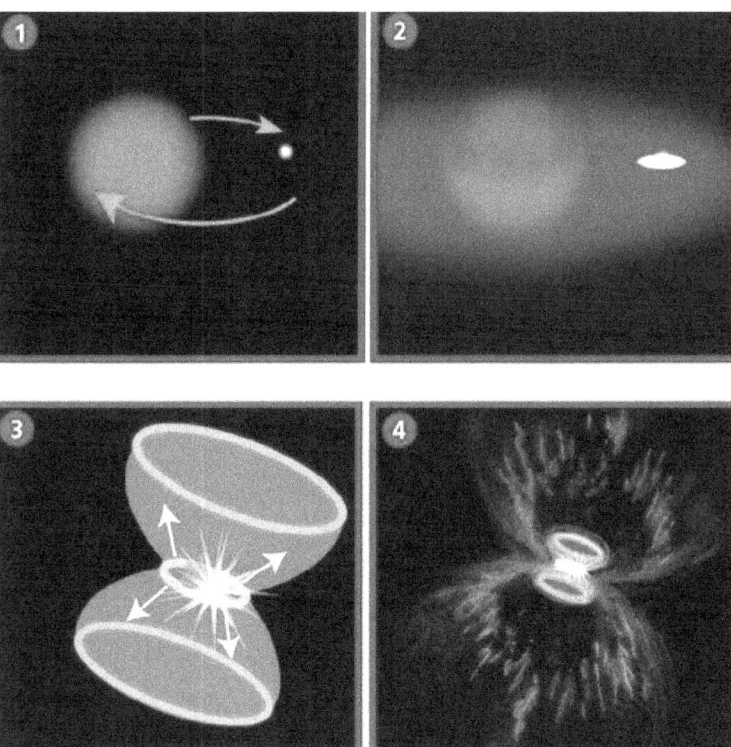

This is not only limited to planets in our solar system. In the Universe, there are giant stars spinning around each other. These stars are binaries, which are also one form of double stars where double stars are another such a form. The difference between the types depends on the distance they remain apart. They keep a certain distance apart and do not collide. In the case of the Sun and its planets, it could be a case that the systems might be too small, or they might be too apart. However, this is not the case with binary stars. They are close, they are big, and they spin around a mean axes called the Roche limit.

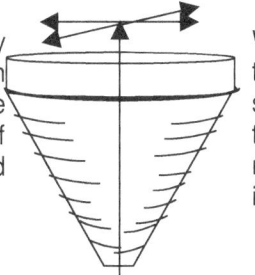

Since occupation may or may not be placing the factor in infinite, the space therefore holds the premier singularity of infinite from which all included in the Universe has come.

When the top starts spinning in a specific position the top merely executed the option to fill the premier singularity at that specific point. When it moves it may take the premier singularity with to the new location it moves through spin or it may fill yet another position in singularity as all is the same.

Because every atom spins, singularity is in every atom surrounding the singularity. With singularity being in every atom, every atom becomes the centre of the Universe holding all outside singularity as space-time and by conclusion of the cosmic splendour, all space-time will dissolve once more back to singularity, bringing about the last atom holding all singularity without any surrounding of space-time left.

The influence immediately above the circle will have the biggest influence and reduce gradually as the value of Π reduces in the leverage that the space has on Π and a gradual but definite change from Π to r will affect the extending of Π progressively more. The decline of Π will follow the same contour of the circle at 7^0.

From there it influences singularity in the triangle flowing through to the half circle. It is an interaction between circular and linear motion as the value of Π continuous past $Π^2$ (at the end of the solid) and every cosmic structure holds an individual and specific singularity. The field where Π extends we call the atmosphere having a value of 21.991 / 7, which is Π.

From this line of reasoning I dismissed the theory of the presence of a force being gravity but rather consider it as a dimensional changing contributed by the spin of the Earth and the spin comes from singularity located in the centre of the Earth.

By claiming the position held by singularity premier as a vacant spot until the arrival of the top, the singularity of the top divides the point flowing from singularity into four sectors holding two half circles

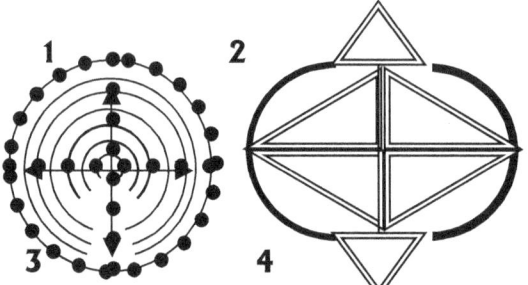

From the star holding a dominant point or most valued point in singularity it affirm all three other structures, each holding singularity individually and in a compliment of 5.

The Roche limit explained in the Newtonian view and showing they have no idea how it functions and by which they clearly don't make much sense about how or why it is there at all.

The Roche limit is:

The region surrounding each star in a binary system, within which any material is gravitationally bound to that particular star. The boundary of the Roche lobes is an equipotential surface, and the lobes touch at the inner Lagrangian point, L_1, through which mass transfer may occur if one of the components expands to fill its lobe. It names after the French mathematician Edouard Albert Roche (1820-83).

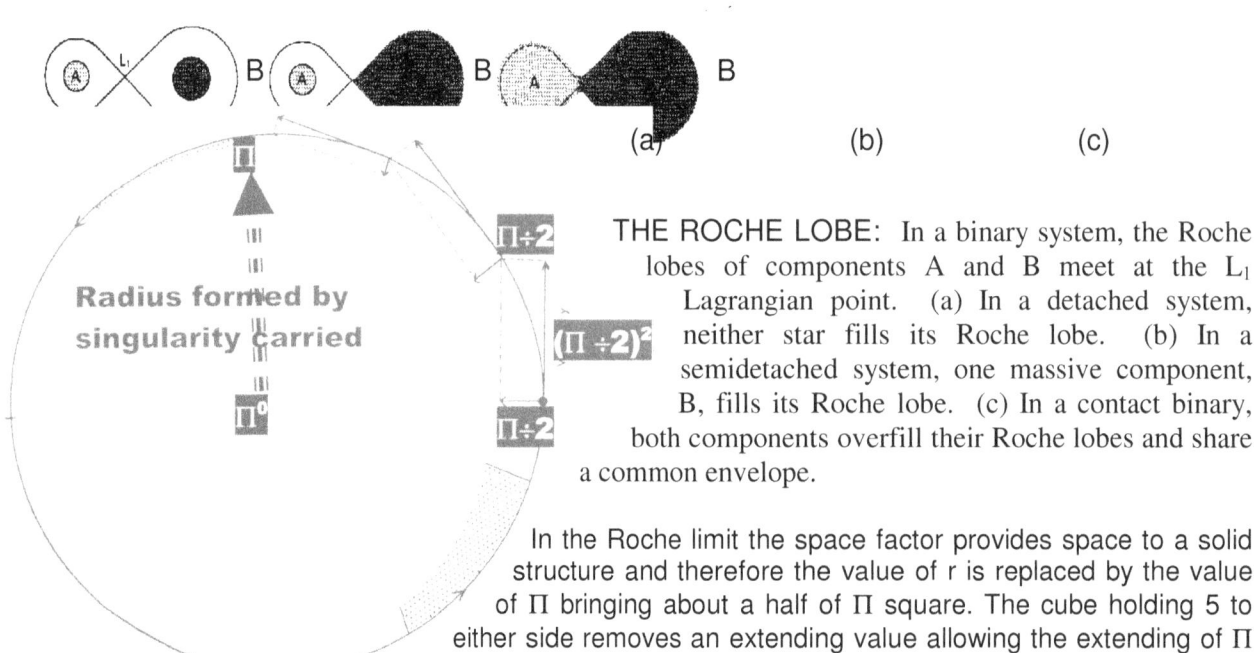

THE ROCHE LOBE: In a binary system, the Roche lobes of components A and B meet at the L_1 Lagrangian point. (a) In a detached system, neither star fills its Roche lobe. (b) In a semidetached system, one massive component, B, fills its Roche lobe. (c) In a contact binary, both components overfill their Roche lobes and share a common envelope.

In the Roche limit the space factor provides space to a solid structure and therefore the value of r is replaced by the value of Π bringing about a half of Π square. The cube holding 5 to either side removes an extending value allowing the extending of Π to indicate position to space by movement and that would be a square. Where Π extends to lock onto the next sphere's extending indicator, Π has to connect to Π forming the square of space and translating that to the half of Π being $(Π/2)^2$.

It begins where the cosmos began and that was when singularity performing as infinity parted from singularity performing as eternity. I am not spending time by going into detail on this mater because of the enormity the subject holds but I have written an entire book explaining how this process came about as it used the four cosmic pillars to generate the first gravity. But as space-time grew, the extending of the influence of the four pillars grew while still holding onto every detail that applied in the first instant time became space. The line that forms is still presenting singularity as a line $Π^0$ and this value supporting singularity runs across and through every atom forming such a line.

At the end where everything is the centralised $k^0 = a^3 / (T^2 k)$ the end holding singularity Π that forms by movement $T^2 = a^3 / k$ that validates space $a^3 = k T^2$ forms a specific relevancy $k = a^3 / T^2$ that goes both ways $k^{-1} = T^2 / a^3$. At such a point Π forms one half reaching into time in order to expand into time while the other half moves down as space or matter pulls it to compress it. This effectively splits Π into two halves and by Π going into a square; it is half of Π going square which is $(Π/2)^2$.

5/2
Five sides divided by two spheres.

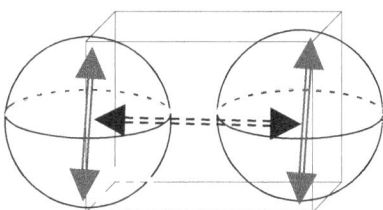

The Roche limit
5/2 = (Π / 2 X Π / 2) = 2.4674

7 Space-to-matter

Explaining the "not understood part" is what is really informing information about the cosmos.

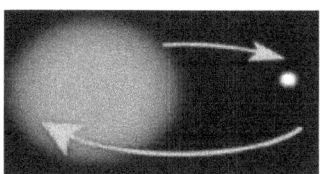 

This is how the Roche limit functions in the practice where governing singularity takes control.

The formula $F = G \dfrac{M_1 M_2}{r^2}$ cannot explain the behaviour of stars going into a melting frenzy such as the pictures indicate because $F = G \dfrac{M_1 M_2}{r^2}$ **presumes that stars has to collide and not come onto battle while the stars are still a large distance apart.**

The obvious occurrence shown in the pictures above indicates clearly how the major star liquefies the minor star and takes control over the singularity dynamics within the minor star. By understanding cosmology principles correctly I can explain what is occurring in this instance and this occurrence connects directly to the Roche limit, as explained above. Not only does the Roche limit explain this phenomenon, but also it ties directly to the Titius Bode principle, but also it rubbishes Newton's formula $F = G \dfrac{M_1 M_2}{r^2}$.

According to the science formula of $F = G \dfrac{M_1 M_2}{r^2}$ the orbiting structures should collide with a bang as r^2 diminishes under the pulling power of the combining force that both objects hold in their mass. Instead they do the tango until one drop, but when dropping it still does not collide with the larger structure, as would the formula used by science, $F = G \dfrac{M_1 M_2}{r^2}$ suggest. The major star liquefies the minor star and when the minor star is in a complete state if liquid, the major star absorbs the minor star by applying gravity as the cosmos uses gravity through the Coanda effect.

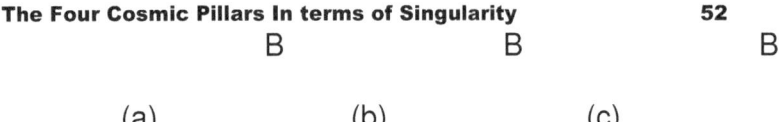

(a)   (b)   (c)

THE ROCHE LOBE: In a binary system, the Roche lobes of components A and B meet at the L_1 Lagrangian point. (a) In a detached system, neither star fills its Roche lobe. (b) In a semidetached system, one massive component, B, fills its Roche lobe. (c) In a contact binary, both components overfill their Roche lobes and share a common envelope.

Once more, this phenomenon should not occur with Newton's presumptions about gravity. These bodies will collide and destruct, without a doubt. When the formula $F = G \frac{M_1 M_2}{r^2}$ applies, there should not be any force, which is able to keep them apart. However, they do exist and what is more, they maintain a certain distance apart. Seen from this view, it is little wonder that the significance of this was lost in the notion that this is yet another "mystery" of the Universe. The scientists of the day (and the past) lost the importance, which this holds for us as Earthly dwellers. Newton used the top to point out why he thought mass was the uttermost important factor that provides gravity. I wish to use the top to point out what Newton missed.

Looking at the top when spinning it seems that the body structure of the top is solid and the air surrounding the top is liquid. It also seems as if it is the air that is supporting the erect stance of the top. The top as a structure composes of solid particles that light cannot penetrate and that material cannot pass through. In that sense it seems to fit all the conditions we set for solidness. The top spins by contracting the air in the direction of the top and it spins through the compressed air that allows the top to spin while being erect seeing that the top air immediately surrounding the top has much more density than the rest of the atmospheric air has.

That which can move is the liquid part of singularity, which forms singularity by the air surrounding the top. This then is that part that moves and that stands related to what cannot move being singularity within the centre of the top. Since everything is singularity everything is immovable but also since everything is singularity that which does not form part of that which can't move forms par of that which can move. It is Π bringing the partition by instituting Π^2 in relation to Π^0.

Every thing outside that which forms the body of the top is liquid with the top forming a solid or so it seems to us. Well yes in a way and not that much either. The top through the atoms that constructs the body of the top is a pump that pumps heat from the outside inwards just like a turbine engine. Every atom that is rotating inside the structure of the top is keeping the centre erect.

The centre is totally motionless because all the atoms in the top are moving with the spinning body while in relation to the centre being uncompromising still as solid the changes their local position and accordance with the centre. With all remaining rigid and still in one the centre singularity, therefore the moving of the top circle goes to the part forming the liquid outside the body of the top since seen from the centre that is the space that changes rapidly with motion. The movement is thus extending the singularity in movement of the top to the outside of the edge of Π forming Π^2 where the top meets eternity. The extending of singularity is holding the air as a liquid and being the liquid the flow of the liquid keeps the top erect and spinning. The spin produces a cold in relation to the hot that the liquid is.

all the atoms are holding a position solid body's atoms never therefore they never move in the atom centres position as witnessed by Singularity also continuing to the value of Π^0

Singularity continuing to the value of Π^0

It is the movement of Π^2 that alters the line of singularity parting material from time Π^0

From this we can see that although it is the top that is moving, as seen from our perspective and seen from our reality the opposite applies as seen from the centre of the top the top. According to singularity coming from the centre of the top it is not spinning because the motion is transferred to the outside of the top. What is moving is liquid and what is not moving is solid. Everything has a reference in relation to another point. That which is capable of relocating is forming a liquid in relation to that which is securing the position of rotation. Everything in the cosmos can move and yet that is only true when there is one point in the cosmos that at that moment in reference are not able to move. The cosmos stands divided between the eternal moving of eternity and the immovability of infinity.

Everything around the top is liquid with the centre being a solid. However the solidness and liquid has cosmic standards and just as it is in the case of hot and cold, big and small, fast and slow, our standards and cosmic standards do not share any measurements. So too does cosmic notions about liquid and solids have a totally different meaning in cosmic terms.

There is a pumping interaction of space-time flowing towards singularity through every point that confirms singularity. Everything in the top that forms the material is also surrounded and partitioned by liquid. By providing motion the matter in the top serves forming the solid connection with the centre as the liquid factor moves towards the centre and in that it extends the space that singularity provides. The structure is composed of atoms. In the atom there is a governing generated singularity around which all sub-atomic material rotate. In the case of the atom all the rotating material forms the heat while the generated centre, which is incapable of rotating, forms the solid factor. Every aspect that is without motion stands in a relation of 1^0 and that which is relatively moving or changing location or finds a new position holds 1^1. These are names I have given to bring some understanding to everything being equal anyway. Everything that is standing still is 1^0 and everything that is moving is 1^1.

Gravity or motion is a constant relation that solids have with heat where heat forms the liquid and solids form space. There is the rotation but part of the rotation is the lateral progressing by rotation to also confirm the generated centre by relocating the entire centre in a straight line. The generation of the line forming singularity is vested in the rotation but the flow towards is the lateral forms another factor motion brings about just as electricity produce a flow of time in relation to space collapsing.

The forming of space-time by measure of gravity is using the same system to do the very same everywhere. Kepler said it to be $a^3 = T^2k$ while Newtonian science portrays it as $F = mv^2$ which when formulated correctly should read $F^3 = mv^2$ and Einstein formulated his version as $E = mC^2$ which also if formulated correctly should read $E^3 = MC^2$. Is there anybody that is brave enough to show how that which followed Kepler was formalised in any other way than Kepler's product given the condition that the mathematical interpretation should read correctly.

There is no substance difference between 1^0 and 1^1 and it is a relation where one moves as the liquid partner and the other stands still as the solid factor. Both are not as much equal as they are precisely the same. Infinity cannot move and eternity cannot stop moving. By parting, infinity had to remain motionless and eternity had to remain moving as it introduced a part of the cycle to one line (point) where it stops moving in relation to the other factor that cannot move but does start moving. Time is a graph that never begins since it never ends and while everything never repeat everything always never remains the same. The factor that shows motion forms the liquid while at that moment the factor that does not show motion forms the solid. The measure of 1^0 is transformed to 1^0 and which ever are 1^1 is passing the extending of space on to 1^0.

Time spins around a centre because everything spins around some centre somewhere in order to secure the centre singularity. But also time moves and in that there is the linear that always are part of cosmic motion. The centre is referred to by heat but heat also secures the centre by reconfirming the centre in the lateral. But in both cases singularity is reinstating singularity by confirming it as it is referring one another. In the manner that 1^1 confirm a position in singularity 1^0 it is supporting 1^1 by generating 1^0. By generating 1^0 it is repositioning and reallocating 1^1 as a position by confirming 1^1.

From all of the above one can deduct that outer space is something viable through which objects travel. It is clear that something filling the space between Jupiter and its first moon because of lightning interaction between the two structures. There is a reference between a structure holding material and the space above as well as the space between two objects. There is electric lightning travelling between Jupiter and its closest planet. If there is lightning there is electricity and electricity means a very distinct interaction that connect the space inside the material structures through the space parting the material structures putting the space in between as a conducting medium which nothing can never do. Understanding this nothing not existing is fundamental in understanding the relevancies applying.

Considering the notion of nothing being in place filling the space between the cosmic structure, electricity needs a conductor to transmit the interaction there are and that disproves the nothing theory and it puts the nothing science places in the Universe as nothing those scientists have between their ears in their understanding the cosmos. It is official that the interaction was detailed as $a^3 = T^2 k$ which is what Kepler found, yet with this information science still do not appreciate the fullest of the implication. I have changed the formula to a^3 (space) $= T^2$ or gravity (time) k (the relevancy applying) and this becomes singularity $\Pi^3 = \Pi^2\Pi$. By applying the atomic value the relevancy changes to $(\Pi^2 + \Pi^2)\Pi^2\Pi 3$ and that relevancy projects to cosmic atoms such as two stars interacting. When two objects come closer than the relevancy would permit, cosmic laws change their application and in this case then becomes $(\Pi^2 + \Pi^2)$ from either side where the three of space changes to singularity Π acting as the influence $(\Pi/2)$ from both sides making that influence a square $(\Pi/2)^2$. All of this I explain in the Cosmic Code and I am not dwelling into this matter at this stage.

There is a battle for position of dominance between closely allocated stars where the rivalry that Binary stars ad in that case "mass" pulling is proven none existing. The Binary systems never collide as Newton's formula $F = G \dfrac{M_1 M_2}{r^2}$ insists on applying through out the cosmos. The closest destruction takes place is when a star is 2.4674 times the diameter of the major star. Any person with the last understanding of mathematics would recognise the value of 2.4674 as $(\Pi/2)^2$. This indicates to the presence of singularity governing the area $\Pi^2/4$ where both object will hold onto their individual singularity while spinning around the mutual point of singularity.

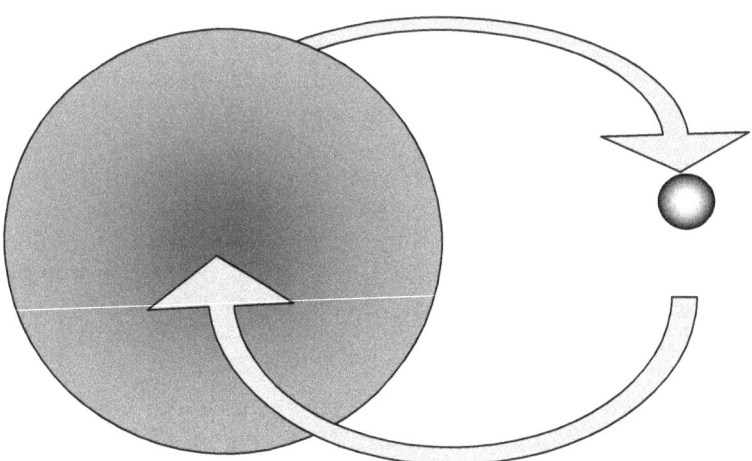

THE ROCHE LIMIT FORMS AT A LINE WHERE THE GOVERNING SINGULARITY OF THE MAJOR STAR CLOSES THE BORDER THAT MAINTAINS THE APPLYING SINGULARITY INTEGRITY.

Where there is two stars in close proximity the one star is unable to defend the singularity governing the structure as to regulate the relevancy of self-conservation, then the major star will take control of the minor stars governing singularity and destroy the singularity by establishing overheating with in the singularity of the minor star. I think NASA refers to pictures depicting this as "blowing bubbles or blowing heat " but do not quote me on that. I just found it amusing at the time that the beast brains in the world would come up with nonsense like that.

To the one object, anything distant from its proton cluster is space, space in what- ever form. By destroying the singularity, the space becomes heat either under its influence or under its control. On the one side, space holds the value of Π and on the other side; it also holds a border of Π. The time however changes to defend the singularity therefore times the

Anything spinning alternates its position in relation to the circle four times. This produces the invert square law. In a developing binary there is one major and one minor component. The radius that forms holds a cosmic relevancy of Π^0 forming Πr^0 that then by movement forms Π^2. The full circle of rotation will be 4 (the 4 directional changing quarters) that are related to Π^2. If there is a structure within the $4\Pi^2$ it is going to interfere with the circle that the major star ahs to maintain in order to rotate. In order to maintain self-preservation in its rotation the major star will clear the circle of $4\Pi^2$.

square of the atmosphere applying then holds position where space normally holds position.

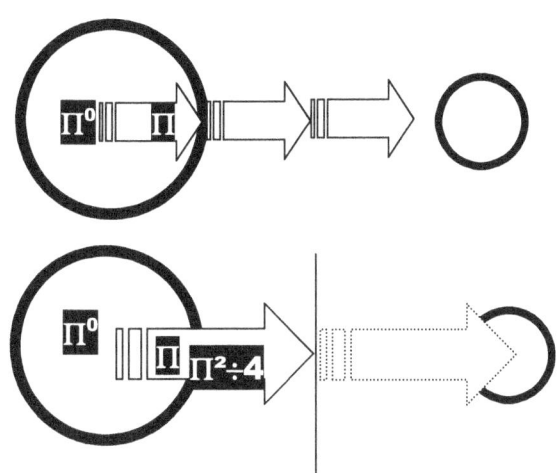

In order to se to the fact that the rotation circle is clear, the major star will take charge of the governing singularity of the minor star and alter the relevancy of the governing singularity ($\Pi^0\Pi$) to fit the relevancy applying of the governing singularity ($\Pi^0\Pi$) within the major star. This will in turn alter the gravity ($\Pi\Pi^2$) of motion of all the atoms within the minor star. By liquefying the lesser star had to point to the interaction there is in the cosmos between solids and liquids. This gave rise to my theory of relevancies swapping and fluids being consumed by solids. What happened during this interaction lead me to believe that the Coanda effect forms gravity and this has nothing to do with having mass or not having mass.

LINE HAVE DUAL OR MUTUAL SINGULARITY

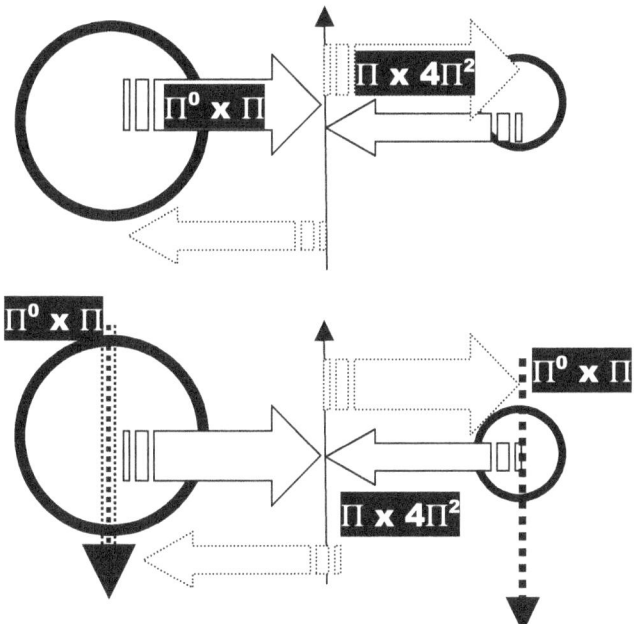

The mutual line of singularity $(\Pi/2)^2$ holds a position referring to each objects individual line of singularity in the value of $(\Pi^2 \div 4)$

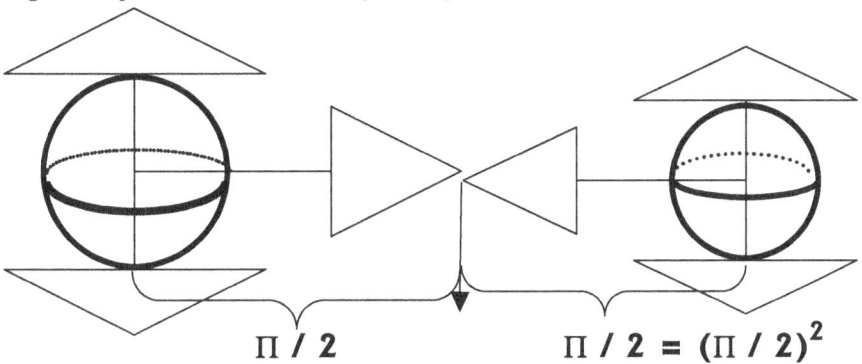

$\Pi/2$ $\Pi/2 = (\Pi/2)^2$

SINGULARITY MEETS AND COMPLIMENTS EACH OTHER.

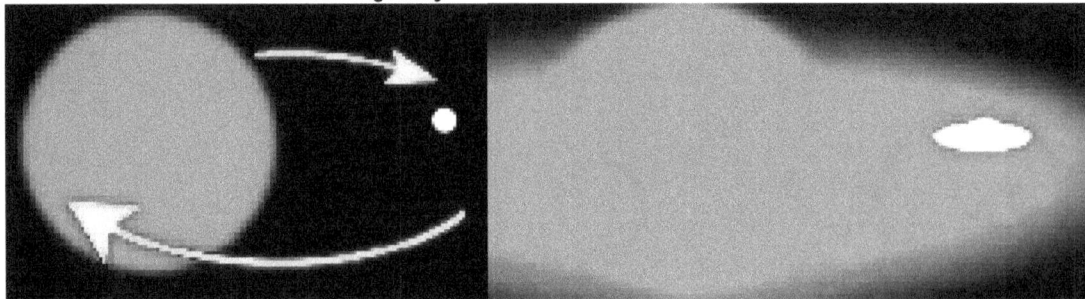

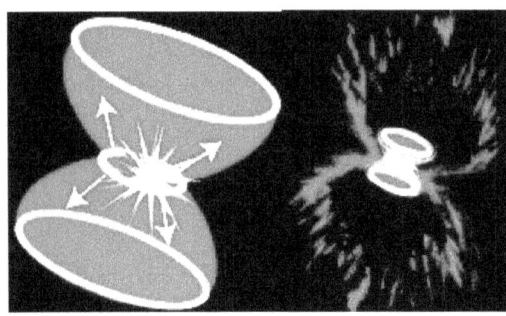

In the struggle for superiority begins between the two points each holding a spot of governing singularity. This is the very same as what takes palce between the spinning top and the Earth or if you wish the Earth and the flying supersonic aircraft where the Earth is the obvious eventual winner and in the final outcome both objects maintain fighting for the claim to singularity, by pushing the space-time occupied to new levels of occupied space-time values.

The result of this is the establishing of an individual singularity dominating because of the unequal ness between the points of singularity and by cosmic law liquid is always present in relation to solid. The minor star is liquefied to bring the cosmic law into effect where at a rotation distance of $(\Pi)^2$ first favouring the on in the matter part and the other in its space part and afterwards turning the points of reference around.

Material moving brings the divide between eternity ands infinity by referring to singularity $k^0 = a^3 \div T^2k$ as Kepler's formula $a^3 = T^2k$ being $k = a^3 \div T^2$ while honouring Newton's 3rd law $k^{-1} = T^2 \div a^3$ forming gravity or time as singularity spinning $T^2 = a^3 \div k$. **Gravity and movement and time is all the same thing. It is repositioning everything in relation to one point holding singularity. That brings us to movement.**

It is Π that sets the margin as to where the material starts and time or outer space begins. It is Π that sets the limit between that which is liquid and that which is solid. It is the positioning of Π by the movement of Π² that part infinity from eternity. It is Π that ends that which can never begin and begin that which can never end. It is Π that establishes a reference point where singularity divert into the eternal and the infinite allowing matter the zone matter can claim space in time, the control of space in time and the influence on space-time, the point of singularity have to reduce the value of singularity on both accounts of the cosmic atoms claiming individual singularity, or enlarge the claim of space by matter away from singularity. This is rather important to understand when arriving at the actual presentation of the formation of the solar system

Any point will be opposing itself within the **rotating of 180°** where it **then change every aspect** of its **previous flowing** characteristics it had or **will once again have** in 360° from there. While in rotation from the view point of a bystander it all may seem static and never changing but to the object in spin every next instant in time will be diverting from every aspect it had every second passing, and the direction it held in relation to the direction it held the previous mille, mille second will totally be incompatible with the direction it holds the very next mille, mille second of rotation. In this fact of an ever-cyclic change going on forever hides all the mysteries about "global warming" and all the misinterpretations persons such as Al Gore and others attach to "global warming". In ever directional changing of 5 to 5 we have the 10 of gravity coming from 2 x 5 as 7 spins through 5 and the change in direction by seven confirms five twice.

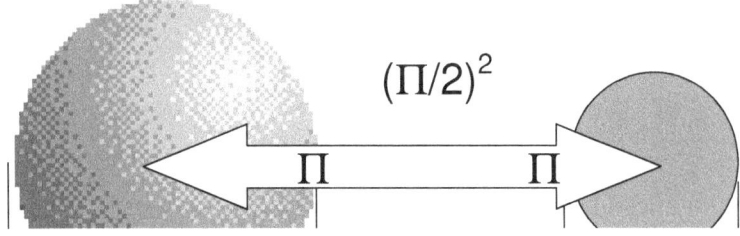

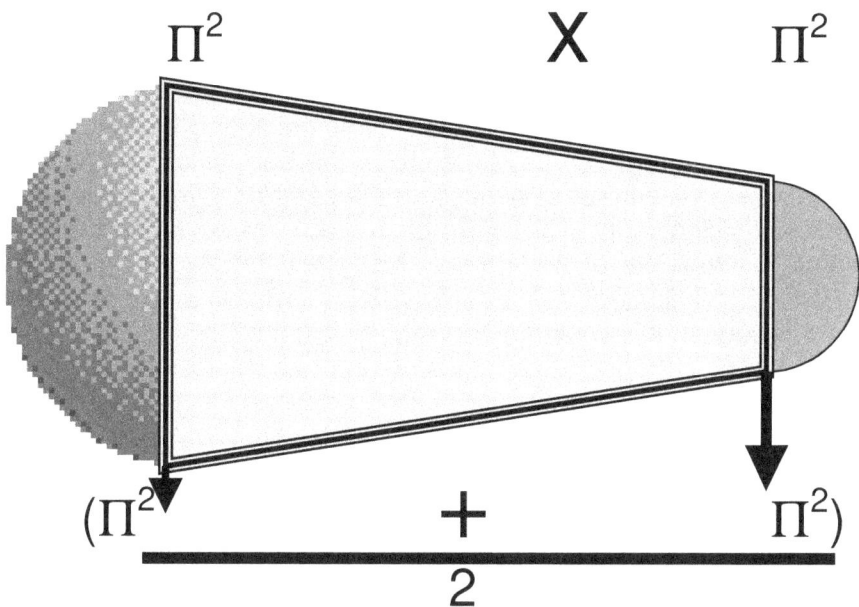

The sectors provide individual singularity a means in sustaining governing singularity by which provision comes through maintaining governing singularity the required spin in maintaining cooling. If this process did not apply, there would be no connecting individual singularity to major singularity.

Because there is a space that may not be occupied by a particle does not exclude the possibility of a particle sometime to the future occupying that space. If the space was nothing all possibility of future occupation will become excluded by the presence of zero that is unable ever to include occupation. I have to be persistent on this fact that zero does not form outer space because my work was rejected on several occasions because science insist on the madness that geodesic outer space holds a firm value of zero.

From the centre of the top runs the premier singularity and as the top starts rotating the top's rotation bring about the sides to singularity, which too was present all the time but filled the being, they're by the rotation of the top. Such rotrion involves four specific quarters.

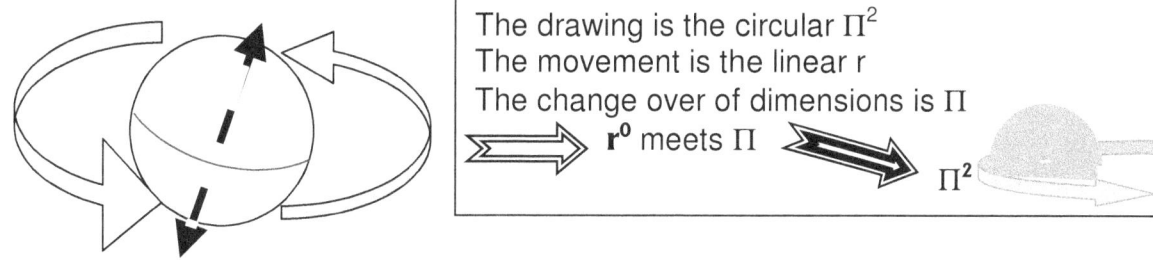

The drawing is the circular Π^2
The movement is the linear r
The change over of dimensions is Π
r^0 meets Π Π^2

In the action of the inseparable drawing closer and moving closer gravity finds the dual value of linear and circular gravity. There is no separation of the two factors acting as one but both have different application and values in the unit. This is the result of singularity having three parts acting as one but giving three distinctions in application.

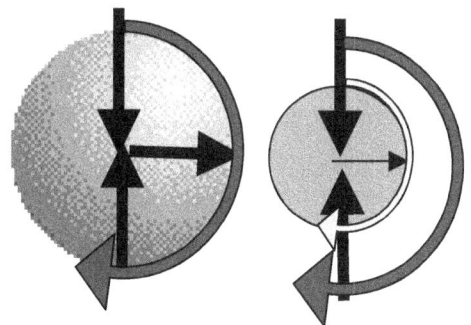
In both stars $(\Pi^2+\Pi^2)(\Pi^2\Pi)3 = 1836$ would apply as a relevancy but the two will apply each according to the gravity within. Within the boundaries of the major star the gravitational relevancy will be $A = (\Pi^2+\Pi^2)(\Pi^2\Pi)3 = 1836$ and within B the same will apply in accordance to that gravitational relevancy applying giving the atomic singularity as $B = (\Pi^2+\Pi^2)(\Pi^2\Pi)3 = 1836$. However since gravitational relevancy is not equal in both stars we have the fact that $\{A = (\Pi^2+\Pi^2)(\Pi^2\Pi)3 = 1836\} \neq [B = (\Pi^2+\Pi^2)(\Pi^2\Pi)3 = 1836]$

The Four Cosmic Pillars In terms of Singularity

Gravity is the dimensional changing of heat holding r as reference to the sphere holding Π as the reference. Heat occupying space has the cube that can apply r, as a straight line bringing about the cube with all its other names than may find attachment to specific form but nevertheless still remains only a six-sided cube with angle changing in some cases. When atoms are within the structure of the Sun, all the atoms form a combining legion that we humans are unable to measure or to witness. In the centre of every atom forms a governing centre singularity. The worth of this are vested within the centre governing singularity of the Earth. This gives the atom's forming the Earth a cosmic code of $(Π^2+Π^2)(Π^2Π)3 = 1836$. I am not explaining this code in this book but I do explain it in the Cosmic Code.

In every star the relevancy of the proton to electron ratio applies differently although the electron to proton ratio seems to equate the same and is the same $(Π^2+Π^2)(Π^2Π)3 = 1836$. That is the atomic displacement difference there is that space-time is reduced from the electron to the proton. I explain this ratio in much more detail in the Comic Code.

Being in the star or planet the ratio will have a different defining ratio because of the gravity of every star holds a different value which is determined by the governing singularity vested in the centre which is derived from the combination of the total of all the singularity centred within every atom forming the line $Π^0$ to $Πr^0$. It is when this ratio can't maintain an atomic equilibrium throughout all the layers in the star that we find a Super Nova blowouts occur in one or more of the layers of such a self destructing star.

It is the presence of the relevancy factor **k** that is responsible for the atomic occupying volumetric size **a³** as well as the gravity **T²** within every star within every atom in each independent star. The differentiation is the result of the governing singularity, which is the result of the combined value of all of the atoms forming the line that makes up $Πr^0$. That is what Kepler's formula o **a³ = T²k**. The curve of Π runs as a divide all along the outer rim of every star and this divide of Π is derived from every value that every atom forms by placing an atomic Π into the equation.

Again I repeat what I so often repeat when I refer to gravity in my own incapable and uneducated way: Whatever gravity is, gravity has to beΠ. If gravity is linked to mass as Newton stated, then mass has to be very closely connected toΠ in order to assume the role of being responsible for gravity. Looking at every aspect that forms gravity, it is formed by a circle that forms a circle in return. Gravity is the instalment of oneΠ initiating the followingΠ. The curve the Earth holds forms aΠ, which is responsible for the circle the Moon follows as the Moon circles by orbit the Earth. The Earth as much as the Sun as much as all stars and galactica holding gravity is round and the roundness areΠ. Taken much further into more philosophy we find the curvature of space-time, the fact that gravity bends light into a curve, this bending comes in the form of a circle that is formed byΠ. The Sun for instance spins around and that is formed byΠ. The Earth holds the Moon captured while the Moon circles around the Earth and the circle is a result ofΠ. If it is with gravity that the Moon circles around the Earth, then in all of this we must locate gravity holding Π as a value. It is this pattern hat I recognised when I formulated the Cosmic Code. I will give one example to indicate the gravitational worth ofΠ. The Universe starts to go three dimensional at a point where Π forms a three dimensional system which forms at a point $7/10 (Π^6) ÷6 = 112.16$ which is where the element table starts forming atoms of various relevancies running from $3((Π^3) = 93$. This I explain in the Cosmic Code.

The Four Cosmic Pillars In terms of Singularity

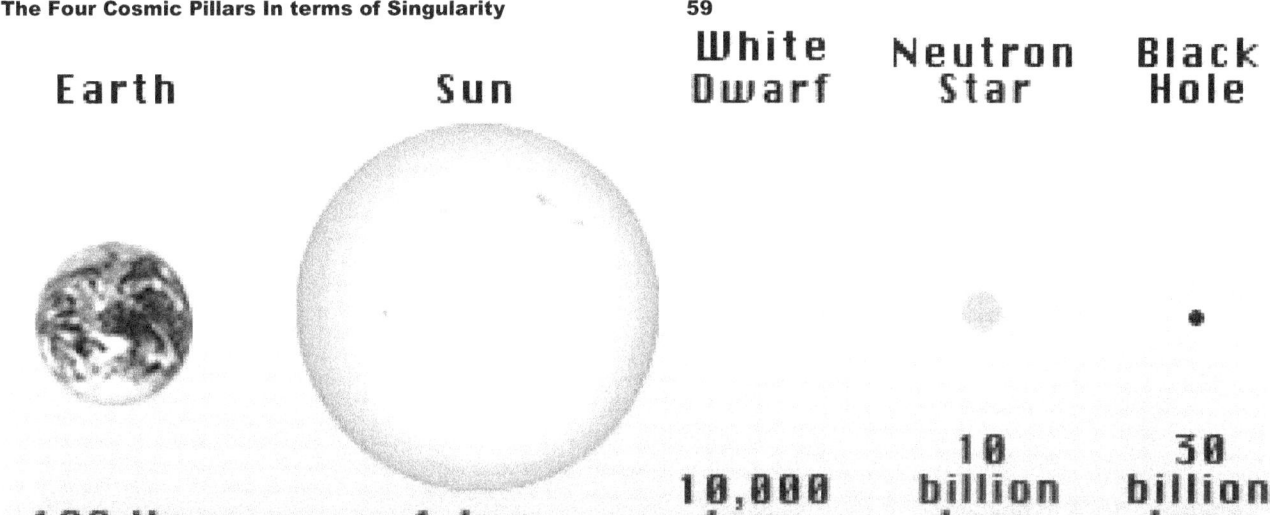

It is clear that the more "massive" the star gets, the denser the star becomes as the star decrease in volumetric space used. The star becomes considerably more compact as its "massiveness" increases, which can only reflect on the situation controlling the atom that controls the stars. The star is its atoms and the atoms control the governing singularity as much as the governing singularity takes charge of the compliment of the atoms in the relevancy value applying as $(\Pi^2+\Pi^2)(\Pi^2\Pi)3 = 1836$. In the more "massive stars" there are more atoms but the increase in atoms forms a reducing of occupied space because the relevancy applying condenses the atom's space-time occupied.

It is the compliment of atoms that form the star and it is the relevancy of the atoms within the star that forms the relevancy of the gravity applying within the specific star meaning that no star ever will have the precise relevancy of any other star. The relevancy applying in the star ($k = a^3 \div T^2$) and ($k^{-1} = T^2 \div a^3$) forms the space –time ($a^3 = T^2k$) that is responsible for the gravity in spin ($T^2 = a^3 \div k$) taking charge and all the while it relates to the governing singularity ($k^0 = a^3 \div T^2k$). It is an interwoven network of interacting relevancies all complying to the governing singularity that is attuned to the controlling singularity. The smaller the circle is formed by the gravity $T^2 = a^3 \div k$, the more compact will the space-time be ($a^3 = T^2k$) and the more control ($k = a^3 \div T^2$) and ($k^{-1} = T^2 \div a^3$) would singularity ($k^0 = a^3 \div T^2k$) have on the star. If this is put correctly in terms of Π applying as it should then it will be as follows: The smaller the circle is formed by the gravity $\Pi^2 = \Pi^3 \div \Pi$ and the more compact will the space-time be ($\Pi^3 = \Pi^2\Pi$) and the more control the relevancy ($\Pi=\Pi^3 \div \Pi^2$) and ($\Pi^{-1} = \Pi^2 \div \Pi$) have by committing the governing singularity ($\Pi = \Pi^3 \div \Pi^2$) and ($\Pi^{-1} = \Pi^2 \div \Pi^3$) have on the star.

The relevancy $(\Pi^2+\Pi^2)(\Pi^2\Pi)3 = 1836$ is valid only as it applies within a star and has a different meaning in relation to what the governing singularity applies. Everyone walks the Earth with the idea that the big stars are those charging "massive" gravity and the small stars has very little gravity because it is "mass" that brings size and the big stars have big sizes and therefore have big gravity fields. That is a load of rubbish and the Hertzsprung-Russell diagram, which is as much rubbish as the idea that mass that produces gravity.

In that there is no mention of mass simply because mass don't feature where gravity is mentioned.

Looking at the Solar system we find that all planets and objects not classified as planets and all things that is just simply forming solar debris has one thing in common…all apply the value of Π in the process where they orbit the Sun, which also uses the formation value of Π to construct the roundness the Sun has. Gravity has much more in common with Π than it will ever have with mass that is producing gravity. Wherever singularity forms gravity, it involves Π which then results in gravity manifesting as some or other form holding Π as a major factor.

However, it is important to note that **k** is not a value but it is a reference. The value of **k** or if you wish then use Π is in determining the gravity factor T^2 or if you wish then use $Π^2$. It is to indicate amongst many things the atomic relevance applying in a specific space a^3.

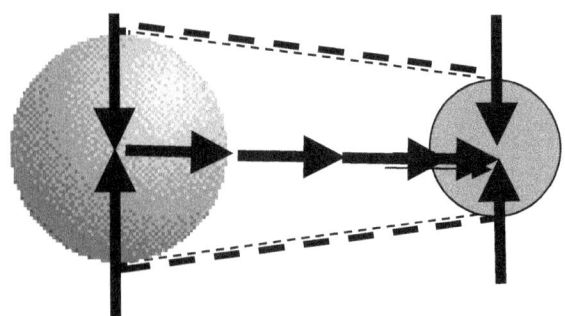

It is clear from the Roche limit applying that the governing singularity influence stretch at least as far as $Π^2 ÷ 4$, which then is 2.4674 times the radius of the star. When observing the Mon we can see the governing singularity influence stretches even further since the Moon uses the Earth's governing singularity as its own axis. We know that $a^3 = T^2 k$ as Kepler introduced space-time. The formula rests on the relevance **k** applying in order to give the atomic ratio the correct relevance.

In the book dealing with the "sound barrier" I indicated the barriers applying on movement and the limitation that the Roche limit brings to such movement at certain heights in the atmosphere. The movement I limits set by the governing singularity capping movement within its sphere of influence. The first sphere comes about at $Π^2 ÷ 2$ where the limit comes as part of creating a surrounding sphere in which the moving craft establishes individual space-time and the second barrier $Π^2 ÷ 4$ shows the outer limit of the atmosphere. Any object that exceeds these limits will become what became to the cosmic debris that destructs at Tunguska in 1908. The Roche limit of $Π^2 ÷ 4$ came about as the comet / meteor / meteorite/ very large piece of cosmic rock which entered the atmosphere at an excessive speed exceeding $7(3Π^2) (Π^2 ÷ 2) (Π^2 ÷ 4) (5Π^0)$ = 12618 km / h that is the Roche limit of the Earth.

The Earth relevance came in place and expanded the atomic relevance that liquefied the atoms of the incoming rock and turned the rock into liquid that then vaporised and turned into expanding heat within the atmosphere's confinement and the rest is history. If any thing enters at a velocity more or less or equal to $7(3Π^2) (Π^2 ÷ 2) (Π^2 ÷ 4) (5Π^0)$ = 12618 km / h or in excess of that limit, the object will turn into liquid. What type of object entered Tunguska is not important because it is not the objects' make up that is of concern but the Roche limit that came into the equation that set the entering object from a solid to a gas that holds the key. This very same phenomenon applies to every spacecraft that enters the atmosphere and in the event where the protective material does not cover the entire spacecraft. The Roche limit vaporised the spacecraft as it did with the space shuttle in 1987 as well as the space shuttle in 200? The protective layer did not cover the entire shuttle and allowed the Roche limit to expand the atomic material forming the shuttle. This had the entire craft vaporise into liquid air that we call flames. However, this also does occur every time a spacecraft enters the Earth's atmosphere and the blanket of liquid heat or flames that covers the craft is the result of the Roche limit coming into effect.

I am not going into detail as to what happens in the atoms because the explanation is comprehensive and I do that in **an Open Letter on Gravity.** There I show what happens inside atom when the Roche limit comes into effect.

In stars with the Roche limit applying the envelope forming Π expands, as the lesser star has to adopt the gravity $Π^2$ of the major star. In this all the atoms within the lesser star has to follow procedure and the atoms start to liquefy as the solid atoms turn into liquid heat. An atom is liquid space-time that spins into a solid structure because the Neutron exceeds the speed of light and the electron speeds heat up to the speed of light. That is the major functions that the atomic particles hold. When the rotation circle widens as result of the interfering of the major star, the speed of spinning slows down and the solid atom's density is brought into question. By not spinning fast enough, the compactness of the atoms in the lesser star expands and overheating takes place, just as it does in the process applying to the nuclear bomb. The atoms in the lesser star goes into a liquid state when those atoms in the lesser star excepts the limit capped by the major star's governing singularity and with the major star's governing singularity in charge of the applying relevance and placing the limit of Π in accordance to the major star's governing singularity, every aspect defining space-time within the lesser star goes array. This brings on the liquefying of the lesser star and by employing the Coanda effect the major star consumes the lesser star as a liquid.

This action I saw happening with the Roche limit then laid the foundation of my theory that everything in the Universe is heat in some or other state of development. There are heat controlled and compressed by rapid motion and there is heat without motion that expands as the controlled heat condenses heat from the expanding movement. The solid atoms consume the liquid space in order to maintain being cold and in that

The Four Cosmic Pillars In terms of Singularity

the relevance of density shifts from the liquid or outer space to the growing solids that increase in size. Nothing in the Universe is able to expand because there is nowhere to where such expanding can go and what we see as expanding has to be a shift of heat in density relevancy. I have written many letters on this matter.

The Roche limit is based on movement related to space within time. There are three forms of singularity that I so far have identified and I got around naming two of the three. The one forms as the structure rotates while honouring its axis and that I named the **Governing Singularity.** Then one forms as the object (for instance the Earth) honours the centre of the Sun and that I named the **Controlling singularity.** It is the movement forward which we see as going straight that takes the circling movement all the way around the Sun and brings the total influence of the entire structure to circle the Sun. Then there is the third one which would either be the atoms bringing about the governing singularity moving around the controlling singularity or it would be the governing singularity taking the controlling singularity one step further as the entire solar system moves around the centre of the Milky Way. This movement will be the third movement of singularity applying a reference. I have to come up with some name in order to describe the function.

The Roche limit is about movement and movement is about displacing material in space through time. At a distance of 12 756 km x $\Pi^2 \div 4$ km the lowest speed one could travel and still meet the equilibrium of gravity is the following formula $7(3\Pi^2)$ $(\Pi^2 \div 2)$ $(\Pi^2 \div 4)$ =2523.6 km / h. The value of Π^2 refers to movement at a specific height formed as Π^2 and being $7(3\Pi^2)$ km / h. Then the next level would be achieved at $(\Pi^2 \div 2)$ and the third level will be valid at $(\Pi^2 \div 4)$.

These borders forming the Roche limit are limits taking space into certain heights and depend on density applying. Also the distances applying in km / h are set to standards applying on Earth and only on Earth. It is taken in ratio of the movement the Earth applies as space moves through time.

However the time then applies to the Earth and only to the Earth. One cannot take the kilometre of the Earth and use it on Mars or on the Sun because the space movement going through time on those structures don't remotely apply as it does on Earth. A kilometre of movement on Earth will become (I guess) a billion trillion of Earth kilometres in relation to what would apply to a dwarf star such as the Sun and immeasurably in comparison to what applies when real dynamic stars are involved. The main concern is that atoms expand in relevancy by accepting the relevance of the major star applying and the movement speed up to a point where the atoms in the minor star just simply liquefies. The process whereby this happens is simple to understand once a person looks at the process of movement carefully.

The Four Cosmic Pillars In terms of Singularity

Everyone in general has this idea that moving an object is shifting the entire object from one location to the next location as if one would push a bicycle from here to there. When looking at accidents and the way a car crumbles and crumples in a collision it is clear that the car arrives at the scene of the accident in stages. The car in reality shifts from one point to the other point by relocating every fragment of every atom in breaking down the atom structure and rebuilding it at the next destiny. In this we find the two types holding singularity-movement in charge.

The governing singularity is taking all the movement around a centre and the controlling singularity shifts the object around a controlling structure as the Earth is to the Moon or the Sun is to the Earth. It is the task of the centre governing singularity to keep the atom in tact (T^2) while the movement of the entirety (a^3) relocates by the process of the controlling singularity and (k) shifts from one to the next location. Remember the governing singularity of the Earth is in relation of keeping the body of the structure of the Earth intact while the controlling singularity is vested in the Sun and is the orbit the Earth holds around the Sun and therefore the Sun holds the influence of the movement in relation to the controlling singularity. In the event of the major star taking control of the centre gravity of the minor star the controlling singularity is not able to sustain the integrity of the atom while the controlling singularity relocates the entire atom.

From the past
$k^1 = a^3 \div T^2$

Going into the present
$k^0 = a^3 \div T^2 k^{-1}$

Onto the future
$k^{-1} = T^2 \div a^3$

Movement cools heat in order to control expanding. With movement the object moving increases the occupied space-time and by increasing its volumetric space in use, it decreases the heat it holds by spreading the heat over a larger used area. Stars are quantities of heat that reduces the heat found in space by the movement it exerts in space. Outer space is heat expanded to the limit and stars are heat compressed to a point as far as the movement will cool the heat levels at that point in the cosmos. The reason why the top can spin erect with individual movement is that the top gained individual time applying within the influencing sphere of the time applying on Earth but the top has a time apart from the Earth.

I don't care much for the way science defines Universal time because when strictly applied as science defines Universal time then the Universe has time starting in Greenwich which is the main naval base for the British fleet and only a Brit will show that weak mentality as to think that the Universe has time starting in the British Navy headquarters. Brits all over think that the game of Cricket start and ends at Lords and Rugby is only played at Twickenham and British parliament rules the Globe…and time starts at Greenwich where the British fleet is based and built… and all this they believe just because they believe Britannia still rules the waves. I might even presume that that is why cosmology is in the Newtonian state it is in and that is bad! A Universe spinning that is as much as it is also holding singularity everywhere and in that everything that is spins around a centre holding singularity still within one hypothetical position represents time. It is eternity spinning around infinity where infinity can be any point but also be only one point leaving eternity to be all of the rest of what would form the entirety of entirety.

The movement of the Sun cools outer space to a point where outer space turns from the cosmic gas it is to a cosmic liquid. The Sun is a liquid fridge notwithstanding the overwhelming idea that says the contrary.

The Four Cosmic Pillars In terms of Singularity

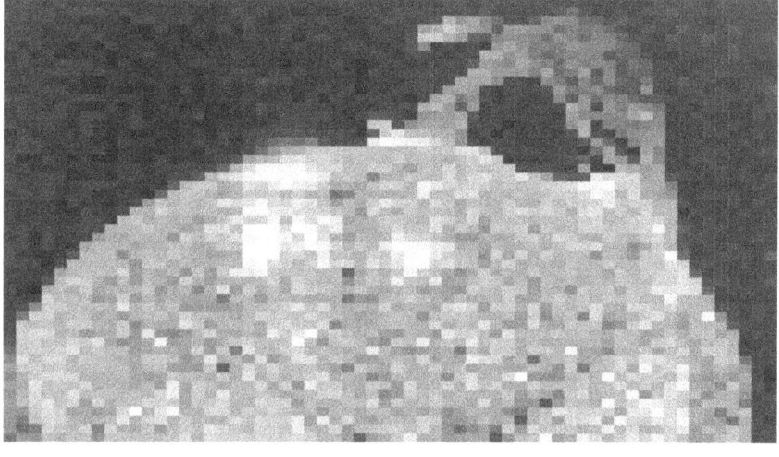

See the fluid push out of a bowl of liquid we think of as the Sun. That we see that is spilling both sides as it falls back into the Sun is not gas but it is liquid. The Prominence is liquid that falls back into a bowl of liquid and the Sun is one spinning sphere filled with liquid. The inside of the Sun is not gas but it is fluid. Stars are material floating in liquid and in that we have either material or fluid, but there is nothing else found in the cosmos than the two variations formed by singularity that is holding singularity in one or the other form.

Every one knows that a gas is one dimension HOTTER than Liquid as liquid again is one dimension HOTTER than being solid. The heat surrounding elements is not a match that fit all and is a separate issue in every case with every element. The heat /space /material relation gives elements their characteristics in form they are recognised by. Such characteristics changes as the heat /space/ material relation changes and alters the presumed characteristics

Hydrogen 1	melts at $-259°$ C,	boils at $-252°$ C,
Helium 2	melts at $-269°$ C	boils at $-268,9°$ C
LITHIUM 3	melts $180°$ C	boils at $1300°$
BERYLLIUM 4	melts at $1287°$C	boils at $2770°$C
BORON 5	melts at $2030°$ C	boils $2550°$ C
Carbon 6	melts at $804°$C	boils at $3470°$ C
Nitrogen 7	melts at $-210°$C	boils at $-195.8°$ C
Oxygen 8	melts at $-218.8°$C	boils at $-183°$ C
Fluorine 9	melts at $-219.6°$ C	boils at $-188.2°$ C
Neon 10	melts at $-248.59°$ C	boils at $-246°$ C

Melts (meaning that the element becomes a liquid) at $-259°$ C. **Hydrogen 1** boils (meaning that the element then becomes a gas) at $-252°$ C**, This does not concern the basic atomic element but only applies to the space on the outside of the atom**

In the Universe there is one substance, which is singularity that I call heat but that can just as well be space. The one falls into the group we think of as material and the other falls into a group we think of as liquid. In the centre and forming $\Pi^0\Pi$ we have heat very much controlled by movement representing the solid part and on the other side of movement we have liquid formed as $\Pi\Pi^2$ that holds no specific form.

The two forms I have just mentioned being gas and liquid are the same and the only difference there is, is the state in which elements can find form being one of the two formed by Singularity. It is a choice between the two forms of having liquid/gas and choosing solid elements. Holding elements we find heat forming a component that material could occupy where the space that forms will allow such a compact substance such as material but still it is heat that is forming space being vacant or filled. Space is one side and being compressed it is liquid which is a denser form of space or gas but it then manifests as the heat we know. There is a liquid which is a denser form of gas and then we have solids, which are a denser form of liquid, but in all it is still singularity that is heat in some form. The two forms holding the liquid I called the cosmic liquid and a cosmic gas where the liquid is the more compressed form of the gas and then the third form is also a much more compact form of heat which in that case we then call the state in which the elements are being a state of solids. The form of being liquid or gas or even solids alternates and the changes come about with more or less heat being part of the density factor. It is the intensity of movement bringing about the density that the space holds. The heat forms space when filling a larger area and then being more (when in the form of gas) or less (when in the form of liquid) or absent (when in the form of solids) establishes and reforms the state in which the elements cluster together or we have heat in a specific formation between atoms. Nevertheless liquid and gas is ranging in heat levels being compressed and dense or expanded and less dense where heat is the opposite of space but also is the very same thing that came about before the Big Bang event took place and represents a period at the time the cosmos was still forming the second form other than atomic solid elements.

By the way I leave just a thought: This could just as well be the "matter and the antimatter" that science is searching for when science claims there was a partition that came about at some point of cosmic development and as they put it the one part started devouring the other part in the matter vs. anti-matter fighting, but this is a solution I offer according to my opinion. Some particles became matter and others became antimatter or plasma or heat or just what you wish to call the by-product that came about from the friction that brought about the heat that led to the expanding of the space. Those forming units we call elements had established characteristics relating to space-time where each one holds an individual identity according to the number of protons the element has in one cluster unit. There are elements. The elements presume in the role of being solid and form as units the solidity of solid materials. Then there are liquids. Amongst those elements mentioned, we regard some of them as being natural "liquids" and others being natural "gasses", however they are solids notwithstanding what our culture call them. There are those we consider to be mainly gas or liquids and only then there are the state of being "frozen gas" (as we regard outer space to be) but such presuming underlines our mistaken culture we have.

In our human culture we swapped the definition of hot and cols around. When something is hot that something expands and what can be more expanded than outer space is. When something is cold that something becomes dense as it contracts and what can be more contracted than what solids inside an atom is. This is a nature law that humans in science missed and that could not be ignored. Within the Solar system what could be more contracted and denser than the atmosphere around the Sun. When that space becomes dense then to our view we have, that space becomes hot because we feel the heat. This is a long and arduous argument I could never complete in a book such as this but I do argue this opinion of mine to lengths in other books, which I have already written. A star is a compacting devise that cools the overheating outer space and in that controls the overheating by applying movement where there is a total lack in movement. In **THE VERACITY OF GRAVIY** I argue this point in much detail.

We can see the star that expanded due to overheating and in the process it shows the star holds liquid containing material. If the star was gas that expanded it would be invisible just as the black part surrounding the star is, but to be able to transmit heat in the form of light it has to be solid enough to be liquid because light is the most solid any cosmic liquid can get. If the star is liquid on the inside holding material, and the liquid evaporates when coming into contact with outer space such as the case is when the prominence squirts out into outer space, then outer space is the hottest, notwithstanding what ever boundaries and values we humans attach to the dimension. When something is compact it freezes by losing space. When something is hot it expands by gaining space. That is a law of physics no one can deny. When dealing with cosmology our human standards have to change to accommodate the rules laid down by the cosmos and not apply our personal interpretation of the cosmos to suit our rules of hot and cold, big and small, near and far. In the case of the Super Nova, something prevented the liquid turning into gas, therefore overheating before the event where the Super Nova took place. The liquid is cooled heat frozen by movement to form a liquid and the movement cooled the heat to become a cosmic lollypop. That movement or gravity spinning the star, which is what gravity is, 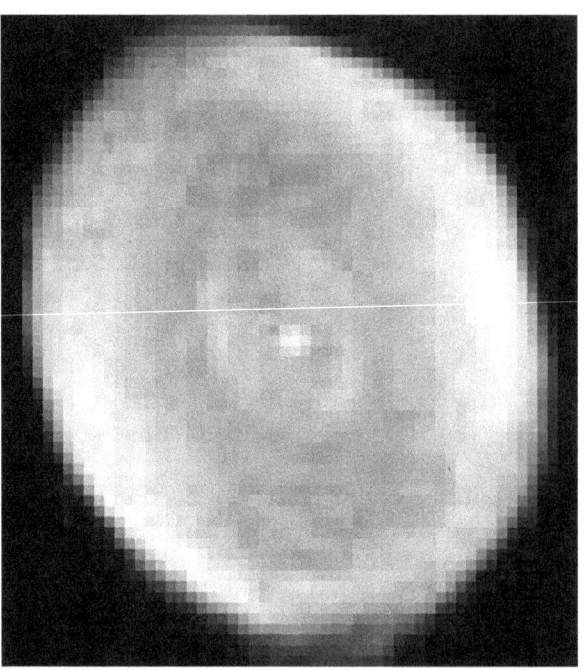 prevents the overheating by turning the layers within the star into frozen identities and the movement cools that star and prevents overheating, therefore it became a liquid outside the star. This turns the star into a miniature galactica, sustaining billions of individual points that represent singularity, because the governing singularity did not destroy, but the singularity of every nature is still in support of one another. From this picture (and others of Super Nova) one can learn a lot, if one is truly interested in applying cosmic law to the picture and not some human response to what we think would apply to an earth-like star that holds gas as an ingredient. A Super Nova is not a star holding gravity that has gone mad because gravity has no intellect to lose. A Super Nova is simply a star that was turning to slow in some or all of its layers and by spinning too slow there was insufficient cooling whereby the layers overheated and expanded. This is the Roche limit that allows the minor star to simply overheat into liquid.

Outer space is heat that overheated to such an extent it is still over heating because it is still increasing volumetric space. Outer space is heat in such a heated state it can expand no more but by the margin time allows it to do so. This statement explaining the process in which this happens is far too complex to discuss in this book and I do so in others that allow much more conversation. But with a slight bit of intellect the feasibility of my arguments are much more acceptable than to think gravity is magic that can go mad in Super Novas. This gives a real explaining to the cosmos and the explaining shows how simple and transparent the cosmos can be. The way science thinks of the cosmos and puts nothing into any equation that renders gravity the possibility of becoming magic and grants the cosmos magical powers is very loony in the least. By not being able to say what forms gravity, then one grants gravity magical powers and that leaves the cosmos in a fairytale with magic running long ad short. By excluding nothing from the equation space becomes something bringing in a value lying inside the realms of the infinite that must form singularity. Applying this logic to the Lagrangian system as well as the other three phenomena and interpreting that information to the law of Pythagoras a clear pattern comes about.

This takes us to the Coanda effect.

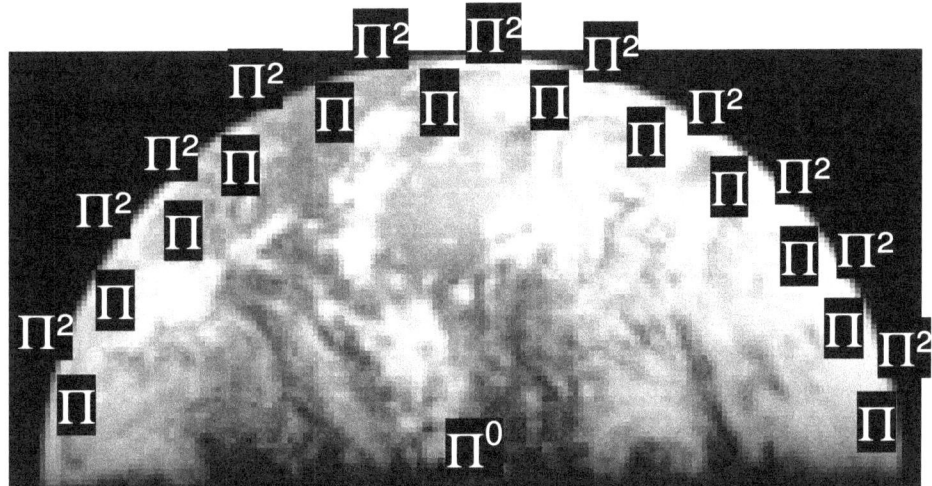

Again I have to press the thought that it is singularity determining space-time To think that matter can be solid liquid or gas in incorrect It is the condition of the space-time derived from singularity that places the form and conditions valuing into a balance the form of the elements between solids and liquid. Hydrogen can be as much a solid as gold can be a gas. Being solid or not depends on a ratio that exists between cosmic liquid and cosmic solids

By denouncing nothing brings about that:

Matter is singularity that turned solid by movement exceeding the speed of light.

Heat is singularity holding a density in which it is liquid that turned singularity to a cosmic fluid

Space is singularity formed the cosmic fluid that turned to a cosmic gas.

All elements are solids that are parted by a substance. The substance might be tiny in quantity, which then allows the atoms to be tightly packed, and this is a hard solid. When the substance parting atoms are more in ratio it forms a fluid and when it is much more in quantifiable volume that the solids are it forms a gas because it is the relation that requires the state of position the substance form in relation to the rest of the cosmos. Hydrogen is as much a solid as gold is a gas because it involves the unoccupied heat surrounding the element to form the state in which the form is that which represents the substance we see as material. This is what becomes clear when investigating an aircraft going past the sound barrier. It is the result of the heat that the structure of the craft builds up when dynamics change in the phase of changing singularity. In the structure one can note from the cracks in the body of the craft.

This the result of the surrounding heat amplifying, as the heat becomes space when the craft slows down in landing and becomes liquid heat again when the craft exceeds the sound barrier and the surrounding air heats again. The cooling and heating process happens in the air when the space parting the atoms forms cracks in the structure of the body of the craft as the heat and cooling thereof becomes permanent space or body cracks after it was heated too many times with too many heat sessions that left scars and traces of the process that applies. It proves that there are dimensional implications all around the body structure of the craft when flying and then landing once more and that the dimensions that changes are valid. The process of heating is the result of the Roche limit applying but it is because through the Coanda effect gravity is pushed to new limits.

Stars only become stars when the stars start to spin and develop its individual time distinguish it from the geodesic, meaning the stars interior remains solid allowing what is beyond the outer edges to become a relevant liquid when this relevancy is comparing to the geodesic of outer space and having outer space going onto a form of relative gas. When looking at the Earth we can see shows how the "atmosphere" of the Earth carries the value of Π^2 in relation to form the liquid as the surface forms the solid being Π. This is why mass becomes a factor of sorts.

The body standing on the surface holds a connection with all the other points serving the governing singularity and becomes part of the line formed between the governing singularity and the point holding Π and since it is part of the line and forms the eventual Π it forms a factor human can use in a measure of calculation. It is the differentiation brought about by movement that is distinguishing the solid space from the liquid cosmos by turning the Earth movement into a solid that puts space, which stands, still in a position of being a relative liquid. The liquid forms part of the rest of the cosmos being a liquid and it is including everything filling space not within the boundaries of the solid Earth.

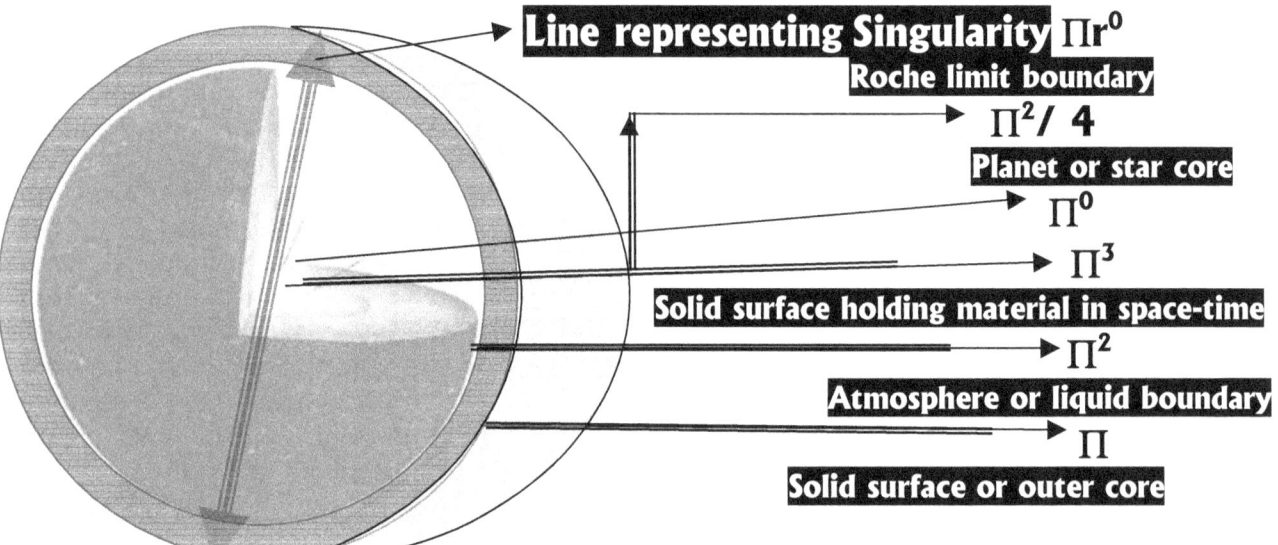

Line representing Singularity Πr^0
Roche limit boundary
$\Pi^2 / 4$
Planet or star core
Π^0
Π^3
Solid surface holding material in space-time
Π^2
Atmosphere or liquid boundary
Π
Solid surface or outer core

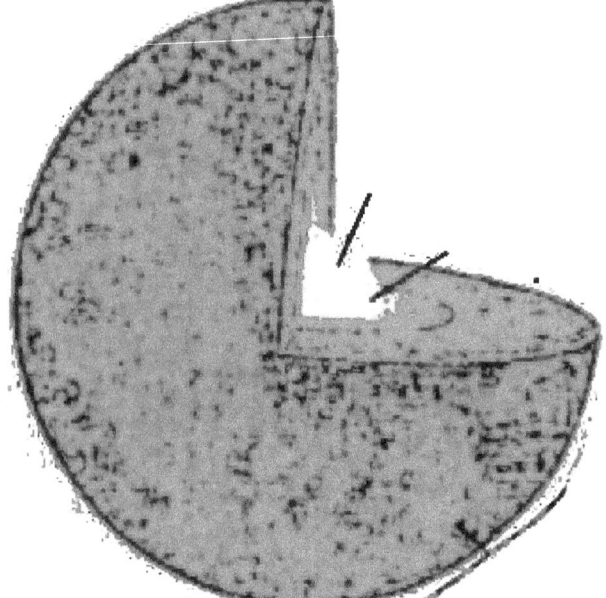

These boundaries are not specifics but relations to certain limits set from the position of singularity outwards. Te boundaries are set in relation to the Roche limit applying and setting such limits.
In the centre of material is singularity forming infinity.
On the outside of material is singularity forming eternity.

Between singularity in infinity and singularity in eternity movement introduces a divide we humans think of as matter or material. That in principle constitutes the Universe which by right is formed by singularity and singularity does not exist but for relevancy applying set by movement and the rest or individual elements of insignificance because size has no meaning in the cosmos.

Every atom is a small microscopic pump that draws heat through the space between all the atoms and mass is the restriction that the flow presents.

The star is a centrifugal pump that pumps heat as a coolant towards the centre of the star and this is dome by the protons in every atom. The protons expand and contracts and in that way it deliberately "sucks" heat from the outside through the electron going and flowing with the liquid neutron towards the proton that then delivers the heat to singularity in order to maintain singularity in temperature.

When a star starts to turn by forming an axis that forms singularity the solid part separates from the liquids where the spinning atoms form the solid and the rest becomes liquid having a fluid atmosphere. Only then can it start performing its duties as a star by converting space (gas) to fluid (heat) which then turns to matter and matter back to singularity. This is achieved by combining protons to denser proton clusters and in that manner serving singularity in a more supportive manner. If the Sun or any other star were using gas, it would not have the ability to generate the means of sustaining fusion because "gas" can't compress well enough to compress solids into bigger chunks of solids. Only heat as liquid in the stars would have the ability to compress material. This is how stars form and that is the basis for galactica being in place.

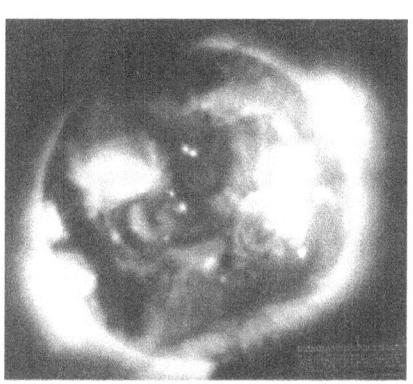

In **the picture of the Sun** we find not withstanding whatever name we attach to the **red liquid substance flowing from the Sun** into space and back to the Sun, **that liquid is heat** in a very direct form. **If outer space was the coldest place in the solar system** the heat **should** immediately **escape to outer space** and **not return to the Sun** as it clearly does because if outer space was the coldest place it will freeze everything and have the lot flow towards the shrinking cold...that it does not so therefore the Sun has to be the coldest because it is shrinking outer space having outer space flow towards the Sun. If **outer space were colder the heat would not return to the sun**. Kepler's figures prove that space within the entire solar system is flowing towards the Sun and that forms the gravity the Sun provides the planets as to keep the planets circling the Sun.

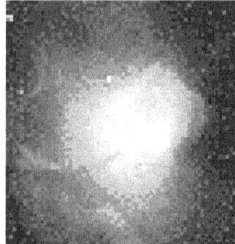

Stars can and stars do **overheat**, sometimes and the **Polar Regions** where **the Titius Bode matter-to-matter applies** holding the square matter (7+7) in relation to the square of space (10) and **other times** in a double relation to the **square of space** 10 to that of matter in a half square (7 /10 or 7/ 10). When saying the above, as I just said, one has to differentiate between heat and overheating because a star represents the coldest space in the Universe and not the hottest space as Newtonian wisdom wish to interpret when using Newtonian standards to read into laws applying in the cosmos. **Heat and cold are relevant dynamics** forming **in appreciation of singularity.**

The Sun is the coldest place in the solar system and that is fact notwithstanding what man tries to tell the cosmos how it should be because that view may fit mans perception. Looking at evidence the Sun provides, it contradict everything science wishes to believe about cold and hot. Science wish to see the cosmos through the eyes of what fits the needs sustaining life on Earth and what benefits in maintaining the surroundings as to support the security of life. Man has this novel idea to look for conditions as one find on Earth whereas life has no part in the cosmos except for being located on parts of the speck of dust we call Earth.

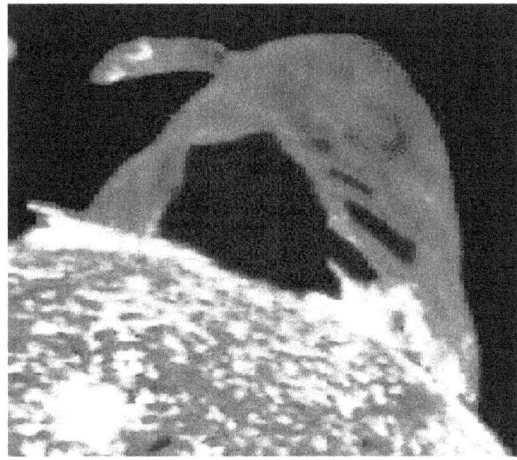

No person with normal eyesight can deny the fact that the prominence squirting from the surface of the Sun is a liquid substance or a fluid. This fluid can only be the purist liquid one may find anywhere. It is cosmic gas that chilled down to form cosmic heat in the face of the cosmic material within the Sun. When a substance such as liquid squirts into space in the manner that it happens within the Sun, this could only be as a result of high temperature differences occurring within the Sun where high volumes of heat makes contact with relative low temperature material within the Sun and the differences causes this violent expanding. If it is liquid squirting from the Sun it can only be liquid the squirting comes from and that makes the Sun not a hydrogen gas bowl but a liquid pool filled with the as pure a form of fluid any place in the cosmos.

Looking at the cosmos in an impartial manner we find that life is not the pinnacle of the cosmos and to the cosmos life doesn't even exist. In terms of the cosmos, life is an afterthought added on and in term of the entire cosmos as far as all evidence goes life is only located on one tiny spot we named Earth with no cosmic significance. If we wish to study the cosmos we have to look for the evidence that supports another view other than holding life as the panicle. Every aspect in **the cosmos is the very opposite of what** science believe it is. The Sun is **not a ball of gas but** a **giant sea of liquid**, frozen **without any** form of **gas or air** in the interior. Having a liquid interior **the Sun** has **no pressure** but has the **very opposite of pressure** to which there is yet no name given. **The liquid comes from singularity freezing** space-time within the atmosphere of **the Sun**, and such is the case with all stars still in the shining phase. **Stars more developed than the Sun is frozen solid causing fusion.**

Since motion in stars forms the condensing of cosmic space to become a liquid cosmic fluid, the movement causes a cosmic gas to go on to form cosmic liquid heat. That is condensation resulting from material spinning. The top spinning condenses the air around the top and that is what keeps the top erect. The spinning causes the differentiation brought about by the movement of material. The density of liquid heat results in much higher thrust and through the drive obtained from this liquid thrust this forms a much higher transfer of power than what a term such as mere pressure would suggest. When an explosion demolishes matter, no force in the cosmos will stop the accompanying destruction. This we find in nuclear destruction

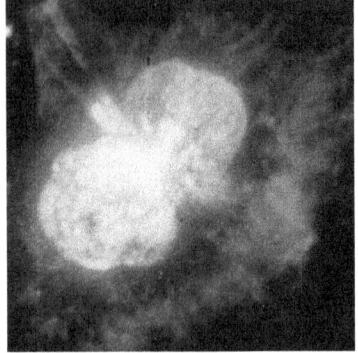

of atomic structures such as we have in Hiroshima and Nagasaki and the Bikini Island atomic test as well as in a limit way what we find present at Chernobyl.

A Super Nova is the same process but going on a scale man still has to invent words to describe the difference. The reaction starts because there is a massive unbalance in the relevancy of space occupied to heat bounded by matter to specific space occupied. The heat levels in the star surges because the movement of the spinning star is not sufficient to cool the star structure and the atoms are not equipped to generate a strong enough governing singularity to produce the gravity (or movement) to maintain the required heat levels and thus produce the required cooling through movement. A sudden super abundance of heat coming available places space occupied in a disadvantage to space available since all the available heat became available space through the process we refer to as an explosion. This process has all to do with motion not creating cooling conditions that will freeze space into a solid condensed liquid state.

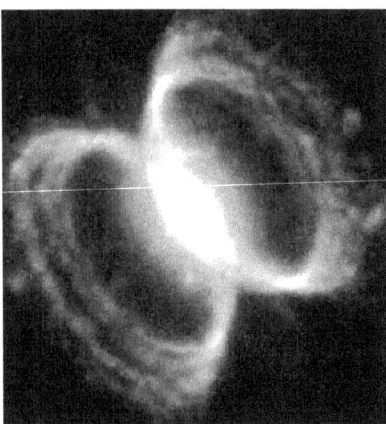

This exploding star is a picture of the release of heat that was previously held within the star. If the star were anything other that one bowl of liquid heat, this picture would not be what it is. It is clear that the released liquid heat tries to maintain a form of Π as the liquid through gravity tries to maintain a connection with the governing singularity. A picture such as this shows that a star is concentrated heat that can overheat and the only way it can overheat is when it's cooling goes array. The cooling comes from a process of movement and the more rapid the movement is the more it freezes everything onto singularity and then the colder and more compact the star is. Gravity is the cooling of space by implicating movement to concentrate heat that surrounds material. **All elements forming matter in** as much as being the heat concentrated by movement exceeding the speed of light when then forming **an atom can be** as much a **liquid as it is a gas and a solid**. With **little heat** in the ratio **all elements are a solid** and with **overwhelming heat** in the ratio **all elements forms a gas. In the cosmos there is no hot as there is no cold. It's about storing energy in space or in heat, which is another cosmic equal being opposing similarities.**

Hot and cold are **relevancies brought about by singularity valuating space-time** and during **the Big Bang** the Universe was **freezing cold** at **three billion trillion zillion degrees C**. The measure of heat is not important because if the one point forming the heated value was exceptionally hot, then the other point forming the cold limit was exceptionally cold. There has to be a zero to have heat finding a measured relevancy between two points forming the limit to hot and the limit to cold. It is the relation matter has with heat that provides the form the particle has at that moment. Say during the Big Bang space was 10^{37} degrees Centigrade in space then he points holding material were zero but if the points holding material were 10^{37} degrees Centigrade, then space was just about minus zero. If there is a scale, then there has to be limits to provide the scale legitimacy. The increasing or decreasing of the heat levels will alter the form of the element. Therefore all elements forming **matter is as much a liquid in composition or not than it is a solid or a gas. It is the space surrounding the atom** that hold the non solid base which provides the form the atom finds its relativity to the rest of the atoms it share space with. Hydrogen is as much a solid as tungsten is a gas and the form depends on the ratio of heat in relation to the solid within the space holding the matter. If the argument in reply is that it is the gravity pulling the heat back to the Sun, then that confirms my theory that gravity is all about reducing space and eventually confining heat to such a small area that gravity is about collecting heat onto matter and with that statement it then declares that outer space is being the hottest and therefore the most expanded place. It is the concentration of heat in space being relevant to form and that come about as cooling reduces heat where any increase in heat will bring about an increase in expanding. When a star is overheating then a star turns its liquid to gas whereby it merely transforms its interior to a relevancy it has from the pre- to post- Big Bang era.

We humans on Earth think that hydrogen is a liquid at − 259° C but that only apply to the Earth. Looking at pictures taken from the Sun we see in such pictures that it clearly shows the **heat in a liquid** flowing **from the Sun** and **back to the Sun**. In the **Sun the hydrogen holds enormous quantities of heat in a liquid at a temperature of 6500° C.** When a star has its singularity secured, the star is bitterly cold because it has heat in a liquid form flowing back to the point of singularity although we may regard the star to be rather on the hot side, however that observation serves human interpretation and is very bias to life. The cosmic truth is that the Sun (for instance) freezes hydrogen to a liquid form at 6500° C. The value of 6500° in terms of hydrogen being in one or another state in terms of human concepts is meaningless because the movement of the Sun and the cooling of conditions on the Sun is so much more influential than what applies on Earth that the drawing of comparisons are just proof we humans are without understanding. If hydrogen remains a liquid at 6500° C, just think how cold it must be as the star's interior approaches the point of singularity. Therefore fusing protons comes from cold and not from heat or pressure and in that way the process becomes sensible. By allowing the singularity to overheat all the atoms in the star must overheat and then the star overheats and heat within the star flows from singularity to outer space freely. In such an event outer space is then colder than the star because the heat releases to outer space with no intention of returning whereas under normal conditions in the Sun it returns as soon as it leaves. There are two ways to reduce heat; one is to bring about expanding space, as the photographs clearly show. The second one is where heat will reduce when an object is in motion by spin. When withholding motion or critically retarding motion then the reduction of movement will bring about that matter will overheat. Gravity is the motion of unoccupied space through the dimensional transformation to occupied space. This comes about as the star moves and therefore duplicates its position in space more rapidly and with that it will spread or distribute the heat over a larger area and in that bring about a drop in heat levels in the entire area.

Motion is cooling which is pace reduction and then to counter this statement not moving of space therefore is the anti-, the opposite, the negative and therefore expanding of space being in contrast to cooling that is the concentration of heat. Condensation of heat is a process of cooling space where this cooling process will rather seem as exaggerating the heat levels by heating the space and heating will lead to expanding which will to us seem as if it is cooling because the higher heat levels are also spread over a bigger area holding space. With singularity overheating the expansion of the singularity drives heat to form more space, creating space to compensate for the levels of overheating. **That is a natural phenomenon.** The only reason why **heat will** rather **flow back** to the star than **escape to outer** space once the star released it into outer space is **if outer space presents more heat than does the star,** because **heat always flows from hot to cold** no matter what influences may arise. Outer space must hold more heat than does the star but the accumulation of space in relation to heat makes it seem colder bringing expanding of heat to become space. <u>**Space and heat directly relates being the one form of the other**</u>.

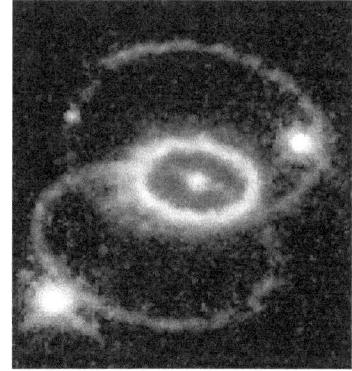

The cosmos is all about **converting space to heat** which we see **as gravity** and **returning heat to space** as a **control mechanism** always **keeping** a very delicate **balance** which we see as **a star shining or being normal.**
The purpose of the converting of space to heat is to supply the core a dose of heat to cool the point holding singularity where singularity is with heat. **It turns space to heat** sustaining matter but sometimes

singularity overheats and then matter converts to heat allowing heat to convert to space. That we call many names amongst others exploding into a Super Nova.

Whatever the names used is less important because the **process rests on space and heat interacting to form energy** but I am reluctant to use the word energy since energy is used for chocolate as much as it is used to describe petrol as much as it is used to describe life's vigour. The concept behind the use of energy in terminology is like a whore. It is everybody's wife while it actually belongs to no one. The release of heat to form space was what **the Big Bang** was and **the Hubble Constant** is all about where space not holding **matter converts heat to space**. I show that **space and heat is the very same thing** and there **is no such a thing as pressure** but releasing **heat produces space** and **concentrating heat reduces space** with the two interacting on singularity demand setting time to space with time being the spin or motion of heat in space. **Heat and space form the second singularity** caused by the **fragmenting of singularity to compensate overheating during the pre-** Big Bang matter forming era. That is what we see as **light and space,** which again is the **same thing and is fragmented** to hold points **serving singularity forming radiation and heat, where the star re-transfers heat back to space due to an overload.**

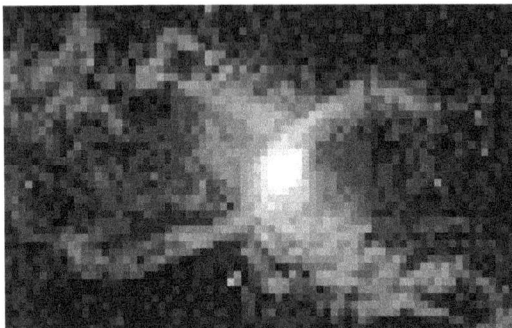

Looking at a star overheating it is obvious how singularity disconnects the control it has and the space once tightly packing the star by spinning movement losing control of the movement and the interior of the star then breaks into heat. By demolishing singularity it means $\Pi^0\Pi$ and $\Pi\Pi^2$ demolishes the very point holding singularity at Π. There is still a centre but the control the centre had is lost as the liquid forming fluid heat spills into outer space. However, take notice of the fact that the 7+7 /10 ratio of the Titius Bode law is still in effect.

The fact that stars overheat is never mentioned and in that the question never asked is why would stars overheat? We can blame pressure, but pressure would not bring about a star disintegrating from the centre, as the star depicted here clearly does. A burst from pressure should blow the sides out.

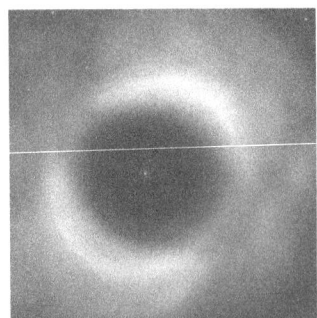

Stars we call Super Nova has blowouts. That man knows since before writing began, but since of late this phenomenon becomes more and more seemingly misunderstood. If stars blow as stars should and as we can clearly see from the picture, then the explosion happening to the star just above this Super Nova surely comes about from other principles. It is very obvious the two occurrences are not a result of the same basic method the Universe uses in destroying stars. When looking carefully to what happened the centre holding the once governing singularity is still present but the heat became space as the gravity became too slow to keep the integrity of the star structurally intact.

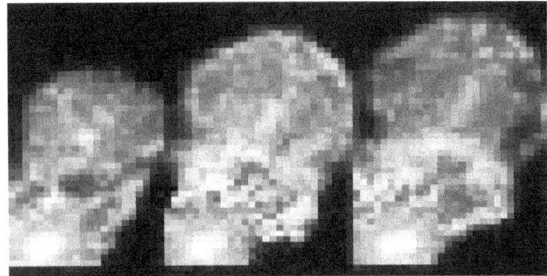

When heat surges and becomes too high, it turns into space. That process we call an explosion but are just heat reforming value as it reshapes concentrated heat into an equal measure in space. It is frequently seen, yet it is never acknowledged by science. When heat reduces by cooling, it relinquishes space in the producing of more concentrated heat, and this process we see is cooling. **On the other hand it is true that the reduction in controlled heat brings about increase in heat that manifests as space.**

Whatever the terms are used to describe a process there must be a recognising of the inter relation between heat and space where the reducing of the one will lead to the increase of the other.

The star does not apply pressure to bring about fusion, it freezes (actually it freezes time but that concept I argue in a far more specific book) and by getting time to stand almost still, the elements compress into fusion. This it does by applying millions and even billions of degrees Celsius. It is our conception of hot and cold bringing total confusion about the principles of cosmology.

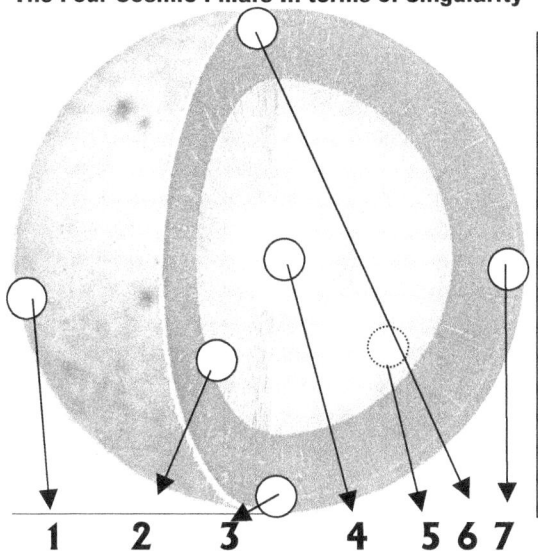

Elements do not determine form because any element can be a solid as much as it can be a liquid or a gas. It is the state of heat between the elements determining the state or form and that comes from singularity governing space-time. The layers in a star stand under separate conditions with space-time. In the sphere Π^0 in singularity holds equal Π in six specific points becoming a total of seven Π and no radius. When removing Π^0 Π in singularity the circle loses some and sometimes even all of its integrity. Where the singularity no longer control space, Π compensates by relinquishing space to r. where Π removed its value r becomes the square and the circle then becomes $\Pi r^0 = 10\Pi$. In the normal tongue we call it the atmosphere.

1   2   3   4   5 6 7

Every star is a Universe holding principles applying only to that star and standing on Earth while viewing the star it is impossible to fathom the values applying within the star because as we stand on Earth we can only value while being under the control of the governing singularity as we experience it apply on Earth. The fact that it can freeze heat to liquid surrounding hydrogen while holding a temperature of 6500^0 C should be an indication it is not what we seem to acknowledge as normal. The Sun is freezing hydrogen to a dense liquid at 6500^0 while space is boiling and over spilling into more space (expanding through overheating according to the Hubble Constant) at -273^0.

Science academics have to review there thoughts on relevancies because what seems to be hot is cold under certain circumstances and what seems to be cold to a point of freezing is boiling hot. The only constant applying is that there is no constant ever applying anywhere. There are no standard issue and fit all through out the Universe. Every point holding singularity attaches different criteria to borders controlling the space-time with in it rule. What fits humans on Earth does not even suit conditions everywhere on Earth let alone conditions applying on the Moon, yet science can't appreciate that Mars applies very different standards to that of every structure and every structure is a cosmos on its own turf, supplying its own turf.

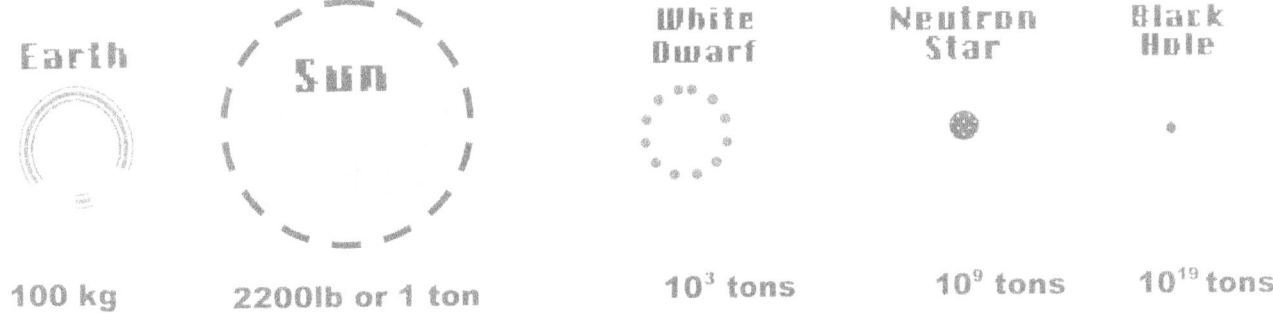

Every structure in the Universe applies $(\Pi^2+\Pi^2)(\Pi^2\Pi)(3) = 1836$ as the atomic relevancy but the range in difference are valid in a different way ranging from a red dwarf to a Black Hole. It is not the specifics that are of importance because the specifics change considerably because this is most apparent when taking into account that hydrogen remains in a fluid-like frozen state at 6500^0 C on the outskirts of the Sun and therefore it is obvious we have to look at other clues to give some indication of what is in process. On Earth in the time we have as a duration we find hydrogen freezing at minus $269\ ^0C$ as where it freezes on the Sun at 6500^0 C, which implicate the reduction of space to an enormous increase in time duration.

In conditions on Earth the rotating velocity of the electron is 3×10^5 km / sec. With conditions being that different it can not nearly be the same in the Sun. As space reduces time increases and this is a fact that is a constant. By having the space reduced to such an extent that it matches near Big Bang relevancies (a period where heat flowed like water and which is the very same conditions we find within the Sun) the time would apply accordingly. We also know that relevancies is all about conditions showing similarities under variables and therefore the space and heat component may seem altogether incompatible but is almost the same given the singularity presence within the Sun and comparing that to the Earth.
What is applying to stars inside the galactica centre is applying to particles inside the Sun.

The Four Cosmic Pillars In terms of Singularity

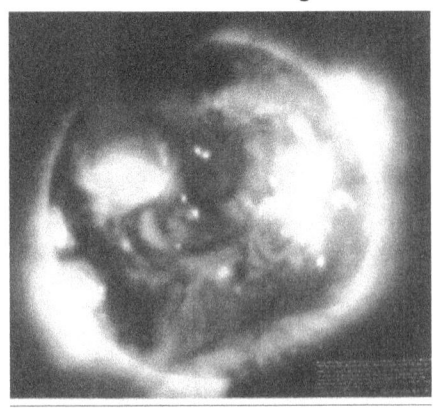

Science sees the nuclear reaction but do not recognise and therefore do not admit that the nuclear reaction is three different phases. One can see that the atom is "canned heat" bottled by spin into a canister. If you break the canister by letting the spin reduce, the heat escapes by going liquid and the liquid turns into gas buy forming gas. That stuff coming out of the atom when exploding reminds very strongly of the liquid coming out of the Sun but that doesn't make the Sun one big atom bomb because then the Sun would reduce by losing its gravity-supplying material. It just says the Sun is what makes the atom container containing what the atoms contain as heat "canned". The Sun helps the atoms within the Sun puts heat into the already existing container by using the four cosmic principles. At the beginning of the nuclear explosion process all the heat is solid, placed in a container by nature and the container has a human name called the atom nucleus. In the atomic explosion there are three ingredients that are distinctly apart. When the solid melts down, it becomes a fluid. The fluid we gave the name of light. There is not enough space to explain the detail of the argument, but light is not a gas, it is a fluid. The first step of the nuclear explosion is converting the solid to liquid. In the liquid state the star does not overheat. The overheating becomes part of the second phase. That phase involves the turning of the heat-fluid to a heat-gas we call space. Space is heat overheated creating space, as heat is space concentrated creating a fluid or liquid not yet correctly named.

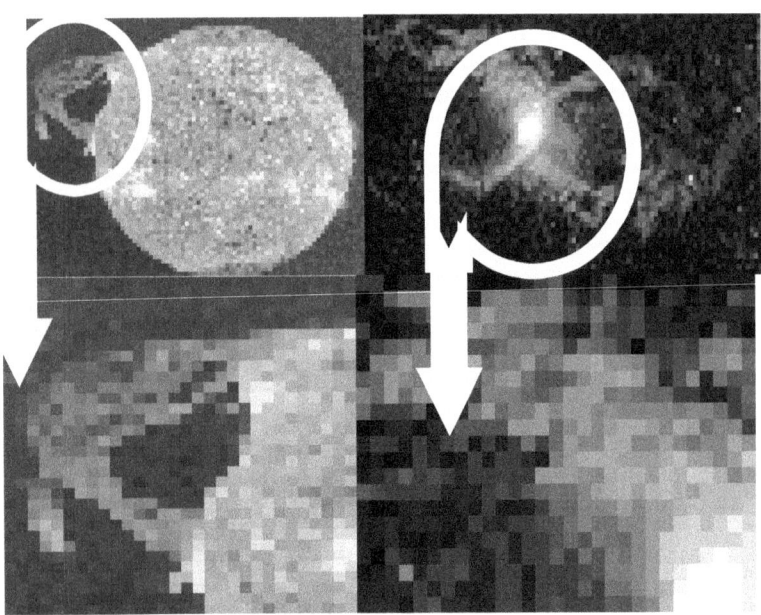

In the pictures to the right the appearance is more than obvious that the overheating of the star at the core is heating the interior to a level where is hotter than it was when movement controlled the structures integrity and a blow out into outer space took place and that brings about that the heat will flow to a colder region. In the one case (the Sun) the star is overheating at the edge and blows liquid in a squirt into outer space. There is an obvious difference between the Super Nova that the governing singularity could no longer protect its integrity and the difference of a star NOT overheating being "normal with liquid pouring from it and then becomes a gas that evaporates and condenses back to Sun. This is more the result of compressing the heat into the Sun and it hits the cold within the Sun. Gravity is the reducing of heat

After establishing the reference point to either singularity reduction or space-time enhancing through allowing matter to grow, the Titius Bode law applies, which I have explained in the pages preceding this page. Total annihilation and destruction of the singularity in one object may result in the object fragmenting to smaller parts where each part will still hold singularity, affected by less matter claiming space.

If mass is responsible for producing gravity and if gravity is responsible for keeping the Universe in check, then mass has to be one part liquid and the other part solid and if not then Newton had it wrong all this time. The cosmos is about liquids interacting with solids and that is what all four phenomena prove. Science can go on to deny this and remain stubbornly witless in their concept as to what goes on in cosmology, or mainstream science can reach a point where they admit that Newton made errors and Newton explained everything in physics, but should be kept out of astrophysics in total.

Every view of the Universe tells a story of liquid being contracted by solids and when movement can't keep the lot together by enlisting sufficient gravity, the lot goes array as liquid expand again and release back into the heated outer space.

The Four Cosmic Pillars In terms of Singularity

When starting to explain the Coanda effect I wish to start again with explaining the forms coming from the divide brought about by $\Pi^0\Pi$ going on to Π^2. The Coanda effect holds its value in the differences there is between singularity represented by liquid and singularity represented by solids. What ever is in the cosmos a visible or otherwise hold it's meaning

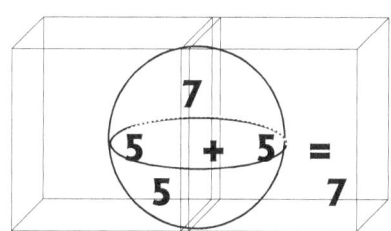

The TITIUS BODE

Principle Inside the sphere
7 / 10
5   = 7 / 10

Space-time is a four dimensional position of the Universe where the position of an object is specified by three coordinates in space and one position in time

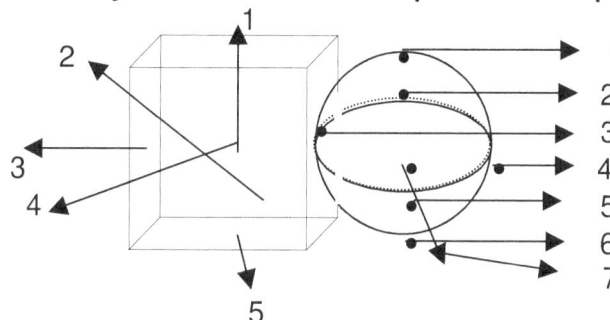

1 As the meeting of r points to a very
2 distinct different r in direction
3 such a point of meeting opposes
4 the other points in meeting and
5 will lead to destruction of the form
6 Π in any the event of any value
7 changes by Π changing Π^2 and r.

By coming into contact with the sphere the cube loses on dimension to the seven dimensions dominating six bringing about that the cube then has 5 sides to the seven of the cube. That is the Lagrangian system with five cosmic atoms holding relevancy to the centre cosmic atom where the centre cosmic atom stands in for seven and the orbiting cosmic atoms standing in for five positions in space. There is a more explicate explanation about this somewhere else in this book.

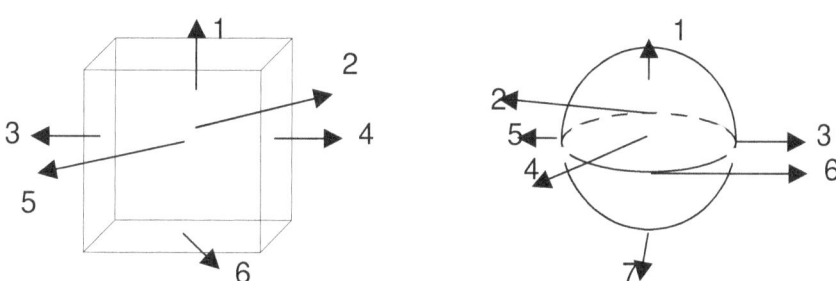

In the cosmic sphere there is no radius but only the extending 1^0 to 1^1 or of Πr^0 running from the centre Π^0 in six opposing directions relating to one another by the square of Π but remaining Π because of the unity the matter holds in relating to space. In every sphere there then are the six $\Pi^0\Pi$ relating in precise dimensional and positional equality to the centre Π^0 as well as to one another by 90^0 and 180^0 implicating the dimensional positioning. Therefore the sphere holds $_7\Pi^0$ and the cube holds $6r^2$

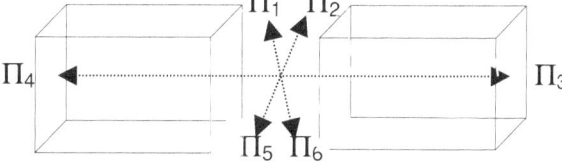

This is where the cosmos starts. This is where the element table has its root value. The Universe turns dimensional coming from being flat at Π^0 going to Πr^0 and then going $7/10$ $(\Pi^6)/6 = 112$. This is where the cosmos becomes a sphere (Π^6) in relation to the cube (outer space) having six sides (/6) in which the sphere turns (7/10). I explain many more such cosmic codes in **The Cosmic Code.**

Seeing our spinning top from the top, there are four quarters opposing each other and by that opposing one another.

It is moreover the individual singularity in maintaining the major singularity, which sustains the governing singularity providing equilibrium in space-time.

In Newtonian science there are more forces flying around than there are ghosts in an old cemetery. They even award forces to principles within the Coanda effect where the water receives its very own force or spook and the glass receives its own personal force or spook and all of this comes on top of the already existing four forces flying all over the show.

Newtonians, would you please believe me when I say there are no spooks, ghosts, fairies and forces and you can start to feel safe at night in the darkness with evil bats flying around in the darkness trying to haunt you because witchcraft is an old wives tale. Gravity is not the result of magic or unseen and unknown forces hiding within the atom and inexplicably trying to pull whatever it sees and then capture it by way of magic. The Newtonian way of looking at science is laughable when considering they are dreaming of something hiding within the star or the atom that grabs the next star billions of kilometres away through a vastness that offers nothing as a valid commodity and the grabbing hooks they call the graviton which hides somewhere in the mass of atom that then gives mass to the star that then gives gravity to the star with which it pulls and so the magic goes on.

The condition for the presence of this spot holding singularity the movement of Π^2 that is initiated by Π that forms a divide between eternity or that which is in the Universe that can never reduce because that is what can never start and singularity which forms eternity which is that part that can never end because that part can never increase.

The Earth spins through space at a value of $(10/7 = 1.42)$ which then represents the interaction between space reducing and the Earth spinning. This is one very crucial part of the Coanda effect in establishing gravity. Secondly the Earth and all objects spin in a double motion where the moving straight represents the controlling singularity puts a spin directional diversion of 7° and the governing singularity puts in another directional diverting spin of 7° forming a total of $(7+7) = 14$. With the atmosphere (space-time) condensing by a value of 7 taken from the movement of the Earth and 10 as a value compressing the atmosphere the 10 forming a relation with 7 brings about space and time or solids (7) and air (10) mixing. I have already given the mathematical multiplication that then produces the movement of Π to form gravity or the numerical value of Π^2. In one dimension space became 10 and in that same dimension matter became seven. In order to separate matter (7) and space (10) through time (the spinning of matter) (7) in space (10) and space (10) spinning the matter (7) the following result came about through the application of the Roche principle $(\Pi/2)^2$. The idea of space compressing is what forms the principle that makes the Coanda effect viable.

If not for the moving of Π to form gravity or Π^2 the value of Π^0 will run across the Universe as wide as it goes and hold singularity stretching into both in infinity as well as eternity with no breaking of the two time limits. As I have shown. The value of singularity holds no s[ace and with no parting of space the entire Universe will fall back into one point holding no space. The entire Universe will form one enormous Black Hole where time will disappear into and the Universe will again be no more because infinity then again locked eternity onto one spot holding time captured in one

The Four Cosmic Pillars In terms of Singularity

position. What forms the entirety called the Universe is the parting that Π brings into the Universe that gives credence to whatever has a measured meaning in the Universe.

Space mingles with time 10/7 and space separates from time 7/10. In this, the result is that matter forms an accumulating movement stretching into space allowing the solid to gain a time value. Time divided into space $(10/7) \div (7/10) = 2{,}04$
$(10/7) = 1{,}4285 \div (7/10)\ 0{,}7 = 2{,}04$

Then also space moves towards matter compressing the cosmic gas to form cosmic liquid and this can become as dense as any solid could be. Light can cut through any material.

Space divided into time $(7/10) \div (10/7) = 0{,}49$
$(7/10)\ 0{,}7 \div (10/7)\ 1{,}4285 = 0{,}49$

Then space multiplied by time $(7/10) \times (10/7) = 2{,}04$
$(7/10)\ 0{,}7 \times (10/7)\ 1{,}4285 = 2{,}04$

That brings about that the Roche Principle
worked both ways (double)   1. $(7/10)\ 2{,}04 \times (\Pi/2)^2 = 5{,}033$
and with multiplication   2. $(7/10) \times (10/7) = 2{,}04\ (\Pi/2)^2 = 5{,}033$

Resulting in the combined value of $5{,}033 + 5{,}033 = 10{,}066$

On the other side the other combined value came to $0{,}49 + 0{,}49 = 0{,}98$

And the result from this product was

$0{,}98 \times 10{,}66$

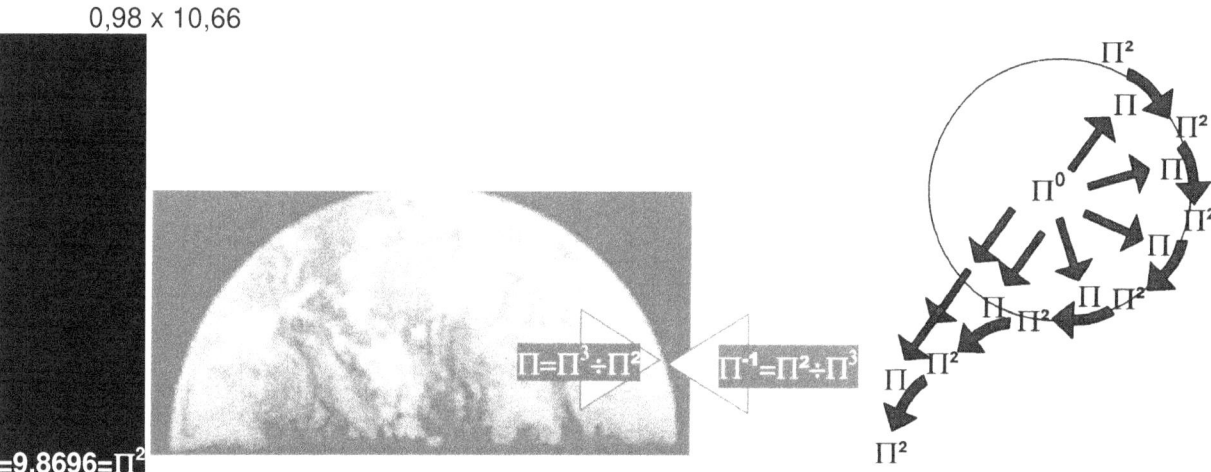

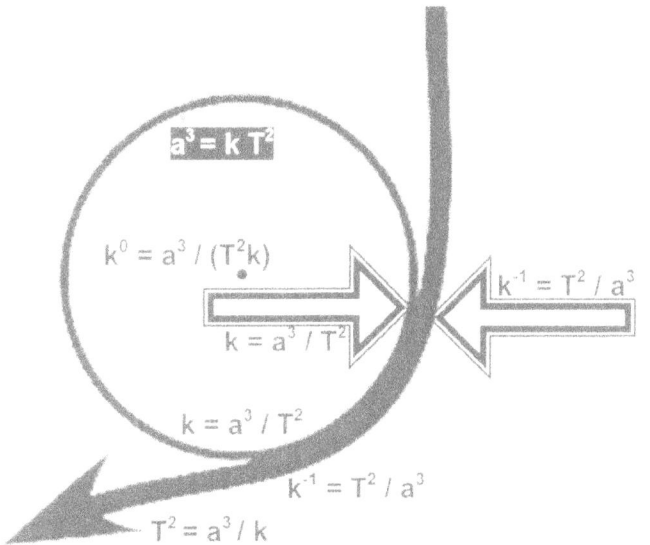

This shows that everything, controls everything and is everything is the centralised $k^0 = a^3/(T^2 k)$ singularity that forms by movement $T^2 = a^3/k$ of space $a^3 = k\ T^2$ in relevancy $k = a^3/T^2$ both ways $k^{-1} = T^2/a^3$ (Newton's 3rd law) thereof.

The only way one could explain the Coanda effect is by investigating the way Kepler interpreted the cosmos through the manner in which Kepler stated his formula. This is done by completely ignoring Newton's complete misconception into the work of Kepler. In this way it is also true that the only way one could explain the cosmos effectively is by investigating the way Kepler interpreted the cosmos through the manner in which Kepler stated his formula. This then too is done by **completely ignoring Newton's complete misconception into the work of Kepler.**

This explains the Coanda effect and the Coanda effect is gravity and gravity "glues" the water to the glass in the same manner as how gravity "glues" air into the Earth or cosmic fluid onto the Sun.!

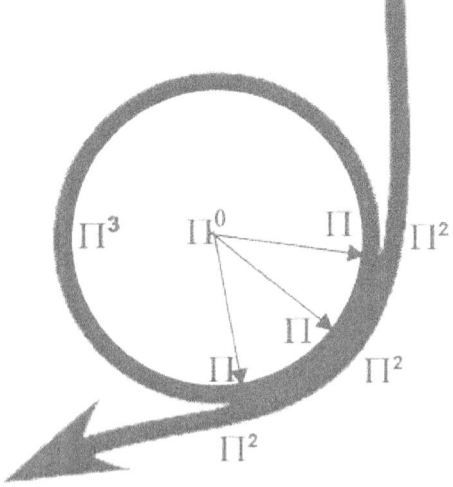

This process happens to all spinning things and as much as it happens to a piston connected to a crankshaft, just as much this will happen to a atom spinning an electron in a similar manner as a the crankshaft is spinning holding a piston connected. This concludes that $\Pi^3 = \Pi^2\Pi$, which when said in terms of mathematics seems very simple, but it explains an entire Universe formed by the movement thereof.

In order to initiate centre singularity Π^0, gravity Π^2 forming movement of space Π^3 is required. To initiate space Π^3 singularity Π extended has to be confirmed by movement Π^2. To secure space Π^3 movement thereof Π^2 must install space-time $\Pi^3 = \Pi^2\Pi$ that confirms singularity Π^0 forming solid Π^3 as well as singularity Π^0 confirming the liquid $\Pi^2\Pi$. This is what Kepler's formula brought to human knowledge and not once did I mention or indicate to mass being prominent or having importance or even being relevant.

This not only defines and confirms the Coanda effect, but it describes what happens with the Coanda effect and the Coanda effect is the result of the other three phenomena uniting and forming the culminating result. The **Coanda effect** is the coming together of the **Roche limit,** the **Titius Bode law** and the **Lagrangian system**, to form gravity not by mass but by movement of space filled by material in relation to fluid.

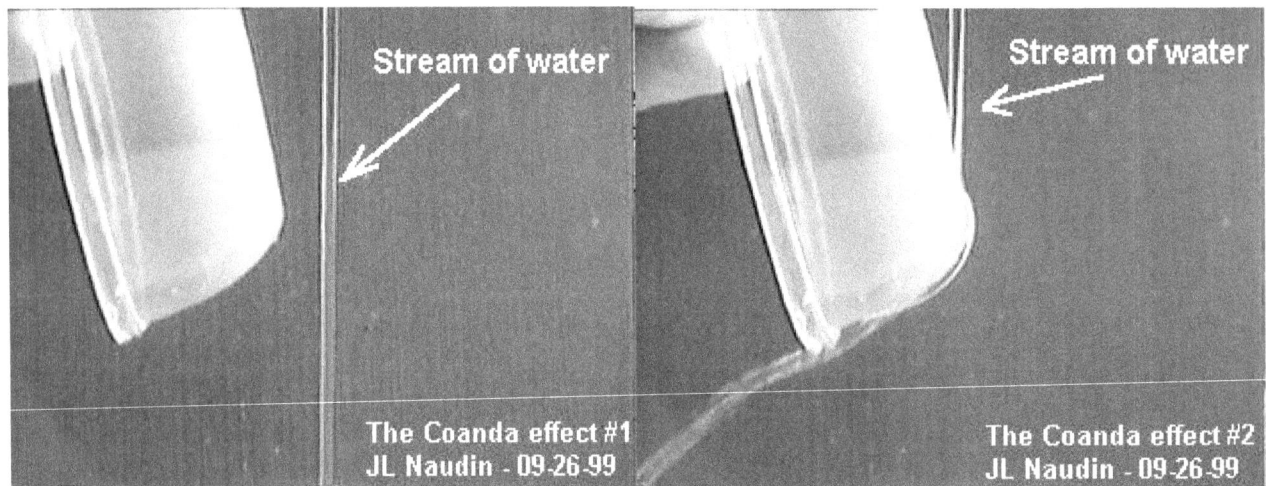

This proves that gravity is the Coanda effect and in another book I prove that the Coanda effect has its origins in Π forming a value and that value forms gravity.

THE LANGRANGIAN FIVE POINTS

In order to understand physics applying in cosmology I had to start by dissecting the set-up forming pi.

At this point I can introduce my theory on the ***Absolute Relevancy of Singularity*** At the point in the centre of the circle line must start. In the beginning when I explained the way I figured how the line start I said a lot of dots has to continue in order to form a line. It would be 1 + 1 + 1 etc. because the line must form by holding singularity after that point does mathematics begin but in the line that forms all factors holds 1. The lie can only form when all the points forming the line have the value of 1 being 1^0. In that conclusion one realises something must separate singularity from all other factors because singularity hosts all other factors but is by own initiative Π. Only when singularity meets the end value can the end value have Π where the final ring of the spinning circle forms Π. That will be the spot of origin forming the relevance in Π. That will hold the eternal spot…the smallest spot ever because all spots that ever can be were secured in a position in the centre of that spot that must continue as a line that forms. Because of the progress singularity follows from the single dimension singularity only allow mathematics a start at Π^0 progressing further too onto Π^0 and from there the line is born as $\Pi^0\Pi^0\Pi^0$ and to $\Pi^0\Pi^0\Pi^0\Pi^0$ etc. where Π^0 then may form the concept and value of r. But the line starts at $\Pi^0 = r^0$. This forms because cosmology is singularity based and the value is $\Pi\Pi^0$. This line $\Pi^0\Pi^0\Pi^0$ of singularity can only continue because every spinning atom preserves Π^0 in the very centre and since $\Pi^0 = \Pi^0 = \Pi^0$ the line is the same without finding conclusion except at the end where it forms mass at Π. At the point where Π forms, the movement Π^2 of the circle defines the space Π^3 of the circle and it confirms the centre Π^0 of the circle through the rotation. Let's call this the solid forming or if you wish, let's call it Kepler's singularity. After that singularity forms a line $\Pi^0 = \Pi^0 = \Pi^0$ where this forms another line again as Newton stipulated it by $\frac{dJ}{dt} = 1^0$. Let's call that the liquid singularity or Newton's singularity and the relevance of singularity having a solid base compared to the singularity holding a liquid base comes about by the movement of gravity.

The cosmos started off with one dot so small eternity met infinity within. Then came one more dot holding eternal heat away from infinite cold, and this parting of heat from cold brought on another and another as the same repeated over and over. This parting of heat from cold was the start of the lot because this process is still ongoing today. I have an entire book written what this process involved as I interpret the interaction of the four phenomena into the process. This parting of the dot from the spot or parting of heat from cold then continued until there were a countless number of dots being as many as we now have that forms the cosmos we see. The accumulative size of the dots were the same size as just one dot was at the time because in the true Universe big and small plays no part of what holds value to the Universe we see. The dots were infinitely small and eternally big at the same time because size is a relevancy and without one the other has no size. So in the true perception, there is no difference in size.

What ever is started with the fact that there is no place or part in with which one may associate zero or nothing. There are no room for a number such as nothing. Every spot in the Universe has an eternal as well as an infinite value, which excludes nothing, or zero totally except in the understanding of the cosmos by Newtonians. Next to the one dot (infinitely close leaving no space to part the two points that represents singularity) one will find the next dot, and if nothing was a factor then that is precisely what one will find the two dots then are. The space parting the two points has no space but for the movement that forms Π where it is Π that grants the points individual characteristics. If it were not for the moving it was a non existing entity, taking up no space, and much more important, no time, therefore the dots are infinitely close to one another, being the same space, while at the same time being eternally big as much as infinitely small. If we as humans cannot find a manner in comprehending this notion, there can be no manner ever understanding the cosmos as much as the start to the cosmos.

When time began every dot was a Universe in its own and the accumulation was a Universe by merit of relevancy applying. The Earth in itself is a Universe as the moon is a Universe, as every atom is a Universe formed by Π^0 connecting Π and Π connecting Π^2 because rules applying on Earth do not apply on the moon and visa versa. When in the ocean another set of rules apply, therefore being in the sea places a body in another Universe. The number of Universal entities is still countless, just as much as it was in the beginning.

Every position in the Universe either holds singularity in a form, or relates to singularity because every spot there is, is also singularity. There can be no position unrelated to singularity therefore every aspect of the cosmos is space-time in various forms under the provision of singularity connecting. Matter cannot be if not being surrounding by singularity and secluding singularity.

Singularity is as close as any spot can ever come to zero BUT IT CANNOT EVER BE ZERO. From singularity diverts space-time and there cannot be space without time as much as there cannot be time without space, not withstanding the size of space or duration of time.

Every dot insignificantly small as it may be, is a part of another Universe as much as it is part of the accumulative Universe and every dot in the infinity holds singularity, which we translate as " nothing" being "darkness or being beyond the noticeable". There cannot be "nothing" just as much as there cannot be "darkness". There cannot be something big or small, but when it is put into relevancy of perception, and then the relativity of perception becomes the question. There cannot be hot as much as there cannot be cold, big or small, far or near, bright or dark because every "boundary" mentioned is just a form of development processes we humans attach to something we will never fathom. The Sun FREEZES hydrogen to a liquid at six and a half thousand degrees Celsius and Universe boils over in the form of the Hubble constant at the temperature (we presume from our vantage point) at minus 273 degrees C. If we Humans cannot or will not abandon our human perception and our manly perspective, we may as well return to astrology for all its worth. Even the atheist places life in the pivot of the Universe while in reality life is on one very small insignificant and unimportant planet spinning around an even less prominent underdeveloped star.

Space-time is a four dimensional position of the Universe where the position of an object is specified by three coordinates in space and one position in time.

With singularity placed in infinity within the centre of every rotating object every atom and its relation to its surroundings including other atoms form space-time diverting from the point holding singularity as far as rotation goes because every object holds three relative positions in as far as where it was, where it is and where it will be in relation to singularity providing time. I elaborate on this else where.

From singularity Π^0 the line runs to both sides forming the edges of singularity Π forming the border through singularity moving Π^2. All aspects of the cosmos hold two halves in four quarters while rotating. One the left side (1) there are two directional changes (2) and on the right side (1+1) there are two more directional changes (4). Each of the halves and all of the quarters are in direct opposition to each other as much as to one another. Within rotating 180° the one point will represent all it had just 180° before in the very opposing way. This is why the Earth's magnetic fields swap every ($\Pi \times 10^3$) year and ice ages change which is an ice desert into heated desert drought cycles and back again. From any point all space is moving from one side of the Universe to the other side of the Universe in quarterly displacement. The time it takes the movement will be the movement from the starting point at a value of Π^0 to an ending point holding that moment of infinity to a value in eternity therefore bringing the square of Π being Π^2 which then forms the value of time. Because time is the movement bringing about the change in quarters from any given point to any other given point nothing can be in two places of the Universe simultaneously

THE LAGRANGIAN SYSTEM.

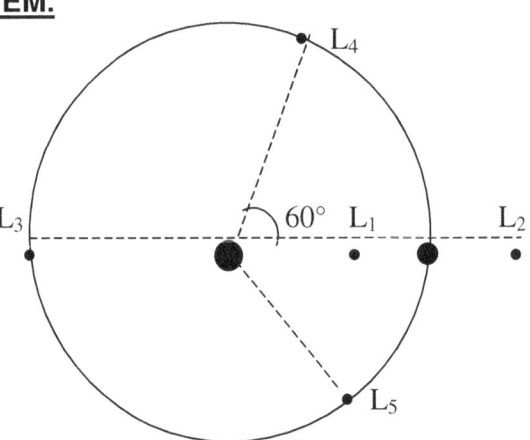

LAGRANGIAN POINT:
The Lagrangian points are five equilibrium points in the orbit of one body around another, such as a planet around the Sun

When going into any one of the four cosmic pillars one has to keep in mind constantly that what applies, applies at a point where singularity turns to space. It happens where mathematics change numerical from holding only form to becoming valid digits If anything was part of the Universe only once that something is part of the Universe forever because then that something has no where to go but to remain in the Universe . At this moment time forms as singularity going into forming space as time moves on and space becomes the visual history of time gone by. But not even time can disappear out of the Universe. Time leaves space behind as the history of time. Newtonians think of mathematics as godly forms by thinking it is mathematics that formed the Universe while the reality is that it is not mathematics that formed the Universe but it is the Universe that formed mathematics. It is clear that using form in the cosmos at the beginning became a principle before mathematics or numbers became valid. This is what the Lagrangian points prove. A triangle, a straight line and a half circle doesn't even approach to be the same in form and yet all three are the same in directional value as 180°. This proves that before mathematics was valid there were other values already in place and from that principle applying a numerical Universe came about. The overbearing reason Newtonians why science can't to an era before the Big Bang is that they do not understand this issue of the Universe being there before mathematics was there. There is a time that proceeded the time Newtonians get stuck at which is when space as heat came about which they call the Big Bang. There was a time when form got to form a Universe when space was still a thought that stretched far into the future. Mathematics proves this as much as that mathematics proves that there was a time when directions were the only valid mathematics used by the Universe. By the way, the name the Big Bang is as silly an idea as the Newtonian idea that mass forms gravity because at the time there was nothing big and sound was just a thought into the future.

It starts as follows:

No line can start at zero because having a starting point of zero there is no line (0 X by what ever reduces whatever to zero). The starting point has to be infinity the shortest any line can be leading to eternity the longest any line can ever be. By having infinity there then has to be a VERTUALL ZERO (not zero) and from that point the rest of the line must start running the other way.

At a point where singularity attaches to space in time the value of the straight lien is equal to the half circle, which is equal to the triangle. In the book **An Open Letter one Gravity** I go into detail as to explain why this occurs but that would take up more space than this entire book permits. To understand the Lagrangian system one has to come to terms with time and what time is. The Newtonian surmising that time can be one and then being 1 time can stand still as depicted in the formula **T = 1 -√ C² - V²** the is a folly as most other ideas seems to be. Time is always carried by the value of three and that is the responsibility of the electron. Time is in the past carried on to the future while moving onto the future.

When one slows the electron down to a point of near standing still one can observe the electron in two positions. That is not the case. What one sees is the electron departing from the past (visible point) going onto the present (point while observing) being in the position of observation while at point three (visible point) departing to the future. Let's slow this down somewhat while remaining relevant to the process. When I see a Super Nova event, the event that is taking place was in my past and therefore it forms part of my past. The light streaming towards me is confirming my past as that is what happened. The light that I see with which I observe the event is what confirms my present as the light is streaming to me by bringing the present. Furthermore the light streaming towards me is confirming my future since the light coming from my past still has to reach me in the future which is also the light that will be at the point where I am in the future and therefore while coming from the past and confirming my presence the light coming from the past is also confirming at the same time my future that I have. That makes me relevant to the Super Nova occurring and that gives meaning to the Super Nova as it catties the present to the future. Therefore, The light is confirming my past (1) (the event that took place) while witnessing the event that took place long ago my witnessing of the event is also confirming and establishes my present (2) (reaffirming what happened by putting me in a time position in relevance to the past) while the light running from my past confirms my future (3) by determining in relation to the past that I have a present since my past becomes my future while waiting for the light from the past to arrive at the present. This account is only the tip of the iceberg because when we start to dissect how light uses time to displace space in order to move the event onto and into the future to where I am the argument really gets intensive and it takes many pages of explaining the complicated issues in hand. This has no bearing on the Lagrangian system but only serves to point out that there is a position where time places space into the Universe in order to have space serve as the history of time. His is the point where the four cosmic pillars serve time in s[ace as to use time to form space. This singularity, which is a point time, places being without space to form a relevancy with space by afterwards becoming space. Singularity being time that is going into space by forming Π is the "white hole" that is in contrast with "the Black Hole" that every one was forever looking for.

The Lagrangian system is part of the four pillars on which time or gravity rests. The Lagrangian system is Kepler's formula put into principle. It is the manifestation of $a^3 = T^2k$. It is the movement (T^2) forming space (a^3) in relation to time (**k**) related to singularity (k^0) where singularity is excluding space-time ($k = a^3/T^2$), including space-time ($a^3=kT^2$) and secluding ($T^2= a^3/k$) space-time by motion of space-time in time forming a negative or opposing curve ($k^{-1} =T^2 / a^3$). One that cannot see this principle in practise as singularity improvise space by replicating time, such a person must be blind or a Newtonian, which makes no difference either way, it is still the same person seeing only mass and nothing.

The Lagrangian points system is the manifestation of time forming space-time by leaving space as delayed or historical time. I am not going to explain this for there is other books that detail what I say. Every time space ends in four, the very next point confirming singularity will be point five where singularity then provides point six and seven as a non movable line. It is the principle of how time rolls into space and the Lagrangian system provides the full range as to how time interprets space as the delayed form of time. The space forms five points (four in time plus one in continuing the alignment) and from that the line shows that the seven points positioning singularity forms the sphere. Time moves through the four sectors of the circle and the very next point in which space confirms s a new point (point five) from where space will grow into a new circle or cycle.

The Lagrangian system is working as the growing part of the Coanda effect where the liquid holds singularity Π^0 extended $\Pi^0 \Rightarrow \Pi$ in relation to forming a liquid in motion Π^2 in relation with the solid that forms as Π^3 confirming the liquid borders. Where $\Pi^0 \Rightarrow \Pi$ in relation to forming a liquid in motion Π^2 to form the attachment of $\Pi^0 \Rightarrow \Pi$ is the indicator of the attaching of fifth point. The liquid holds solids as markers in five locations and the liquid spins around the solid as the solid confirms the liquid by singularity extending to form the border confirming space-time. The Lagrangian system is one part in three other that forms the Coanda effect as the cosmos interprets gravity forming and the Lagrangian system is the improvisation for gravity by way of the Coanda effect.

But most of all, the Lagrangian system delivers every proof of how singularity forms space-time. It shows how singularity Π^0 places the relevancy of Π in relation to gravity moving Π^2 and in this movement the space forming contains the Universe that singularity provides to secure and define that space-time it preserves as Π^3.

The Lagrangian system proves the singularity places the straight line (180^0) equal to the half circle, each being (180^0) and that is put in equal relation to three triangles, each being (180^0). The Lagrangian system indicates the manner which singularity uses to put in place the characteristics of space-time relating to singularity. It is the straight-line (180^0) connecting the (two) half circle(s) (180^0) to the (thee) triangle(s) (180^0).

But moreover it is the manifestation of the atom as singularity improvises the connecting of space-time contraction. Every star is doubling as a cosmic atom and every star is just as much an atom as being the accumulative product of all the atoms combined.

LAGRANGIAN POINT:

Singularity presenting the triangle with 3 markers each only on one side of singularity.

Singularity presenting the half circle with two marking points

Singularity presenting the straight line with one point forming part of the circle

Triangle holds three, the circle holds two and the line also holds two points it shares with the circle by bringing a dividing half sharing points as singularity infinite.

It all forms part of singularity as a unit by three, two and one holding five in total and with space being the result from matter dismissing Π to favour r space must either join matter by becoming Π and dismissing r or maintain r and hold a maximum of five points to singularity at greatest value. With the Universe always in division by singularity the singularity holding seven position to Π will relate to the two singularities affecting the position of a cosmic atom. That will form as double points to space where five then multiplies with the two aspects of singularity divide and form the value of 10.

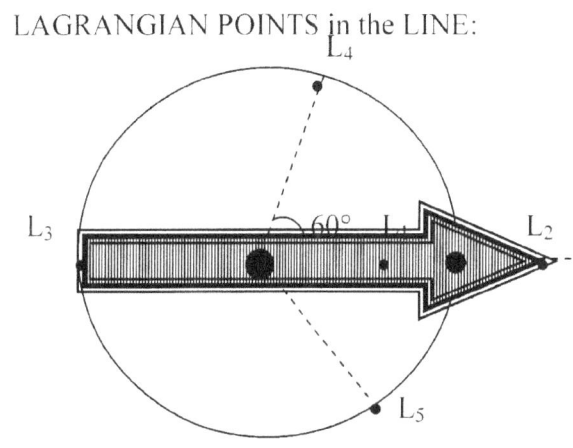

LAGRANGIAN POINTS in the LINE:

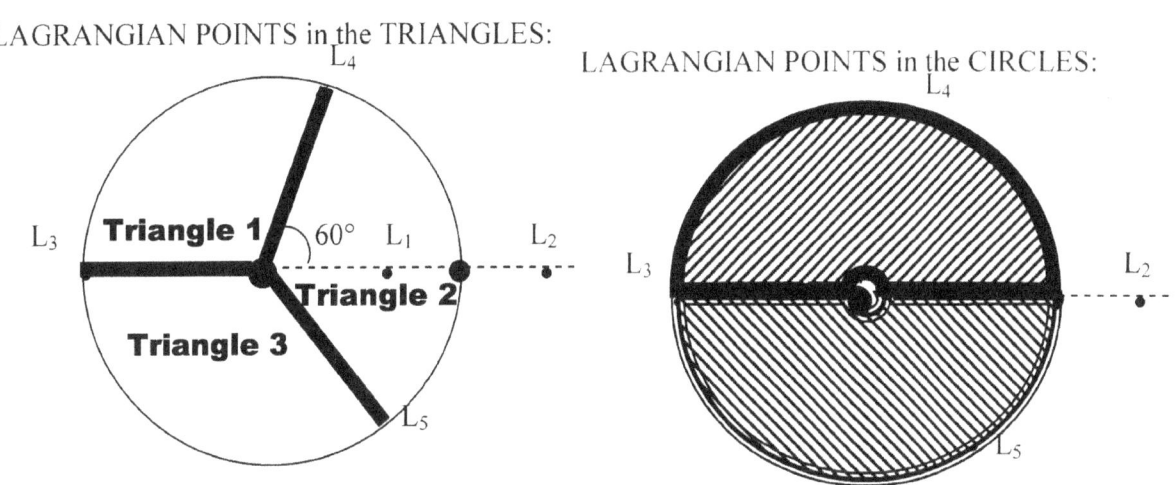

LAGRANGIAN POINTS in the TRIANGLES:

LAGRANGIAN POINTS in the CIRCLES:

1 Half circle   = 180°   L₃ L₄ L₅
2 Triangle 1   = 180°   L₃ L₄ L₅
3 Triangle 2   = 180°   L₃ L₄ L₅
4 Straight Line = 180°   Singularity
5 Double Circle = 720°   Sphere

Singularity can only meet space by measure of a straight line that is forming is running to both sides forming the edges Π of singularity starting at Π⁰ the border Π of singularity Π⁰. This means that all aspects of the cosmos hold two halves in four quarters. Each of the halves is formed by Π meeting space on either side of the Universe (two equal circles forming by the measure of Π). The movement forming Π holds a relation to a position (point) that forms (one digit) in relation to the other side having two reference point By having the three points forming three quarters that hold two halves that are separated by one straight line five pints in all form. In the full rotation that comes about as Π moves to form Π² of the quarters form that is in direct opposition to each other as much as to one another. The four quarters hold a value of fifty each where in the Pythagoras triangle 7² + 1² forms the fifty. Since space holds a value of 10 in relation to the 7 that gravity brings about, the dividing of 10 into 50 puts 5 places in a dimensional alliance with the 10 of space to form the fifty that From any point all space is moving from one side of the Universe to the other side of the Universe in quarterly displacement.

Then forming Π as a full rotating circle the 4 directional changes holds 5 points each that positions 20 points in the full circle. This is how the value of Π or 3.14159 comes about. Having singularity in development (.991 or put into context with the circle Π - 3 = 0.1416 and 0.1416 x 7 = .991 which means that singularity expanding into space holds 1.) On the other side the three movement of gravity forms 7 x 3 = 21 points and with singularity growing as time another 0.991 adds to 3.14159 or Π. That brings about the proof that Π is the way gravity forms a continues decline in the density of outer space as well as an increase in volumetric occupation of space within matter and the fact of gravity is vested in the value Π carries. In **The Veracity of Gravity** there is a lot more explaining as to how this comes about. The time it takes the movement will be the movement from the starting point at a value of Π to an ending point holding that moment of infinity to a value in eternity therefore bringing the square of Π² to the value of time. Because time is the movement bringing about the change in quarters from any given point to any other given point nothing can be in two places of the Universe simultaneously.

From singularity extending into space there comes three values each holding 180° and this fact science is familiar with. The values predate mathematical numerical and formula values and prove the most basic of mathematics was in place long before numerical values start applying. From this the numerical formation came about and I wrote a book dealing with such a matter of development. As everything grows from singularity the straight line is always a potential triangle with on side apparent and the other side in infinity.

There is no zero from where a line can start or grow and because of the absence of such a point mathematics brought about a diversion to escape the zero mark not existing. This abolishes the century-old idea that a line springs from zero or that a graph implicates zero as a divide. If the straight line did cross zero it would not be one line but it would be no line since the one line will discontinue cancelling the line at one point. All lines have to have a start and end at a single point without space therefore no line can be half a line. Therefore every line that forms is also a circle just as much as the circle includes a straight line to represent space just as Kepler said it does in $a^3 = T^2 k$

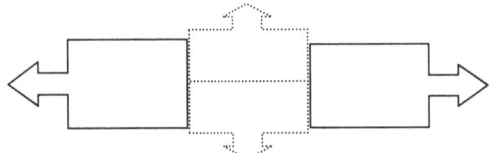

In order to overcome such a problem the straight line holds another line Π on both sides Π² as a point in infinity Π⁰ to half the line as to enable the line diverting from 1⁰ and grow to Π.

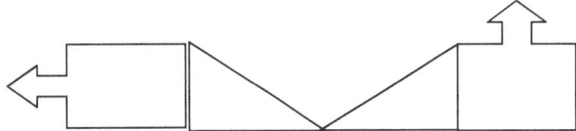

Because each line represents the other side of singularity dividing singularity by half the square of such a halves represent singularity two the half of a circle thus bringing total of the two halves would match the other half of singularity in half the circle. This is very important when considering the way that light travels in time through space.

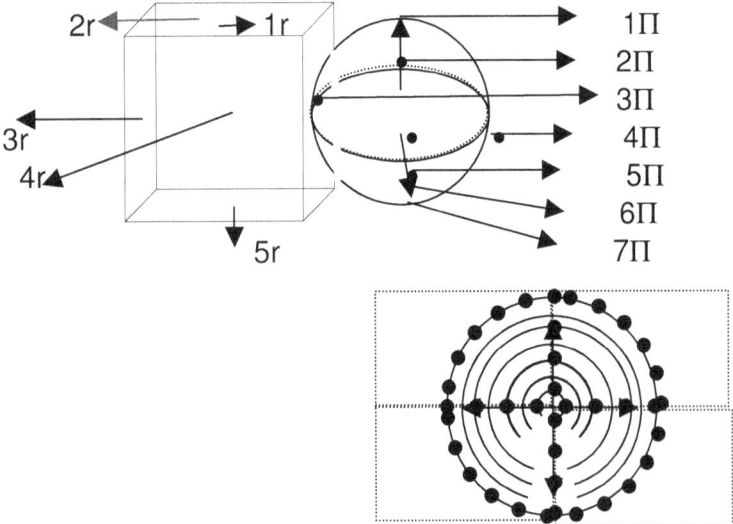

From space the cube holds the value of 5 times four quarters in relation to singularity forming the four five points of the square in the cube.

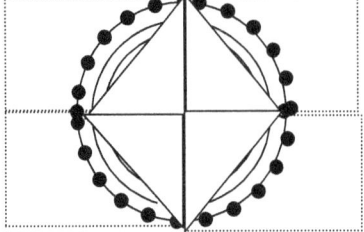

From singularity holding the relevancy the five sides in the cube as a square holds four triangles to two circles.

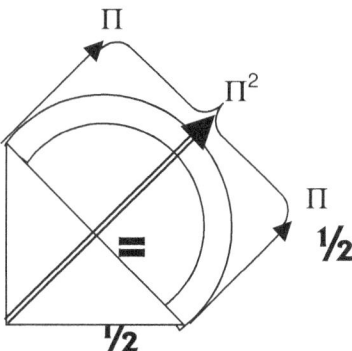

$½ + ½ = \sqrt{1}$

$360° / 5 = 72 / \Pi = 22.91$ which is 1 more than 21.991

What this proves is that when the circle is completed singularity bringing the future (10) is already in place because $360° / 5 = 72 / \Pi = 22.91$ shows a value one point more than Π because Π is 21.991 in relation to 7.

1 is singularity on the one side of the Universe and 0.991 is singularity on the other side of the divide.
10 from singularity on the one side of the Universe and
10 from singularity on the other side bringing about the
Π that holds 21.91 to 7
Since the sphere is double the circle and half the circle represents singularity by the square, half the square of the triangle is a straight line diverting singularity the law of Pythagoras is valid.

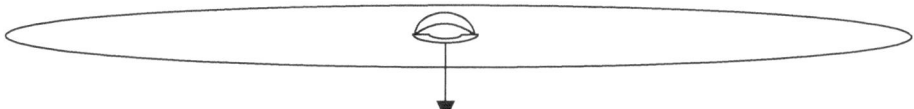

The divide bringing about the two sides of the Universe where the one (1) to singularity depicts the one side of the Universe and the other side depicts the other side holding space from singularity (.991) bringing about the singularity value of 1.91

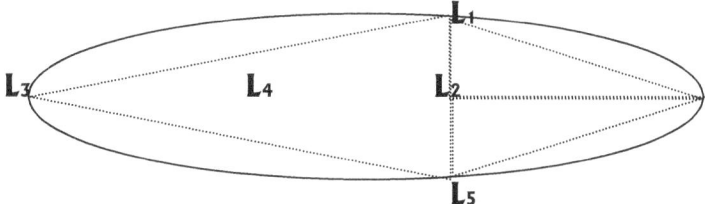

Dividing the four fives singularity holds a centre line (1.991) with one on one side and 0.991 on the other side but since it is a space relating to a sphere only one of the quarters on either side of the divide relates to a specific therefore unlike the sphere where the full value of Π relates to four fives, bringing about Π as the dominant the space separating the sphere from the points in space holds a combined value of one cube in line with the divide singularity supplies having five points.

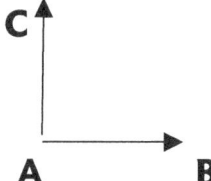

It takes any line time to relate to space and only nothing is instantly and nothing is what it is it is the absence of anything, therefore moving the line through the space it covers is the square of time in as much as Π^2. The time factor stands in all cases in relation to the space factor in the square because time being single dimensional develop space as three-dimensions while the space is in the sphere that spins in the cube and that is a law of cosmology. To that reason is why I reintroduced Kepler's formula as to explain the cosmos by the formula $a^3 \div T^2 k = 1$ because this is the way the cosmos gave information to Kepler. Using $a^3 \div T^2 k$ puts a relevancy between time $T^2 k$ in the line and space a^3 to the line.

Since the triangle in singularity are on both sides of the divide of singularity and the circle holds the time aspect relating to space in the square, therefore the triangle then must relate to space in a square in order not to duplex singularity in the divide.

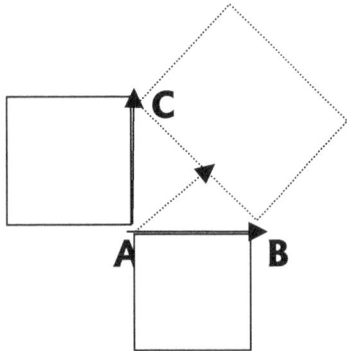

The time affecting the space of AC will relate equal to the time the line AB relates to time and space and where time is always in the square the lines will be the square of the triangle forming in relation to the square existing in the total of the time to the space relation forming between the lines.

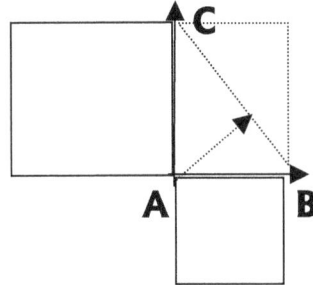 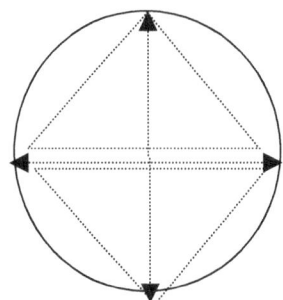

Since the triangle forms on both sides of the divide and all things concerning singularity is in duplication the double triangle will be the square. In the same way the circle represents both sides of the divide of singularity that forms and then also has the value of 2×180^0 as does the double triangle triangles. That brings about that Π relates to the square Π^2 to form three-dimensional space and from this fact mathematics can substitute Π by using r^0. However that is not the case in the dimensional aspect and as mathematics gets its queue from the cosmos and not the other way around the substation may apply to singularity as space in the sphere but not as space as the sphere. Heat is concentrated space and space is expanded heat. Gravity and electricity is the very same thing where electricity is a concentration of heat demolishing space in a very specific location and gravity is the concentration of heat in a less dynamic form but acting in a much broader space. In both instances it is the polarized motion of iron $_{56}$ linking time directly to singularity through the conducting of heat surrounding elements. To this effect I wish to point out that no element is either a liquid, a gas or a solid as all elements are all three forms and it is only the state of the relation that apply to an element at a very specific position in space-time that will allow the element to act in either of the conditions that the heat or space which is the same thing will allow. Gravity is the dimensional destructing of space to the concentrating of heat in that space by increasing the time duration through the Titius Bode principle or when matter holds less space to the normal allowing of matter occupying space, matter will produce the Roche principle in guarding its individual singularity of the mutual singularity between objects.

This principle also is the only difference of notoriety between electricity on Earth and gravity on earth. On Earth Π has a very slightly higher value than space in as much as space is 3 and the atmosphere is Π. In a cosmic midget as the Sun are the relevancy changes considerably and the space to atmosphere can be 10 Π. There is no chance of generating electricity in the atmosphere of the Sun because the atmosphere of the Sun is electricity in as much as the gravity being 10 Π to the Π of the earth. That explains the fact that the Sun liquefies heat to a watery substance. With the heat in a liquid the Sun becomes a sea of heat.

By matter applying Π as the reference there is little man can do to change that. In the case of electricity using r to form the value C we can change that because as r relate to space and we are part of space as space above the Earth in the neutron zone we can change the space holding r as value. When looking at the Sun applying gravity and relating that we see to what we find in electricity there is hardly any thing to recognise a similarity. But when in view of dimensional dynamics it is the same thing because in the Sun even r holding heat becomes Π holding matter as heat becomes liquid and that stands in between matter and gas. With this view in mind it would be worthwhile to have another look at the way we see how creation started and bring heat in as related to r and matter being related to Π.

Gravity is the transformation of space to heat in one specific dimension changing that particular dimension in relation to the other five dimensions. That brings the reason why the Lagrangian system can only allow five positions and allowing any more will destroy any form of dimensional implication between object relating to one another while sharing space occupying in time duration. The Big Bang had its massive motion brought on by first implementing the Roche factor of $(\Pi/2)^2$ after which when matter had a larger claim to space and space broadened the Roche factor adjusted to $\Pi2/2$ and then implementing the Titius Bode principle very much later on to Π.

I also prove that gravity is the result of four cosmic phenomena interacting to form the value of Π which by movement becomes the value of gravity Π^2 and gravity is equal to cosmic time applying in other books also available form LULU.

1) *Absolute Relevancy of Singularity in Relation with Applying Physics*

1) The location, the position and the value of **singularity** as a factor forming space-time
2) Finding **space-time** by dissecting Kepler's formula in relation to valuing singularity
3) Finding space-time, **proving space-time** and **aligning space-time** with gravity.
This is part of the *Absolute Relevancy of Singularity in Relation with Applying Physics*

2) *Absolute Relevancy of Singularity in Relation the Four Phenomena*

4) Finding the **Roche limit**, and explaining the resulting of a law coming about from singularity.
5) Finding the **Lagrangian system**, how and why that becomes the building form of the Universe.
6) Finding the **Titius Bode law** and I show mathematically how gravity comes about from that
7) Finding the **Coanda effect** and the producing of gravity through reproducing space-time.
As part of the *Absolute Relevancy of Singularity in Relation the Four Phenomena*

3) *Absolute Relevancy of Singularity by Explaining the "Sound Barrier"*

9) Proving the phenomenon known as the "**sound barrier**" by proving it **is gravity** generated **by motion** in space becoming independent where motion creates independence. Breaking the sound barrier is the motion in space duplicating space by crossing over gravity borders. It is $a^3 = kT^2$ where ($k \leq T^2$) or ($k > T^2$)
As part of the *Absolute Relevancy of Singularity by explaining the "Sound Barrier"*

4) *Absolute Relevancy of Singularity in Relation with Explaining the Cosmic Code.*

4) Finding the **working principals** behind and manifesting **of gravity** as a cosmic occurrence and introduce it as part of the *Absolute Relevancy of Singularity in Relation with Explaining the Cosmic Code.*

For more information visit www.gravitysveracity.com

The Cosmic Code

WRITTEN BY PEET SCHUTTE   ISBN 0-9584410-8-1
KOSMOLOGIESE EN ASTRONOMIESE TEGNIKA

EXPLAINING
THE
"COSMIC CODE"
IN TERMS OF
PHYSICS
BY
PUTTING
SINGULARITY
IN

THE ABSOLUTE RELEVANCY SINGULARITY

ISBN   0-9584410-8-1

All rights are reserved.
No part, parts or the entirety of this book may be reproduced by publishing, electronically copied, duplicated by whatever means that form reproduction or duplication of any description, without the prior written consent of the copy rite owner.

WRITTEN BY PEET SCHUTTE
© KOSMOLOGIESE EN ASTRONOMIESE TEGNIKA

I do find much pride in my status as being Afrikaner and would like to have my names used by pronouncing it in the manner Afrikaans dictates…therefore I would sincerely appreciate the courtesy when readers will take note that my name and last name are pronounced in Afrikaans, which is originally from Dutch and must be pronounced that way. Peet one would pronounce "here" which is the closest English to the pronouncing of the "ee". The "Sch" in Schutte is pronounced exactly as school is where both actually are pronounced Skutte or "skool". By pronouncing my name in Afrikaans you do me the utmost courtesy any one can.

Being an Afrikaner is what I am most proud of. I submit article to well known physics magazines but my articles are rejected on the most unappeasable grounds and for the most outrageously ridiculous reasons the Newtonians can think of. I explain how gravity forms but I am rejected because they are of the opinion that my work does not meet. One such an article I may use because I said I was going to use the material as an open letter I gladly show.

This book was done with a $25 ^{oo} scanner and a $35 ^{oo} printer and the reason I explain inside. For the same reason this book was not edited or linguistically checked. I could not because that does not work because I am in the writing business and not the spelling business and while I check spelling the writing gets more and so does the spelling and grammar errors. I had a choice; doing the books with no funding or not doing it at all because while I rubbish Newtonian science and show it is the fake it is, they will never publish my work because I trash Newton.

Not having funds and trying to fight science for the truth with the truth was a fight that physically broke my health and still I am not published except in this manner. I apologise for the spelling and language but in poverty that was the best I could do under the prevailing circumstances in which I find myself…. This book is a first in every sense… it unites science and religion because science and religion was separated by human stupidity

Please take note that I sell information and not words or books and therefore the information takes priority and not the spelling or words used to inform the readers. This represents the work of God and not the word of God and so there is no interpretations applying and versus you can learn and sound intellectual but only cold facts you will have to understand.

WHOM IT MAY CONCERN,

I do find much pride in my status as being Afrikaner and would like to have my names used by pronouncing it in the manner Afrikaans dictates...therefore I would sincerely appreciate the courtesy when readers will take note that my name and last name are pronounced in Afrikaans, which is originally from Dutch and must be pronounced that way. Peet one would pronounce "here" which is the closest English to the pronouncing of the "ee". The "Sch" in Schutte is pronounced exactly as school is where both actually are pronounced Skutte or "skool". By pronouncing my name in Afrikaans you do me the utmost courtesy any one can. Being an Afrikaner is what I am most proud of. Another point I wish to highlight is that I feel compiled to produce this work in a comic-like format. I have found that the more intellectual and the more educated Academics are, the less they understand the most primitive or classical mistakes in science as well as physics.

As I said my mother tongue is Afrikaans and my second language is English. I have per suiting this theory that I partly present in this book, of which the investigating research was done the past thirty years. Then I compiled my presentation thereof for the past nine years on full time basis whereby I was tying to introduce my findings to many academics without much joy. This past nine years saw me go without any income as I tried to get my theorem recognised. Going without a steady income left me almost destitute and in order to find a manner to get my theory across to the attention of influential readers, I decided to publish these books electronically as to try and get around the stranglehold of Newtonian bias controlling science at present worldwide. I decided to publish these articles through LULU.com which I saw as way the only manner whereby I could generate funding by which I would be able to have the twenty seven books I already wrote linguistically edited and then to have the books published on a Print-On-Demand basis. With my first language not being English and the books not linguistically checked by an expert there are bound to be language errors that readers will notice. In the past I tried to check my work myself but after checking say one hundred and fifty pages for language corrections, instead of having corrected work I ended instead having four hundred pages of new written information which is still not language corrected but holds a lot more information. This is because my priorities lie elsewhere. I aim to spend money on correcting the work as far as language goes, as I receive money and in the hope that I will receive money. I will have all my work including the one you are reading edited professionally and corrected as I find money to do so. . .

In the book that deals with gravity there are just too many and numerously wide ranging facts that form the complete picture as a whole, which leaves me unable to include a full introduction in a space as small as that which page will allow. The explaining include for instance those phenomena, which I call the four cosmic pillars, but wise as you are, you would not believe me at this point that I have cracked the coconut because I guess in your vast experience you have seen too many idle explanations in the past proving to be senseless and little impressive, therefore my mentioning my success would not matter much either way. The proof I bring is true about gravity being formed as a result of these phenomena, **1)** the Lagrangian system **2)** the Roche limit **3)** the Titius Bode law **4)** the Coanda affect, which I explain by delivering mathematical proof as to how they fit into the overall picture of gravity and which I mention just below. I prove the fact that every individual one of those phenomena is forming a unit that is in total being what we think of as gravity. The phenomena altogether constitutes a unit that forms the process working as gravity. Nevertheless my mentioning these facts will be just completely unbelievable to you without you reading the book, because I guess you have heard some attempt to explain the phenomena before but when I say you have not heard it in the context I put it, you might still be most sceptical because you have never heard it in the correct manner that I explain it and that poses the difference. Still you may not be convinced about my claims and although my explaining the phenomena is correct, does not change the fact that you don't believe me. The phenomena form an intergraded unit that results in gravity forming where each forms a part of gravity. You may still be you would be sceptical ...but convince yourself that I did manage to:

1) Find the location, position of singularity as a factor forming space-time
2) Finding space-time by dissecting Kepler's formula in relation to valuing singularity
3) Finding and proving space-time and aligning space-time with gravity
4) Find the working principals behind gravity as a cosmic occurrence.
5) Find the reason for the Roche limit and explaining the resulting of gravity from that.
6) Find out why the Lagrangian system, becomes the building form of the Universe.
7) Find why the Titius Bode law mathematically provides the foundation of gravity
By proving that the Coanda affect is gravity through activating space-time
By using the above the four cosmic pillars, it enable me to present the proof where I now can explain what conditions bring on the sound barrier. By proving it is gravity that the individual structure generates motion above and beyond the gravity the Earth provide is what is producing individual motion that the independent object earned within the sphere of motion that the Earth's gravity provides where the independent and individual motion put the relevance that gravity has beyond the conserving means gravity has where **the space** that is serving the **independent object** is

independently in motion. The adding to the independence on top of the normal structural independence is creating more individualism by the independent motion of the individual structure being apart from the motion that the gravity of the Earth provides. The fact every one misses is that any structure that is not part of the Earth's crust has an independent gravity and the form this gravity applies is stronger than the Earth's gravity which is why the structure maintains its form and this provides the independent individuality the structure has giving the unique structural space. The gravity of the Earth strives to incorporate everything into the Earth's sphere and into the Earth's structure and therefore the fact that the object is not incorporated into the Earth shows defiance and individuality, which gives it, mass.

By applying individual motion on top of the structural individuality that increases by the motion that the Earth provides, the independence of the individual object is becoming further exaggerated by having independent motion, which is further defying the incorporation the Earth strives to achieve. As the motion of the independent object grows more independent by applying more excessive motion to such an extent **where motion creates almost the ultimate independence that may free the individual object with independence from the motion the Earth creates** is what is breaking the restraint gravity has on all objects with independence formed by their structure. The structure show independence at all times by not forming part of the structure of the Earth within the sphere of the Earth's gravity. Moving about shows even more reluctance on the part of the top when spinning allows the top to eventually become part of the Earth. **Breaking the sound barrier is the motion** in space duplicating space by crossing over gravity borders, which is the limit to what constraint the Earth may produce in accordance with what full independence would allow.

These are the definitions underwriting cosmology and while my work is that much ignored; let's see how far I stray from these definitions in comparison of how much Mainstream science underwrites these definitions by bringing indisputable proof in presenting unwavering hardcore facts.
Quoted directly from the Oxford dictionary of Astronomy the following:
The definition of space-time is as follows:
Space-time is a four dimensional position of the Universe where the position of an object is specified by three coordinates in space and one position in time. According to the theory of special relativity there is no absolute time, which can be measured independently of the observer, so events that are simultaneous as seen from one observer occur at different times when seen from a different place. Time must therefore be measured in a relative manner as are positions in three-dimensional Euclidean space, and this is achieved through the concept of space-time. The trajectory of an object in space-time is called world line. General relativity relates to curvature of space-time to the positions and motions of particles of matter.
The definition of singularity is as follows:
Singularity: a mathematical point at which certain physical quantities reach infinite values for example, according to the general relativity the curvature of space-time becomes infinite in a black hole. In the big bang theory the Universe was born from singularity in which the density and temperature of matter were infinite.
The Oxford dictionary of Astronomy defines gravitation as follows
Gravitation is the force of attraction that operates between all bodies. The size of the attraction depends on the masses of the bodies and the distance between them; gravitational force diminishes by the square of the distance apart according to the inverse square law. Gravitation is the weakest of the four fundamental forces in nature. I. Newton formulated the laws of gravitational attraction and showed that a body behaves as though all its mass were concentrated at its centre of gravity. Hence the gravitational force acts along a joining of the centres of gravity of the two masses. In the general theory of relativity gravitation is interpreted as the distortion of space. Gravitational forces are significant between large masses such as stars planets and satellites, and it is this force, which is responsible for holding together the major components of the Universe. However on the atomic scale the gravitational force is about 10^{40} times weaker than the force of electromagnetic attraction
I have to give potential readers this fair warning that *The Cosmic Code as the Absolute Relevancy of Singularity* **requires a somewhat higher level of understanding and needs a greater degree of insight that the other books in this series does namely**

1 Explaining Physics in terms of the Absolute Relevancy of Singularity,

2 Explaining the Sound Barrier in terms of The Absolute Relevancy of Singularity,

3 Explaining the Four Cosmic Phenomena in terms The Absolute Relevancy of Singularity and

4 Explaining the Cosmic Code in terms The Absolute Relevancy of Singularity

Which all are also available from Lulu.com.

The Cosmic Calendar is part of the Cosmic Code but only portrays cosmic development from the initial development and not the method of operation. This is not the entire Cosmic Code but only forms a very small part of the Cosmic Code; however this indicates the stages in which the cosmos allowed stars as well as galactica and the atoms forming stars and galactica to develop by applying relevancies through gravity which enables the cosmos to grow in time and through space. From every value one can read about relevancy that applied, which brought about cosmic form that predates the Big Bang, explains the Big Bang and indicates the forming of the cosmos after the Big Bang took place. It shows how atoms formed.

$\Pi \times \Pi^2 \times \Pi^3 / 5 = 192$ ⟵⟶ $\Pi^2 \times \Pi^2 \times \Pi^2 / 5 = 192$

$10/7\pi^2/2(\pi^2+\pi^2)=139$	$7(\pi^2+\pi^2)=138$	$7/10\ \pi^2(\pi^2+\pi^2)=136$
$2\pi(\pi^2+\pi^2)=124.$	$2(3)(\pi^2+\pi^2)=118.$	$10 \div 7(4(\pi^2+\pi^2))=112.$
$\$T = \pi^2(\pi^2+\pi^2)=107.$	$3\pi((\pi^2+\pi^2)=102$	$3^2(\pi^2+\pi^2)=98.$
$10/7\pi(\pi^2+\pi^2)=88.$	$10/7(3(\Pi^2+\Pi^2)=84.$	$4(\Pi^2+\Pi^2)=78.$
$10/7(\pi/2)^2(\pi^2+\pi^2)=69$	$7/10(4((\Pi^2+\Pi^2)=55.$	$2((\Pi^2+\Pi^2)=39.$
$10/7(2((\Pi^2)=28.$	$10/7(\Pi^2+\Pi)=18$	$7/10(2(\Pi^2+\Pi)=14$
$2(\Pi^2)\ 19$	$10/7(\Pi^2)=14.$	$7/10((\Pi+\Pi)=5.$
$10/7(\Pi)=2.$	$(\Pi/2)=1.57$	$7/10(\Pi/2)=1$

I am not going to attempt to explain the Cosmic Calendar as it stands here because trying to do that would require an entire book. The explaining is very complicated and effort is tedious. I will however explain one line in order to have the readers see how finding the Cosmic Code enabled me to read the status of cosmic displacement. I will explain the situation by dissecting the formula and reading then from the formula what could be interpreted and how it could be interpreted. Let me explain the following period:

$10/7\pi^2/2(\pi^2+\pi^2)=139$ $7(\pi^2+\pi^2)=138$ $7/10\ \pi^2(\pi^2+\pi^2)=136.$

The number 139, 138, 136 points to heat expanded that is displaced to be confined into singularity by the movement of gravity. This number is normally associated with the number of protons within one atomic cluster and the number of neutrons helping the flow of heat to such confinement. I know there is no atom cluster capable of holding 139 protons, but this development comes at a point where the first defining of space came about and was in place long before atoms was formed. Let's do $7(\pi^2+\pi^2)=138$ to explain the formula as one could read from it the information and then see how that will explain the Cosmic Code. These three values indicate a point where galactica formed with the displacement value forming in the galactica centre as the governing singularity. Every value has an identifiable code by which we can read what event took place, how did it stand in relevance to the rest of the Cosmic development and in what way did it affect the rest of cosmic development where this precedes the element table that started forming atoms as a sphere at $7/10\ (\Pi^6)/6$. 7/10 is the electron spin; (Π^6) is the atomic sphere and 6 is the cube of cosmic liquid in which the solid atomic sphere spins as it performs in the gravity structure of the Coanda effect. At $7/10\ (\Pi^6)/6$ the atomic-solid sphere forming six sides by Π started to spin in the six sides cube.

$7(\pi^2+\pi^2)=138$

7 The seven indicates that the material in bonding found a round shape we now associate with the Universe and the Universe founded roundness by movement, we now call gravity but is time moving space.
$(\pi^2+\pi^2)$ The double gravity indicates singularity forming associations or relevancies using time to form space in relation to the diverting of the roundness of spinning movement.
=138. The cosmic displacement of space-time required achieving this status. In the atom we think of the proton number that will bring about such movement but the proton and the atom developed much later where the atom became a unit as the neutron became a unit and later the electron became a unit that then formed the atom to become a Universe within a Universe setting conditions and standards required to form a discipline within that which we all now think of as the Universe that we live in. **The Universe is within the atom that forms a cosmic unit holding singularity as much as it secures singularity and every atom forms a Universe standing apart, parted by time from all other atoms by the spin produced. Every Universe starts in infinity and ends where each atom's spin is forming relevancy between where that Universe starts and ends. All atoms are a Universe formed within the space   that time puts between infinity and eternity. All atoms are stitched together by an invisible, unseen singularity - string that is present while also being absent and links everything the Universe is throughout the entirety.**

I have no chance that what I state as my theory on **The Absolute Relevancy of Singularity** will be read, or much less that it will be seriously considered and I have not a snowballs hope in hell that it will be accepted by those with the authority to change physics principles. The theory I introduce here and now would never be accepted in my lifetime because science in the Newtonian way is bent on believing in the marvellous, the facts bordering the supernatural, the outrageously inconceivable and the magic of what can never be explained, although they claim to use facts. It is **the marvellous** to think that mass can create gravity. It is **bordering the supernatural** to think that with nothing between stars, yet by the magic of mass, mass has an unexplainable ability to attract another star many astronomical units away. It is the **outrageously inconceivable** to argue that life started on Mars, then overcame the Quite impossible to escape the gravity that Mars holds on all things, and after overcoming the unthinkable, then made a dive for the Earth just to come and evolve over here. Science think they my have the ability to create a Black Hole in a Manmade atom-accelerator because science thinks of the Black Hole as **the magic of what can never be explained** and therefore that proves that science has no idea of what a Black Hole is and I can prove what a Black Hole is. That fact that I can explain what a Black hole is, that the Wizards of Oz will never allow the explaining I present to be done in as simple manner as I am about to explain the Cosmic Code. However, when I prove what a Black Hole is I am going to destroy the fantasy world everyone makes believe as physics. To science a Black Hole is a world of magic where gravity has the ability to go mad and a Black Hole is something that man could manufacture by creating an atomic accelerator tunnel, or so science thinks. In other words the best science at present can do to explain the gravity in a Black hole is to give gravity a level of superior intellect and then take it away (by allowing gravity to go mad as it seemingly does in Super Novas and in Black Holes). Why can I prove what a Black Hole is…it is because I can prove what gravity is and believe me that is one thing science this far could never get around in proving. The facts they use is as much fiction as Little Red Riding Hood's talking wolf…when it comes to explaining the integrating details of how gravity comes about. In science, when following my theory, everything can be explained by using physics, but using my explanation will make all present science become fiction, make all present science look like a fairy tale and make all present science seem to be good bedtime stories deprived of truth…and the money spent on Newtonian fiction-science will never allow me to have success because that would be too costly for the industry money-wise. Why would I call science a fairy tale…well this is just one of many, many reasons. Science wishes to promote something as impossible as time travel, which I show, is impossible. Science believes in travelling at speeds unlimited that could exceed the speed of light. I prove all such thoughts are impossible because I show that gravity and time is the very same thing. No one can beat gravity because gravity as time maintains the structural integrity of the Universe. In beating gravity one wishes to beat the cosmos that hold us secured. That is why time can manifest as what is known as the Hubble constant. Time is the redeploying of space by extending the absolute relevancy of singularity and that is only one of several factors that serve as time. Every time I declare Newton was mistaken and therefore science is wrong in presenting the most basics of physics, the workings of gravity, I am barraged by rejection and silent ridicule. Every time I challenge the Members of science to either prove Newton correct or to prove me wrong, I am ignored…my challenge goes unmet, so please forgive me for showing much antagonism…it is a result of Mainstream Science rejecting my efforts unfairly for many years. What I write is undeniably and undisputedly correct, but the instant science admits to my work being correct, that admission demotes most of the work science has accepted in the past as correct to the level of science fiction. It will destroy the groundwork of mainstream science and demote what is accepted to become fairy tales, which is what most Newtonian based theories are. Let Newtonian science explain what the cosmic purpose or the function is of a star…of a galactica…of an atom…of gravity…they have no idea. By the time you have finished this book you would have found answers to all the above questions in detail.

Mainstream science has so little idea of what a Black Hole is or what could cause a Black Hole that they devised a "Mini Black Hole" to suit there marvellous misinterpretations of gravity. That is a form of fantasy that fairy tale writers can't compete with. Science is so misguided in understanding life that they put life in all places throughout the Universe without ever finding one shred of evidence of the presence of life. Yet they say they work only with proven facts alone. They hold the opinion that life could have come from Mars but fail miserably in explaining how it will be possible for life to escape the gravity of Mars and then fly all the way, ever so precisely guided; directly to the Earth. How would it be possible for life to escape the gravity of Mars without them when explaining such a possibility by employing realistic physics, going into so much fantasy it leaves the story of the three pigs and the blowing wolf seem real. Science has the explaining of the exploding Super Nova down to the last detail where they explain that a Super Nova is gravity that has gone mad without ever proving how gravity can go mad because the truth of the matter is that gravity has no intellect to "go mad" in any way. Mainstream science always places new object found where their findings prove that the newly found object is on "the edge of the Universe", meaning where the Universe ends by forming an edge. This fantasy they dish up to anyone willing to believe him or her without ever telling what is beyond that edge. All they can see is an end of the Universe but in reality where there is an end there has to be a beginning of something else…this is physics. The Universe I show can't have an edge because I show where the point is that could never start and I show where the point is that could never end. I show that which can go no smaller and I show that which can go no bigger. I am about to introduce a Universe that mathematically can never start and the same Universe can mathematically never end.

I have been on a self-teaching mission that lasted thirty years and now that I have the answers and from which I have drawn the conclusions, I now find so much resistance from mainstream science in getting the findings my research uncovers out in the open. I offer tot academics many books in which I use diagrams, sketches, mathematical explanations and cosmic photos including other tools I employ to promote the required understanding needed to bring the ideas across that I wish to promote. However, publishing in this manner is very costly and money is one thing I do not have and therefore sending it to academics with no reply is an expense I cannot endure. Any academic feeling confronted by my accusation, please show how you prove $F = G \frac{M_1 M_2}{r^2}$ is applicable and is true. Show how the use of the formula could be applied meaningfully to present an answer worth of anything. Use the Newton's formula to show when the Moon is going to hit the Earth as the mass of the Earth pulls on the mass of the Moon. Better still, prove that mass does contract to create gravity and then explain how this is done...and please leave out the graviton because that is a joke! The idea that mass draws mass closer $F = G \frac{M_1 M_2}{r^2}$ is mathematically proven as an untruth, which means it is not true. What is the truth? ...when you have completed this introduction you will have had a peeping view, a tiny glimpse of the truth...but as little as you would gain from reading this introduction alone, when put in comparison to what any person can gain from reading all of my work in total, you will gain endlessly more than what science is to explain about the truth, because what you then have gained by reading this document is much more than what science know about the truth. What I try to convey is that there is a good reason why academics block any and all publishing of my work, and when finishing this book, in comparison to what I offer, you have not even opened a first page of what I offer as new information when judging what my other work uncovers. Still, your effort in reading this document allows you to discover so much more of true science than what previously was known If you think I am boasting I challenge you to show where any of my explaining gravity requires superior intellect to understand... however in my simplistic approach to gravity I prove everything I say by applying the simplest mathematics there is.

The effort that this book represents the informing about an entire new way of cosmic appreciation meant to show that there are grounds for concern in the way science thinks and this book does not even bring all such arguments indicating concern in full. That one can only find when reading the first ten letters forming books named as with a title beginning with **Open Letters...**and those titles are included as books which I mention on my website, having the same name as this book namely www.gravitysveracity.com.

I am about to prove that gravity is **the Coanda effect** and gravity comes about from four cosmic phenomena never yet understood since it was never yet explained. Science doesn't believe there is something such as **the Titius Bode law** but science does believe that mass would generate gravity. Science has no clue about **the Roche limit** but science believes in spite of the Roche limit that big craters on Earth are reminders of massive asteroids that hit the Earth in giant collisions. With the Roche limit in place these crates are the result of something else because it can't be from asteroids colliding with the Earth. We all know how the bicycle rides and we all think we understand how the bicycle rides but having the bicycle ride on two wheels have little to do with balance and everything to do with the Coanda effect.

The bicycle rides forward when peddled but also the bicycle rides downwards when peddled and the two are both linked to gravity. I am going to prove that the Coanda effect forms gravity. I am going to prove that the **Coanda effect** comes as a result of the **Titius Bode law**, **the Roche limit** and the **Lagrangian positioning system** but most of all how these are related to singularity. That means I am going to prove that mass has no effect on gravity but mass comes as a result of gravity. I am going to prove what singularity is and that there are two types of singularity that in the end is only one type of singularity.

Teaching ever since time began forms a pillar on which memory and remembering what you are taught is the most prevalent part of tutoring. One is expected to remember what those coming before and which are tutoring you, wish you to remember. The Tutor lays a foundation by ensuring that everything known and accepted coming from the past are well and truly founded in the mind of the student. In that there is no problem. The problem arises where the information studied is flawed and no one ever realised that. Fortunately this does not occur regularly, but if and when it does, notwithstanding the exceptional par it forms, ten becomes a major problem to deal with. Therefore what comes form the past are carried on into the future as unblemished truth and no person meddles with the thoughts called information given as study material. However, as unlikely as it could be, this did happen and it is part of the basis of physics. When the student is taught, the student is expected to accept without argument. What comes from the past are considered to be tested beyond suspicion of doubt! One can only start to think and through arguing set by reason new thoughts, after the learning by memory process is well established and it then forms a solid base for everything the student knows. This mostly takes about all the time one lifetime presents. Well, what happens when that everything that everybody believed in the present, inherited by all from the past, was totally flawed? It has happened to physics and no one in physics yet realised it. Then the mistakes will carry on forming the past, carried over as flaws into the future for as many generations as it takes to realise the mistake and could continue indefinitely, if there is no clear minds working the recognise and correct what needs to be corrected. I ask of you not to judge me for I fall short. Judge what I present you with, for then you will realise with all my shortcomings, I present you with a truth that exposes short fallings in the basics of physics.

Now show your academic worth and your educated dignity and accept the challenge I make to you and to all of your kind: I challenge one and all: **PROVE ME INCORRECT IN ANYTHING I SAY!**

Their method of accepting Newton and his idea of mass producing gravity proves the lot were and are unable to think If not they would have realised that gravity connects far closer to Π that gravity has any connection to mass. Every aspect in the cosmos uses Π to produce gravity by spin but the planets arrange their allocated position while completely ignoring mass. They are arranged by a means of totally ignoring size in the range to indicate allotted positions or mass in arrangement. This should tell someone something somewhere somehow…and yet in three hundred years it did not tell anyone anything this far.

Whatever gravity is, gravity has to be Π. If gravity is linked to mass as Newton stated, then mass has to be very closely connected to Π or else have no implication on gravity. Looking at every aspect that forms gravity, it is formed by a circle that runs in a circle that runs by connecting to another circle. The Earth as much as the Sun as much as all stars and galactica holding gravity is round and the roundness is Π. The curvature of space-time, the fact that gravity bends light into a curve, this bending comes in the form of a circle that is formed by Π. The Sun for instance spins around and that is formed by Π. The Earth holds the Moon captured while the Moon circles around the Earth and the circle is a result of Π. The Moon is captured by Π to the extent the Moon does not even serve its own axis but honours Π by adhering to the centre of thee Earth. If it is with gravity that the Moon circles around the Earth, then in all of this we must locate gravity holding Π as a value.

Looking at the Solar system we find that all planets and objects not classified as planets and all things that is just simply forming solar debris has one thing in common…all apply the value of Π in the process where they orbit the Sun, which also uses the formation value of Π to construct the roundness the Sun has. Gravity has much more in common with Π than it will ever have with mass that is producing gravity. Yet with all this in mind I have not been able this far to persuade any one with my logic I try to use in my arguments.

Kepler gave his formula symbols $a^3 = T^2 k$ that do not quite represent gravity in its true symbolic nature and that then was the reason why I came on the idea that gravity has to link to Π more that any other value or symbol. It is because everything holding gravity or representing gravity (not mass) is round. Gravity connects by the use of Π. We have to part what mass does and what gravity does. Mass is where the object connects to one point on Earth and being at that point with mass the Earth does the moving by spinning. The spinning of the Earth then represents the movement or the intention to move because the Earth spins by Π. This movement gives mass its qualities because mass does not possess the influential value of Π since mass is a quantity representative of the amount of atoms and not the spin of the atoms within the mass quantity. If we look at the way the Moon connects to the Earth, it is done by committing movement in a circle. That represents Π. When we look at the way the solar system connects to the Sun in circles every planet holds an individual symbolic value to Π circling in relation to the Sun. If we look at the roundness of galactica, the formation represents Π. The connection gravity has is not by mass but it is by Π. When we go in search of a cosmic resolve to find gravity, we better start looking for the influence Π has on the subject or leave the entire subject alone because the gateway in understanding gravity goes by the meaning of Π.

If we look at the value of the Roche limit we see the value of Π used extensively. When looking at the "sound barrier" we find that what we see is only Π bringing any explanation. The Earth spins at a relative value of 7(ΠΠ²), which is 217 km / h and that would be the highest take off speed where objects star to rise into the air. The value of 7(ΠΠ²) is gravity and that I explain in **Absolute Relevancy of Singularity in Relation the Four Phenomena.** The "sound barrier" comes about at 7(ΠΠ²) (Π² ÷2) =1071 km / h which I explain in **Absolute Relevancy of Singularity by Explaining the "Sound Barrier"** In the **Absolute Relevancy of Singularity in Relation with Explaining the Cosmic Code** I prove that the code that the cosmos was built by is absolutely written with the use of gravity which is Π² and is the only cosmic code because it explicitly uses one value and that is Π.

In gravity there are always three aspects of gravity interacting as one. That is what the cosmic code is all about. Gravity isΠ. Gravity is the value ofΠ. When we look at Π we find that three times seven plus o.991 isΠ. I am not going into the value of 0.991 at this stage but as gravity is Π we can see that 7° by 3 methods of spin formingΠ. Gravity is the change in directional movement we named spin and that puts a directional change in relation to a centre that controls all dynamics and aspects of such a spin. Having Π as 3 x 3 we can see that to form Π there has to be 3 ways of spin to forms the total cosmic value of Π. The one we can easily see as the turning or spin the Earth produces as it spins around the Earth's axis. The Earth spinning and changing direction implicates one 7° change that formΠ. The next as obvious 7 that produce the measure of Π is the spinning of the Earth around the axis of the Sun. This would brig an orbit and the orbit is directional change on a wider scale, which is seasonal, but as the Titius Bode law implicate it holds a gravitational relevance and produces the second 7°. Having Π there is a third, which could be the atoms spinning inside the Earth and could just a well be the Sun taking the Earth with the entire solar system on a journey around the Sun. This is because the fact of movement is a relative indicator of what is applying and does not prove a specific value. It is gravity applying and there are always three indicating factors working in tandem to bring a specific movement about. What does always apply is that whatever gravity is. It is an interaction between liquid space and solid material.

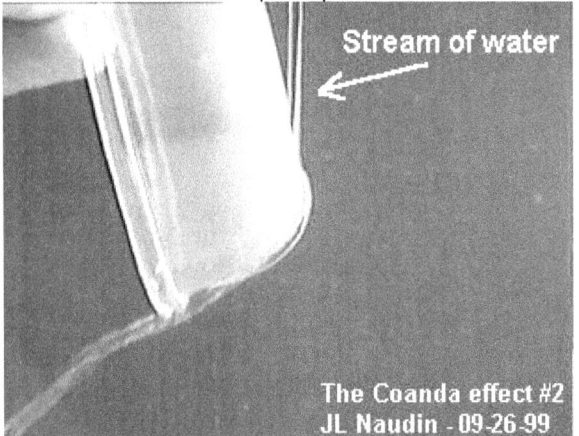
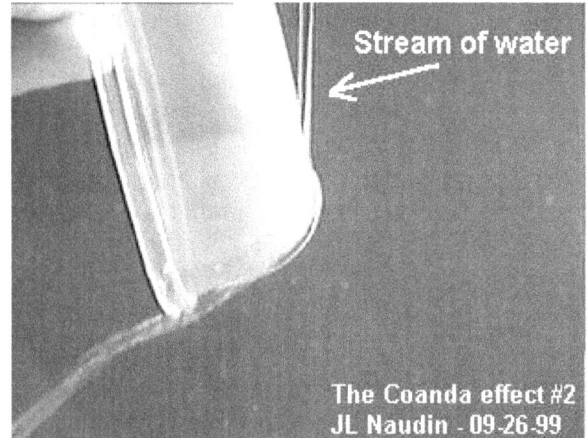

I say this phenomenon called the Coanda effect is gravity. I say mass is a product of gravity whereas Mainstream Science has been saying for centuries that gravity is a product of mass. Science says that gravity is due to mass establishing gravity while not one person could ever explain the least detail as to how it is done. I went on to research Kepler and I discovered gravity through discovering Kepler. I concluded that gravity is the movement of material through space. By following Kepler's guide as Kepler formulated the process in introducing the equation four centuries ago being $a^3 = T^2k$ he gave us an explanation to what gravity is...if only Newton took notice of this important document. This says material holding space moves through space and proves that gravity has nothing to do with mass while mass is the product of space moving.

First you should decide what belongs to the gravity factor and what forms part of mass. Newtonian science is of the opinion that when a body is floating up in outer space the body has micro gravity…that just can't be the case. Newtonian scientists confuse the factors being responsible for mass and for gravity because if not, then please explain which is gravity, the part that tries to move the body to the centre of the Earth, or the preventing thereof? We have to see that mass is created by the pushing of an object onto the Earth and from the pushing (not pulling) comes mass while gravity is what is doing the pushing. While resisting further movement, mass comes into the picture and while moving towards the Earth or intending to move towards the centre of the Earth, that movement constitutes as gravity while stopping the movement leaves the object with having mass. Mainstream science loves to confuse the two issues because Mainstream science love to confuse everyone because Mainstream science is completely confused about the science they say they are the Masters of.

Is gravity that factor, which makes all bodies fall to the centre of the Earth, or is gravity that which prevents the further moving of bodies having gravity to fall further down to the centre of the Earth and then by restricting the movement, then forms weight or mass? By restricting movement towards the Earth a mass factor comes about which gives weight! It is presumed that the body has micro gravity because the body is weightless in outer space. This prompts me to ask the question underlying what has never been decided… what is gravity and what is mass. A body floating in outer space has maximum movement because when it moves slower, it starts to fall to the Earth. Being in outer space that body has little mass because it has a micro weight.

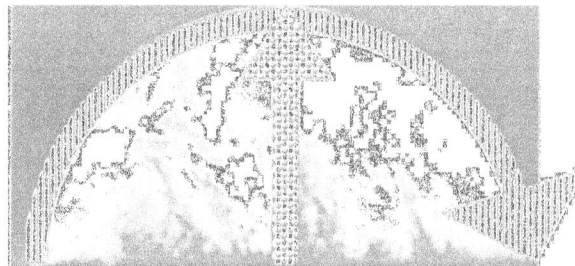

At that point when floating in outer space the mass (measured as weight / kg) is indefinably small while the movement that is applying is at a maximum in maintaining orbit. However, that is speed measured by distance (meters) travelled in time (seconds). Mass has a value, which is measured in the same currency in which weight is, and then mass is weighed as much as weight is and therefore, undeniably and in contrast to the logic of mainstream science's confusion and frenzy trying to confuse what can't confuse any further, mass and weight is connected as the same way of measurement and using the same measuring tools while gravity is movement notwithstanding mainstream science trying to put mass and weight far apart. If mass was equal to movement as gravity is, then mass must be measured in meters / second. Instead mass has the value which is the same as weight which is measured in grams.

When any object is in a state of having mass the object has to be standing still and being secured in a position on the Earth at that point of having mass. The object has to be in a position of absolute rest while it is on the Earth. At a point of standing still in relation to the Earth while excepting only the movement the Earth allows any object to form mass and it is where at that point that the object with mass is resting while all the rotational movement is equal to the movement the Earth delivers where the Earth is rotating. Rotating at the speed the Earth dictates form the factor science call mass. When the object leaves the surface of the Earth such an object will have to move much faster than the Earth moves or have less density than is required to maintain a steady position on the Earth.

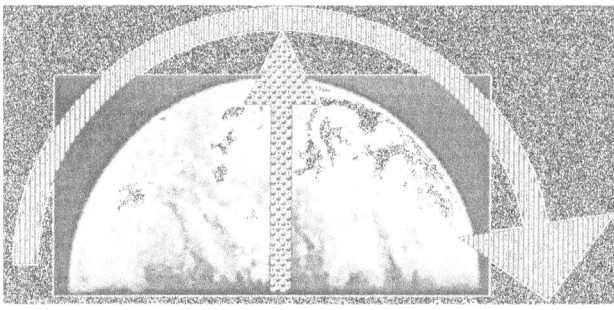

When any object is standing still in mass on the surface of the Earth, an object has micro gravity because the individual gravity left to the object in mass is infinitive small and is left to become an indication of attempting further movement towards the centre of the Earth while the Earth's material blocks the micro gravity to move and hence apply mass in doing so. Mass is not something inherent of the object but is the annexing of the object given mass by the Earth to secure the position of the object to ensure the object becomes part of the Earth structure. Having micro mass (not micro gravity) is where the body in rotational movement extends beyond the limit at the point where the Earth surface would award a mass factor. The movement speed goes beyond the speed required by the Earth at the Earth limit where rotation velocity secures mass as a factor. By exceeding the rotational velocity at a higher rate, such movement would exceed the movement or gravity of the Earth that is required in order to grant a mass value.

This issue is of cardinal importance and could deliberately be altered to hide the misinterpretation science wishes to connect to mass in order to hide the fact that mass does not bring on gravity but it is gravity that brings on mass. Mass is achieved when the object is resting motionless on the surface of the Earth while it is gravity that is still attempting to obtain movement as to try and move the object down to the centre of the Earth. This movement consists of two parts where one part is following the curve of the Earth while the Earth is rotating and the other part of the same movement is the thrusting of the object educing the object to move to the centre of the Earth. Mass is the result of gravity and not the other way around. Gravity brings on mass and mass depends on gravity to have any value or function.

A person that acquired the skills of peddling while staying upright on the bicycle has achieved the method of rearranging gravity within singularity. Without motion the bicycle falls on the spot it holds. When the bicycle is put in motion the bicycle can maintain the upright stance as long as the motion applies. When the motion stops the bicycle drops. To introduce motion to the bicycle the motion brings about a stable unsupported upright stance where balance can result from the motion the Earth enforces to the balance coming about by the bicycle using independence gained from motion of the space holding the bicycle. The space that the stationary bicycle holds is the direct result of the Earth providing the motion.

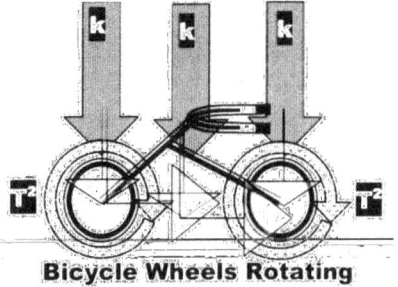
Bicycle Wheels Rotating

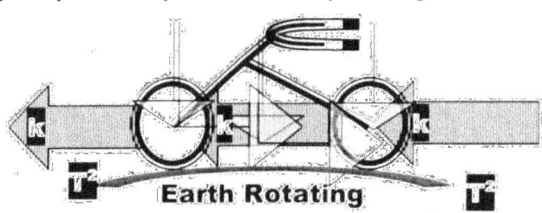
Earth Rotating

Gravity has nothing to do with mass although mass is a by-product of gravity. Gravity is the compressing of the space surrounding any spinning object and all things filling the entire Universe forming a solid structure spins in some or other way. By spinning the object contracts the space towards the centre of the object. Let's refer to the object as the Earth. The Earth draws the surrounding space towards the Earth and by doing so it contracts the space into ever more compacted layers of heat. By spinning the Earth diverts directional movement by the measure of 7 and later on I am going to show how this forms Π. Since it is Π that forms and it therefore is Π that moves from point Π_1 to Π_2 Π then becomes Π^2. This is as simple as gravity is and that is what I prove without finding any audience because science wants to have the absolute breathtaking marvellous filling their need for the magical mind boggling. The super intellectual wants to find that which would challenge the absolute brilliance of such a genius person. Because I am rather utterly simple by nature and not extremely intellectual, I chose to stick to my explanations by using the top and the bicycle as mediums to explain what I try to convey.

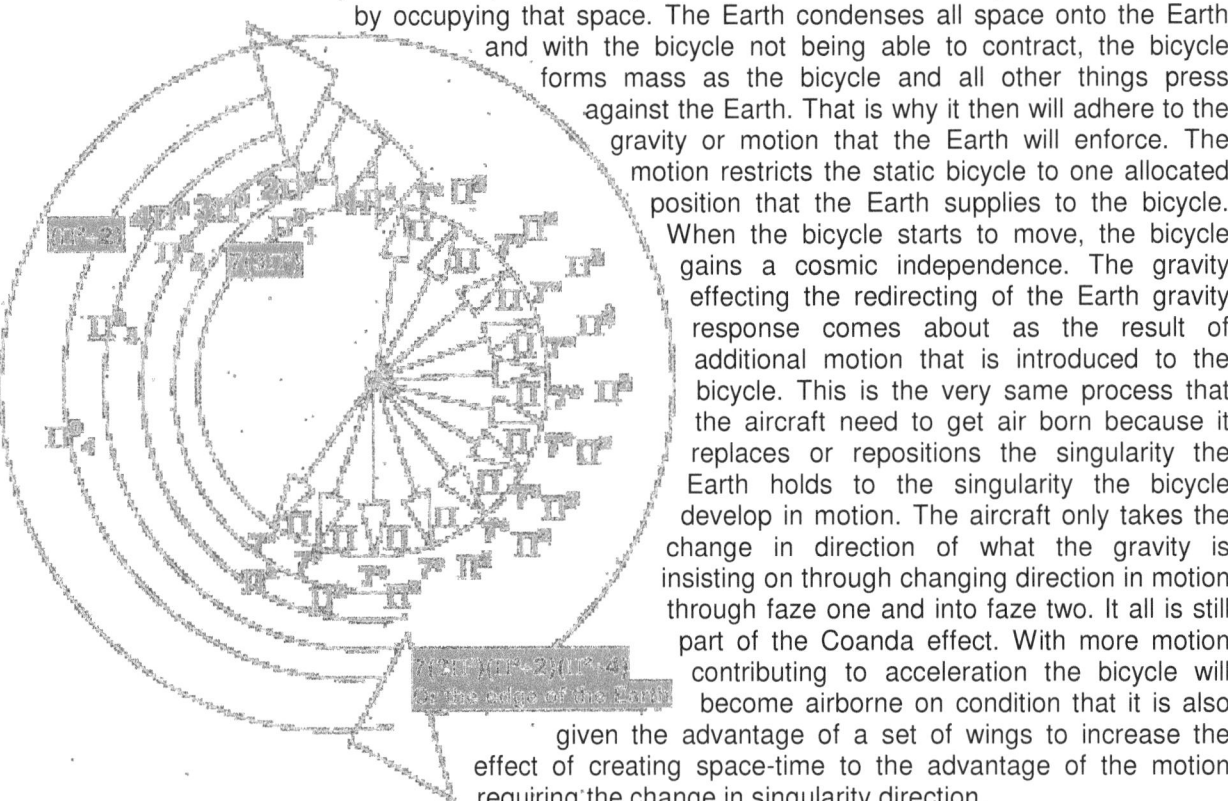

The Earth contracts not the bicycle but the space the bicycle finds itself within and the space the bicycle fills by occupying that space. The Earth condenses all space onto the Earth and with the bicycle not being able to contract, the bicycle forms mass as the bicycle and all other things press against the Earth. That is why it then will adhere to the gravity or motion that the Earth will enforce. The motion restricts the static bicycle to one allocated position that the Earth supplies to the bicycle. When the bicycle starts to move, the bicycle gains a cosmic independence. The gravity effecting the redirecting of the Earth gravity response comes about as the result of additional motion that is introduced to the bicycle. This is the very same process that the aircraft need to get air born because it replaces or repositions the singularity the Earth holds to the singularity the bicycle develop in motion. The aircraft only takes the change in direction of what the gravity is insisting on through changing direction in motion through faze one and into faze two. It all is still part of the Coanda effect. With more motion contributing to acceleration the bicycle will become airborne on condition that it is also given the advantage of a set of wings to increase the effect of creating space-time to the advantage of the motion requiring the change in singularity direction.

I have specifically chose to use a bicycle in my explaining because the bicycle is the object that rely the most on singularity achieving the required balance in which to operate. It is singularity, which puts space in balance of the time the space uses to duplicate. The singularity create space-time and such space-time results in a balance of space and time $a^3 = T^2k$. It is through the Coanda effect that marries the motion to the space that gives the balance that keeps the bicycle up right while it is singularity that allows the bicycle to move or duplicate the material by relocation through time. It is an act of balancing singularity that gets the bicycle as a machine working properly. It is also the next best thing to illustrate how singularity by motion provides gravity in addition to that which the Earth already produces. In the Coanda principle there are two factors where one is motion and the other is space and the two provide both duplication as well as contraction of space-time.

We think of the bicycle moving in a mono dimension where the bicycle as one unit moves from one position to the next position. That is not true because accidents show otherwise as the bicycle fragment when having collisions. The bicycle re-affirms its position changes as it moves by displacing the body atom by atom and the movement is as much part of the bicycle as the bicycle is part of the cosmos. Every movement is a new association with whatever forms the bicycle's surrounding and the movement is as much part of gravity as the bicycle moving independent. It is very important to see the bicycle moving as not forming part of the cosmos but as a result of life manipulating the cosmos by forcing the bicycle to move. Without the direct involvement of life the movement of the bicycle will never come about and therefore the bicycle receives independence as a cosmic structure not through the cosmos energising the bicycle but because life intervenes and life is not a natural part of the cosmos. The bicycle stays upright and riding because life puts it there.

In the time of Newton steam was the Rocket science of the day…and that was literal because the first steam engine was named "The Rocket". The concept of anything being drawn or pulled or powered by any source outside life was still in a concept form and to place much emphasis on movement or to think about what drove what whereto was in its infancy. Everything was either wind driven or animal driven or powered by slaves…yes slaves was still part of the daily practise of getting things done. That is how far back Newton's ideas go. Yet what Newton introduced has never changed, has never been changed and has been excepted without thought of changing or reviewing the concept Newton introduced notwithstanding the fact hat the entire idea Newton brought about was proven completely inadequate and incompetent.

Back then as it still is every one was and is sharing the Newtonian vision of a contracting Universe where the lot would one day again come together and Creation will end where Creation started some time ago which is what was presumed at the time Newton go wise to cosmic principles. Newton introduced the concept that the Universe has mass that is pulling mass towards one another and we are in the centre of an ever shrinking Universe. That is what the lot of us can see… we are forming the centre of the ever contracting cosmos where every Newtonian can vividly see with his or her eyes through any telescope that all Newtonians minded scientists are sharing the centre stage of the ever collapsing Universe. The Universe is about to end where all mass contracts into one huge lump of material. This was and still is the Newtonian mentality.

Then along came a man that had a good look at the Universe through big lenses. He looked at the sky and came to a conclusion the lot was not shrinking but everything that was thought to reduce was expanding. There were no arguments to have because any one that would look through his eyepiece could clearly see the lot was not shrinking. The lot was growing apart. In some cases he said the lot was racing apart. The Universe was growing by miles and not shrinking into nothing. The main discover had a name distant to for fame, and the man had a position of seniority, of education and of status and by having all that no one would dare to go against him even if it meant that he went against Newton. With the mark of such importance it prevented others from pushing his opinion aside. The man was E.P. Hubble. Through his telescope any one could see that the Universe was expanding and the expansion was most rapid.

This unleashed a problem the world had no name for. Everything previously thought of being known to science was at that point devastatingly unknown to science. The contracting Universe that science was familiar with was expanding and not contracting which made the Universe quite wrong in the attitude the Universe had. It is impossible for anyone or anything to have any vision about Newton being wrong and with an expanding Universe and not a contracting Universe by the mass thereof, this pointed to Newton being wrong. Newton could never be wrong because Newton was never wrong yet…so if the Universe insisted on Newton being wrong then the Universe is out of step with science that then forced science to correct such blatant misgiving as to prove Newton wrong. They would prove Newton correct even if that meant cheating facts by defrauding the public by creating an abnormality. They concocted a mathematical joke using mathematics and by implementing shear fraud mainstream science found a way to defraud the unsuspecting public. This postponed the unavoidable correcting of facts that the Universe had to comply with since the Universe owed the Master Newton some apology. The Universe had to start contracting and stop this silly notion of expanding because Newton has to be correct or science will unleash the tormenting truth on the Universe. Science will have Albert Einstein correct the Universal expanding even if it meant that dark matter was to be invented to do just that. Did the Universe not know that he whom never can be wrong is in a person going by the name Isaac Newton! In order to correct the Universe's incorrectness they had to find a manner in which to set the Universe straight as to prevent this rebellion of the Universe against the laws of Newton. Newton can never be wrong and therefore the Universe better stop this illegal expanding and start to contract as Newton insisted it does. In order to help the Universe set its path straight and follow the righteous road Newton laid out for it, the scientific world set Albert Einstein to task to count every particle he could see as to determine the mass in the Universe. If he did not count all the particles down to the last one, how would he measure how much mass there is in space and therefore I conclude he counted every atom he could see or couldn't see because he determined what the mass was in the entire Universe. Their devious deception they gave a code name so that all the public in the world would be none the wiser about the Universe's failings to uphold Newton's laws of mass pulling gravity to force contraction. To stop this leaking out to the general public about the Universe indulging in criminal behaviour the code name given was the critical density. Einstein was ordered to find the density where the Universe will start to honour the religiosity of Newton as to begin to stop expanding and contract in the way Newton said mass is going contracting with force. Albert Einstein only being human and not a god like Newton found not enough mass that will bring about the contraction Newton ordered the Universe to have although I am sure Einstein did cheat somewhat because as far as I can see there a lot of mass to count and Einstein did not count that long to come up with a number about all the mass being scattered throughout the Universe...and he did not even have the Hubble telescope to help him with his task by measuring all the atoms throughout the entire Universe. But the tricksters did not run out of tricks to find a way to clear the name of the honourable Newton. They started looking for dark matter that no one would see because if I no one sees the dark matter then also no one can prove the dark matter is there and therefore no one can prove the dark matter is not there. This shows the brilliance of the deception and the length to which the Newtonian members will go to defend their god called Isaac Newton. If no one can prove it is not there no one can claim they are indulging in fraud! How marvellous this idea was and it proved to be foolproof this time!

The question never answered is this: If the dark matter is there the dark matter has to have mass or the dark matter cannot be there. If it is mass pulling on gravity and the dark matter is there, what then is stopping the dark matter with mass from pulling at this very moment to bring about the required pulling that will lead to the contraction f the entire Universe coming about just as Newton ordered the Universe to do? What is postponing the pulling of the mass that the mass must evoke, if the mass is there and if mass is evoking the pulling of gravity! What the hell has the visibility got to do with mass pulling by gravity…or in fact having mass or not having mass in the first place? Why will the mass not pull now and produce gravity if the dark matter with mass is there along? What has the fact of visibility or no visibility got to do with the mass being there and activating the gravity at the present moment? Or is it that those clever one's were not that clever when trying to further the deception Newton started? Again I repeat the question I found no answer for as yet: If the mass is there dark or not dark, then the mass has to force gravity to pull…that is law on the condition that it is mass doing the pulling by forcing gravity to comply! Why look for dark matter that is not pulling and not presenting enough gravity to get the cosmos to contract because if dark matter is present dark matter has to pull because then dark matter ahs to have gravity because dark matter has to have mass in order to be present! …And those in charge of physics get annoyed when I call this fraud!

The shrinking radius will increase the effectiveness of the influence of the gravity that the mass can produce by the margin of the shrinking of the radius. During the Big Bang the Radius was infinite at that point, and in that event then that means the gravity was eternal. With the entire Universe being as big as a Neutron, the Universe was the size of an atom. If the Universe were the size of an atom and the mass within that Universal atom could not prevent the Universe exploding into immeasurable atoms, then it would not be able to retract all the atoms into one unit again. If there was not enough mass to start the contraction, there can be no contraction of mass that is producing the gravity at this stage. If the gravity is of such a nature that it allows a continuous growth of the radius, then the radius firstly cannot be zero as Newton suggested and the extending of the radius proves there is no contraction in the way Newton had everyone to believe. If Newton's idea of mass contracting mass is true, then on the other hand it must have resulted in an implosion as that which can never repeat again. With the Universe being one atom there would never in the future be a better instant to implode by using the gravity the mass would present at that point. With Newton's formula of $F = \dfrac{M_1 M}{r^2} G$ forming gravity, then the Big Bang is just not possible because from that formula the Big Crunch must respond. I agree that there is a connection between the Earth and the Moon and between the Sun and objects circling the Sun from as far as Kuiper belt and even much further than that...but it has nothing to do with mass because objects floating in space have nothing to do with mass, just because they are floating while positioned in random to size. If there is any person that is of the opinion about the fact that mass is being used by the cosmos to position planets in accordance with mass then please be so kind as to explain where does the solar system employ mass in order to arrange the planets' status in positioning the solar system by mass providing an indication of the layout.

The smallest and the largest planets are forming the ends at the closest orbit and at the most outer orbit. The first four are part of the smallest but they also are the densest and then it is followed by one giant of a planet going by the name of Jupiter. If mass was a factor this planet Jupiter should be closest or most outward in relation to the Sun but since mass has no influence on location, mass plays no role in the allotted alignment of the solar system.

If one person out there tries to show that mass is critical or influential or of any minor significance, then please show where and how does mass play a part in the cosmos except being something born in the imagination of Newton and incorrectly kept in prominence by his followers. All the planets orbit at the same rate and there too, large and small plays no part. The only clear evidence of anything playing any part in the solar system is the Titius Bode law, and that science put down as being coincidental because it does not apply to or prove Newton's mythical mass brings gravity story.

Locating and finding the presence of singularity

$k^0 = a^3 / T^2 k$ states that whatever is, is also spinning in order to be present.

What is in the Universe is spinning around a centre point. In the **precise middle** of all **objects in rotation** forms a precise centre point that is dividing the object in sectors from where the spin will initiate Π and from that line all **the spinning initiation starts** forming that centre point by continuously moving a circling Πr^0 away from the centre. Thus, the spinning object **will have a middle point**, a very specific **centre point that does not spin** and only holds Π as a specific value because no radius can apply. The point forming holds no space but controls all space spinning. But also the one value such a line **cannot have is zero** because the line **is there and holds contact** to the rest of the material bringing about that **zero does not start any** line and therefore the **value of the line must be infinite**, just as described in **accordance** and by **the definition of singularity**

As I am introducing a very new idea, I wish to explain in better detail what I try to convey.

While the toy top is spinning one will find singularity by moving the rotating line or radius progressively to the middle by reducing the length the line has from the edge to the middle. At one point all further reducing must end but the ending cannot include zero or nothing because the rest of the line still attach the rest of the top.

As the rotating direction moves inwards, the rings will become smaller and smaller.

That point albeit hypothetical, is also as much a reality none the less and is placed where that point **must be standing still** because every line **running from that point** in **opposing directions** is also **in opposing directional spin the other or opposing side.**

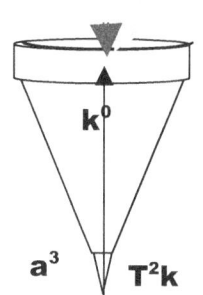

In considering the spinning motion in the fraction of time in the detailed instant every aspect of rotation will turn in every instant of change in time. Although the points had the same characteristics only one instant before, they oppose the characteristics it had just before and just after the very instant in which they are and to which they relate by similar points also in rotation. The fact of the graph proves my point in quarterly opposing dimensions and values.

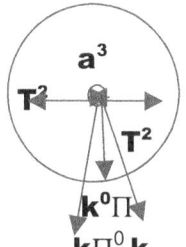

In dimensional terms, which I explain later on the value of **2k** relates to T^2. That relation extends to the next value where T^2 relates to **k**, which relates to T^2. The first space in the circle will then be T^2 **k**. From the centre being in infinity one can realise by applying mental power the single dimension factor not seen but present all the same. Extending that into the 3D comes six **k** and any one of the six will further extend to form a seventh point as T^2 All this is a multiplying of $k^0 = a^3 / (T^2 k) = 7$

In dimensional terms, which I explain later on the value of **2k** relates to T^2. That relation extends to the next value where T^2 relates to **k**, which relates to T^2. The first space in the circle will then be T^2 **k**. From the centre being in infinity one can realize by applying mental power the single dimension factor not seen but present all the same. Extending that into the 3D comes six **k** and any one of the six will further extend to form a seventh point as T^2 All this is a multiplying of $k^0 = a^3 / (T^2 k) = 7$

The very first instance brought the developing of **k** that was equal to T^2 time and a^3 space.

But motion came about on this side as well as the other side of the Universe where the Universe was $a^3 = T^2 k$ and on this side was a^3 while a^3 was becoming $T^2 k$ on the other side and the other side $T^2 k$ was becoming a^3. Then the second in creation came about but the first instant plated a veneer of material in the form of opposing singularity destructing as liquid heat due to a lack of motion onto the singularity applying motion, which brought about specific markings onto singularity. This act represented destroying of some less protected centre governing singularity and the maintaining of other singularity remaining in form and cool. This would see as a flicker from another point holding singularity. From the other side the flicker would seem to cut the duration it took the motion to fill the time half as long while in truth it cut the space in half by doubling the time. Then four flickers came about and eight flickers came about and the action time provided brought about more space as space doubled every time but it cut time duration by half every time.

There is a connection between the Earth and the Moon and between the Sun and objects circling the Sun from as far as Kuiper belt. This is what I named the **governing singularity** and this point could be playing a role either as the **governing singularity** or as the **controlling singularity**, which I am about to explain.

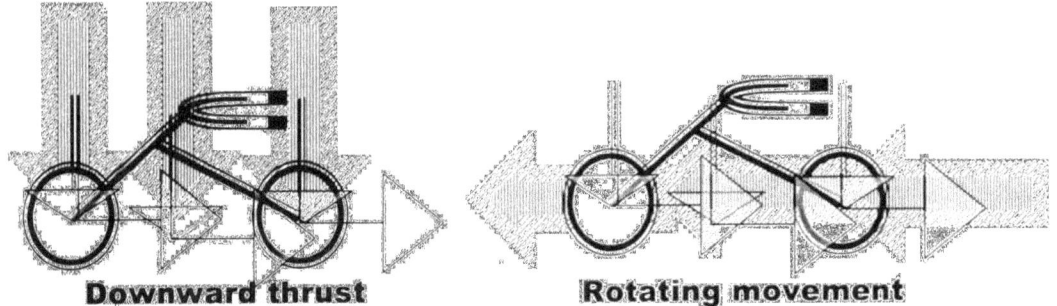

In gravity there is never just one movement but there are always two movements combining as gravity.

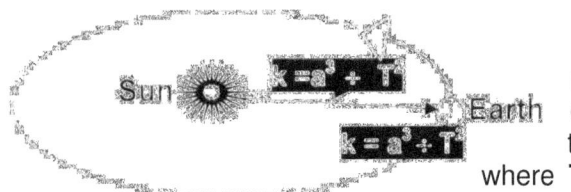

Taking the bicycle as an example we have the downwards thrust of gravity as well as the forward moving of the peddling cyclist that bring about movement. Under normal conditions and without acrobatics the forward peddling is required to keep the bicycle upright but just as much is the downward thrust required to keep the bicycle peddling.

Try as anyone may, the cyclist will not be able to keep the bicycle upright in outer space where the downward thrust is not present. Using Kepler's formula of $a^3 = T^2 k$, it is the body of the bicycle filling the space the bicycle claims a^3, that will move forward T^2 on condition that the bicycle is thrust down k. The downward thrust can only help if the forward movement T^2 grants the space the bicycle holds a^3 independence of the Earth making the moving singularity the bicycle then claims become the governing singularity while the movement demotes the Earth's normal governing singularity to become the controlling singularity. Space to grant a body independence from the Earth requires two directional movements acting as one movement. The body can only be independent a^3 in space if the movement is backed = by rotational movement T^2 of the wheels as well as having k become the directional movement. The value of k is in the reference the object holds in relation to the direction of movement.

Let us take it from a point where the Sun provides a centre as one starting edge of k giving k a directional relevancy, then that centre k will provide a line from the centre and the line k will provide three spots in a formation that produces a structure by the square T^2 of the dimension where T^2 becomes the spin of the Earth. Not once did Kepler indicate size as a contributing factor to a^3. That means every single point that k indicates there are three positions a^3 implicating sides of a double dimension. In the same manner is k not limited to distance or does T^2 lesser by size declining. $a^3 = T^2 k$…that is what Kepler said. There are three dimensions forming space a^3 by movement where two is between any two points T^2 flowing as time from the centre of the sun, which is indicated by the line the relevancy factor k indicates.

From the Sun there are three points moving between two points from one point to two other points giving the six dimensions we find in space. It is space in time or space converting space through the movement of time. It is a location of a point in the third dimension a^3 that will move according to the second dimension T^2 that will implicate k as a reference in the first dimension. It is about dimensions in reference to one another.

The Absolute Relevancy of Singularity — used to explain THE COSMIC CODE

When an object falls to the Earth the body moves in the direction of the Earth k. However at that point the Earth is also rotating away from the point where the falling object is heading to the Earth, which is indicated by the symbol T^2. The space in which the body is a^3 is falling k to the spinning T^2 Earth $a^3 = T^2 k$ all the while the body is descending (falling) $k = a^3 / T^2$ as the Earth is rotating $T^2 = a^3 / k$. It is in this that Kepler's formula comes to prominence. As the body moves towards the Earth k^{-1} the Earth shifts T^2 and the falling body is re-aligning with the Earth by associating with the position the Earth has that then changes. The falling body is declining in space represented as $k^{-1} = T^2 \div a^3$ while the Earth is rotating $T^2 = a^3 \div k$. At this time k shifts from straight down to slanted because the reference point relocated the reference position.

Having to re-align its position of reference as the Earth changes position in rotation, the object changes direction by implementing the triangle of Pythagoras because the Earth moves to the side by 7° and the object fall by 7° and in that the triangle that forms a right angle triangle (both sides equal to a change of 7° and therefore equal forms establishing a hypotenuse where the hypotenuse forms gravity. That forms the basis of what all that gravity is about and gravity is about Π coming into space-time by movement Π^2. Mass comes into question when the body falling has no further space in which to fall but are then obliged to stay still on the Earth surface and form a part (mass) with the Earth. While it is falling, it is in the space surrounding the Earth that contracts by the margin of Pythagoras and the body that s filling the space is also contracting with the space. The body is merely moving down with the space in which it is and it is the space in which the falling body is that is moving down with all the surrounding space also moving down. Being solid the body restrains the contracting that reduces the size of the body but still the space surrounding the body is becoming denser and compacter. In the Universe as in cosmology there is no possibility of the presence of nothing and therefore outer space cannot be "*nothing"* but is cosmic fluid that has the ability to become denser by contracting. There are two forms of matter formed by one cosmic substance, which is singularity. There is singularity formed as a liquid and there is singularity that is forming material. That is it...there is no possibility of anything else and all substances are composed of singularity being controlled by movement (material or matter) positioned in relation to singularity not controlled by movement or forming a liquid. Everything is heat that is forming space but movement, which we call elements and the rest of space, is heat forming cosmic space, which is uncontrolled and is therefore not dense but totally expanded, controls some heat. We think of this substance as outer space. Thewre3 are elements and there are heat covering and surrounding the elements as space forming a liquid in which elements of all sizes float. That is the cosmos. That is material floating in cosmic liquid having no mass because it has buoyancy by movement. That is why not one of the planets indicates any positional arrangement by virtue of mass because the buoyancy of any cosmic object excludes mass as forming a factor.

Always being a solid	Becomes a fluid	Becomes a gas
Hydrogen 1	melts at –259° C,	boils at –252° C,
Helium 2	melts at –269° C	boils at -268,9° C
LITHIUM 3	melts 180° C	boils at 1300°
BERYLLIUM 4	melts at 1287°C	boils at 2770°C
BORON 5	melts at 2030° C	boils 2550° C
Carbon 6	melts at 804 °C	boils at 3470° C
Nitrogen 7	melts at -210°C	boils at –195.8° C
Oxygen 8	melts at –218.8 °C	boils at -183° C
Fluorine 9	melts at –219.6° C	boils at –188.2° C
Neon 10	melts at –248.59° C	boils at –246° C
Sodium 11	melts at 97.85° C	boils at 892° C
Magnesium 12	melts at 650° C	boils at 1107°
Aluminum 13	melts at 660° C	boils at 2450°

When an element freezes it is solid notwithstanding… because then there is much less heat in between the solids. The ratio of cosmic liquid to solids favours the solid overwhelmingly.

When an element melts it becomes a liquid and that means there is just more heat in between the solids.

The Absolute Relevancy of Singularity — Page 102 — used to explain THE COSMIC CODE

When an element boils it is a gas again notwithstanding…and a lot of heat (cosmic liquid) is added where the ratio of cosmic liquid to solids favours the liquid ratio overwhelmingly.

In all of nature there is no **NATURAL GAS** as a natural element as much as there is no **NATURAL SOLID** as a natural element. No element is either a gas or is a fluid but all of the elements forming material are a solid. This solidness comes about because the atoms spin and the spin provide a density that cosmic liquid or heat lacks. We arrange the elements in such a manner, but that is only applying to the situation the Earth grants the elements a status to be thought of as a "natural fluid" or a "natural gas" where in fact even the hydrogen atom is a "natural solid" that boils at a very low temperature. In outer space all elements will classify as a solid because all elements freeze under those (to our thinking) "extreme conditions"

It is the value of Π that puts solids apart from fluids or liquids. Solids spin within liquids at the point where singularity forms space by the measure of Π. The cosmos are either solids or liquids and both are formed by singularity but movement applying or not applying brings control and control brings the definition to where the assortment belongs.

I wish to explain the location of singularity very briefly since shall return to the explaining about singularity later on but trying to exclude much confusion that may arise I wish to explain singularity very briefly. In the centre of all things are a spot that forms a centre, which has no space. When spinning that centre spot forms a line that has no space. If the line had space, the space had to choose sides and since the line forms the divide between that which spins to opposing sides, the line in place can't have space. It is the movement that brings about the relevancy **k** and with **k** in place the singularity inside the centre forms the space, which becomes the top. Putting the relevancy **k** in place brings about the validity of the space a^3 by the movement of the spin T^2.

By starting to spin the line forming singularity in the centre $k^0 = a^3 \div kT^2$ and that is what mathematical significance Kepler's formula indicates. As to why things spin has to do with heat and cooling but I will deal with that argument as the book progresses.

Everyone call this line that forms the axis. Everyone knows about the axis and yet through so many thousands of years no even ever thought to scrutinise the axis. The axis controls all particles spinning around the axis while the axis in itself represents no particles because the axis represents no space.

Having no space would mean occupying no space which means forming no part of the Universe filled with space and yet it controls all the space as wide as the mind can imagine. Without space it does not form a part of the cosmos, but forms the cosmos as wide and as deep as the cosmos goes. The axis could not be seen but with applying intelligence the axis could be witnessed. Having no part in the cosmos in space, the axis could only be understood and never be seen. The axis could be proven but never be shown. The axis is what controls the Universe from end to end because when there is no end there the axis provides one end to what never can have another end and the axis governs whatever spins in relation to such a line.

Again I wish to press the issue to form clarity. The line is without space and only holds form, and therefore the line represents a point not having any dimensions while it still is there without ever being there. If ever there is a concept I have to introduce, then it is the concept of how important the axis is and how science up to now missed the biggest issue driving the cosmos. The line forming the axis is there but only intelligence will ever form the concept whereby one can realise where the line is. Anyone unable to understand this concept can never see the validity of religion because here too, is a something that is there but only intelligence and not a lack thereof can bring understanding to the importance thereof.

The Absolute Relevancy of Singularity
Locating and finding Singularity

In the **precise middle** of all **objects in rotation** is a precise centre dividing the object in sectors that will **start the spinning initiation** from that centre point. Thus, the spinning object **will have a middle point**, a very specific **centre point that does not spin** which only holds Π as a specific value. One value such a line **cannot have is zero** because **zero does not start any** line and therefore the **value of the line must be infinite**, just as described in **accordance** and by **the definition of singularity.**

What goes much beyond my intellectual understanding is that science this far missed the point that when something has the measured value of 1 it has to form singularity. What other number can singularity have as a value of worth than the value of 1? If the value one received from any calculation is one, then it indicates to a point serving singularity, as we find in $k^0 = a^3 \div kT^2$

When the top is not spinning, there is no such line detected, but when the top spins, the line forms from the top to the bottom running all along the centre. This line divides the top into four directional sectors that opposes in direction of rotation. The line's worth has gone undetected since Newton announced gravity and whereas the line is the most important aspect concerning physics, yet the existence of the line was previously never noticed.

That point, albeit hypothetical, is also as much a reality none the less and is placed where that point **must be standing still** because every line **running from** that point in **opposing directions** are also in **opposing** directional spin to the other or opposing side.

Spinning or movement inside the line would **be zero,** but the line, although **being without space,** also **can't be zero** since the line is there for all to see. The movement in the line **is zero** and the space the line uses **is zero** and the line holding a value in size might **be zero,** but the line as **a cosmic reality** just can't **be zero** since the line controls all the spinning taking place.

From this centre line that is only theoretical definable, but is still there all the same, an opposing value always form from a previous turning position to the next turning position that becomes real and distinct when rotating, but loses its distinction when not rotating because then all traces of the line that is not there is lost as the line disappears.

As the line disappears the value of the line not being there changes from being noticeable to zero, and as the line removes from having a notice ability in securing a value by spin to then when not spinning have a value of zero, where this zero value then replaces the most original value it had. When not rotating, zero removes the line from a position it never held in space previously. When rotation begins, the line forms and is only backed in value by having only a hypothetical position claiming zero in spin and in space but not in presence. Being without space doesn't make that the line is not less distinct but the line is more distinct than any other part that in reality does hold space and therefore participate in spin because from that point every rotating piece of what ever is, then will spin around this line that is not there to start with and such spinning will clearly carry from where the line only has a distinction value in the singularity to carry on with a value of Π implicating rotation. The line forming holds 1 in singularity and from where such a line ends, only there does the circle value of Π start.

If the spinning top is all the evidence any one needs to come to such a conclusion that will bring any proof that the singularity governing the top connects to anything anyway, we will then find it when studying the spinning behaviour the top represents. Placing singularity in a location not being present in the Universe is fair and fine, but what will the evidence be in proving its activeness as part of the creation at large?

There are solids that form elements and by the control of singularity producing movement that confirms the structural integrity and discipline of singularity and is contracted to solidity by movement. That then is a solid forming various atoms of all sorts that we know as elements or as material. The elements in atomic cocoons are formed by singularity but the immense fast spinning contracts the singularity to a solid substance. Then we have liquids that are singularity that is unattached and are loosely connected and will accommodate any solid spinning where the forms of solids are in need of occupying that space for any duration of time. The liquids are able to accommodate or house the solids without being affected in way. There are cosmic liquid accommodating cosmic solids where the solids Newtonian science does recognise but the cosmic liquid Newtonian science fail to recognise. Newtonian science calls the solids elements but the cosmic liquid they call "nothing" and then give "nothing" a measurable value. Lately Newtonian science came to think about "dark matter" that has to keep the Universe in tact and I suppose in a way this forms the "dark" or invisible matter Newtonian science so desperately needs. The only viable conclusion about what keeps the Universe in tact would be the idea of a liquid in relation to a solid. Should there be any one that disagrees with my statement about the cosmos formed by liquid, then please tell yourself why would the atmosphere be a liquid with a density of about six hundred time less than water and where does the atmosphere (not density differences) stop and the nothing filling the atmosphere starts.

.
Elements are solid and that which house elements can't be nothing but has to be a fluid / gas/ call it what you like. Hydrogen is as much a liquid as iron is a gas and neon is just as much a solid. In fact all material (atomic compositions) is solids and the ratio of heat in between the atoms determines whether it forms a liquid or a gas. It depends on the element relating to the space/heat in the circumstances surrounding the substance at that very precise instant in time. We have to stop telling the cosmos to show us what we wish to find and start accepting what the cosmos is telling us to find. The culture that I am referring to is all about **nothing.** At present we find that there is something we think of as nothing in outer space. Because nothing is what we wish to find and nothing is precisely what we are getting because we think of outer space as nothing. If you accept the cosmos to be nothing, then please define nothing to yourself and find the definition in the cosmos. What we think of as forming a gas / liquid is when the mixture of cosmic liquid becomes more in ratio than what the solid (atomic element) is and when the substance "freezes" there is less of the cosmic liquid than there is when the mixture turns to gas or liquid. We confuse water with what is a liquid since water tends to mimic liquid because water is very adaptable to changes in form when being in a fluid state. However, being fluid like does not change the substance of water since water forms with the combining values of material and material is a solid notwithstanding human connections to the idea.

This brings us back to the importance of Kepler's relevancy, which Newton got rid of so easily. The value of Kepler's space he indicated as a third dimension a^3 does depend on indicating a structure a^3 that is in rotation T^2 but also needs one position having a constant of some sorts in relevancy to singularity. Any point where **k** may indicate a position one will find a value matching a^3 and the matching location will fit T^2 at that point on the condition that T^2 forms the margins of the specifics of **k**. That is the relation there is in the solar system between all planets and the Sun. The Sun always indicates the centre and the planets always indicate the rotation. But $a^3 = T^2 k$ is only producing a relevancy of three dimensions that is equal to two plus one dimension. That indicates the space a^3 is in place by the movement T^2 thereof in relation **k** to singularity k^0.

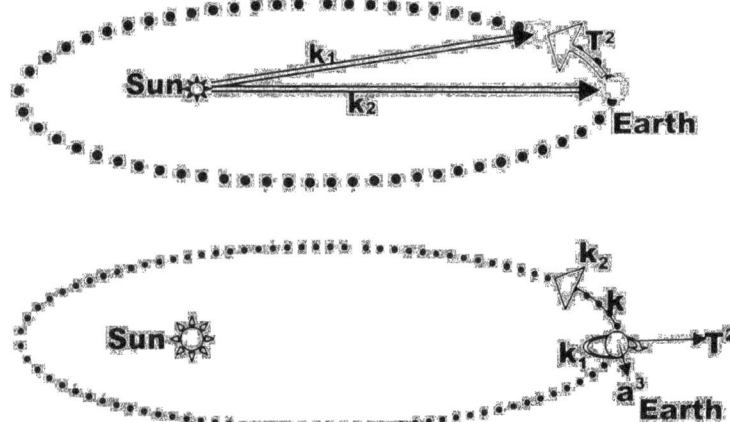

In order to argue this idea that outer space is zero let us return to the sketch and take it from a point where the Sun provides a centre as one starting edge of **k** then that centre **k** will provide a line from the centre and the line **k** will provide three spots in a formation that produces a structure by the square T^2 of the dimension. Not once did Kepler indicate size as a contributing factor to a^3. That means every single point that **k** indicates there are three positions a^3 implicating sides of a double dimension. In the same manner is **k** not limited to distance or is T^2 lesser by size. $k = a^3 / T^2$ That is what Kepler said.

There are three dimensions a^3 between any two points T^2 flowing as time from the centre of the sun, which is indicated by the line **k**.

The value of **k** is not to put a measured value in place, but to bring a reference to singularity $k^0 = a^3 / (T^2 k)$ applying as to place a specific singularity in as the **governing singularity** and another in place as the **controlling singularity** because there always has to be a controlling singularity determining the orbit while there has to be a governing singularity determining the spin of the body in relevance performing as the space a^3 in question in the formula $a^3 = T^2k$ where in that formula k determines the relevance of k^0 as in $k^0 = a^3 / (T^2k)$. However, this burden k forever with the responsibility of forming a line and a line is what places the Universe in place. Every space a^3 in question puts singularity k^0 in position by the motion T^2 in relation k the position allocated in the Universe.

The implication of the relevancy produced by the use of the formula $k = a^3 / T^2$ brings about that when dividing T^2 into a^3 there is k left. The fact is that a^3 is a three dimension (^{3}) of single k (^{1}) showing one or T^2 is two dimensions of k being the one dimension it means that k is a part of space a^3 or T^2 which is time. It is the same thing in a double dimension or space being a triple of k then k is one factor and k cannot show a position of zero. If $k = 0$ then there is no possibility of $k = a^3 / T^2$ because $k = 0$ then $0^3 / 0^2 = 0$. That does not make sense. Mathematically space cannot be zero because those being of the opinion of space could be zero or nothing must first prove mathematically that space is zero. Moreover they then must prove mathematically how zero grows through the Hubble constant. By translating Newton's vision of the circle in completing a cycle would become zero through rotation…well that does not count because in the use of the formula when calculating a^3 please replace any factor of $a^3 = 4/3\ \Pi\ r^3$ with zero and calculate the end result. If k cannot be zero then k could not start from zero. With $k = a^3 / T^2$ no point can be zero because k shows space $a^3 = k\ T^2$ is no reference to the volumetric mathematical formula used to calculate $a^3 = 4/3\ \Pi\ r^3$. Nor does it show the use of the circle in the second dimension being $a^2 = \Pi\ r^2$. In the case of the Kepler formula the circle factor becomes the square as indicated by the duration of the time T^2. The factor standing in for the line which normally would be r and then be the square value is in the case of Kepler's formula not the value indicating the square. That means Kepler never indicated a circle of mathematical procedure but said mathematically the distance of the planet from the Sun k holds space a^3 in relation to time T^2

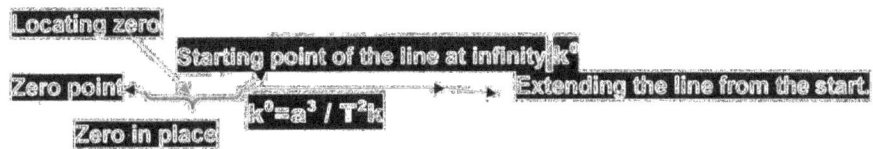

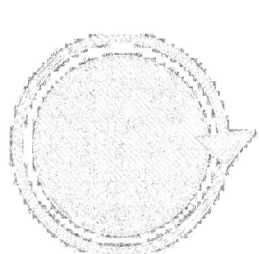

Lines mathematically cannot start at zero because there is no evidence of zero as a factor in mathematics. Should you disagree with my statement the question in need of answering is this: **What will the length of the shortest hypothetical line imaginable be and moreover, what would the total overall length be in that case?**

The shortest line that can be valid is a line having the start of the line and the end of the line holding the same spot. That points to singularity forming 1. There can be no line shorter than that and a line having zero as a start must have no start in order to qualify as being zero. The Universe is lines that form from every angle possible and with zero removed from the Universe we find that the Universe starts with singularity 1^0, 1^1, 1^2, 1^3, 1^4 and so on. With zero excluded from mathematics we can return to gravity.

When an object has gravity the objects is following a line that will form a direction of movement. Every conceivable object in the entire Universe is spinning while also moving in a specific direction, which will result in forming a circle. Everything in the Universe is submitted to movement and all that holds space moves. When an object is standing still on Earth with mass, that object is moving. This is because time is the movement of all things in relation to specific point.

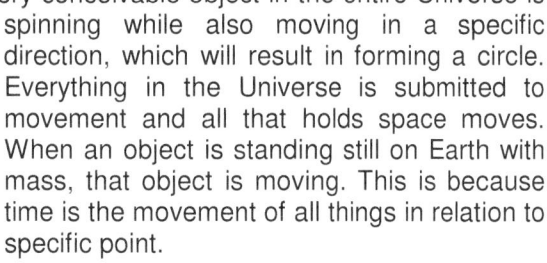

Gravity is a product of movement and not a product of the influence of mass. By orbiting at a specific distance, the distance from the Earth is determined by the rotational speed the object encounters. When the object reduces the orbital rotation (circular velocity), the gravity by slowing down will bring the object to start moving towards the Earth, which is falling and which is what everyone knows is to be gravity. One then must accept that mass is having an object being in a point of only moving with the Earth while gravity is the movement or inclination it shows to produce what is required to further move towards the centre of the Earth or the inclination of forming movement towards the centre of the Earth.

The inclination to move to the centre of the Earth is gravity while stopping such movement is forming mass. That is the difference science never finally concluded…gravity is movement or having the inclination to move while mass comes into play when that which moves is standing still in reference to the Earth while having the Earth move.

Anything that is spinning, by spinning shows contraction in moving surrounding space towards the centre of that which spins by the centrifugal forces pulling space to start to decline. Everything is moving towards the centre and I have indicated that resting in the centre we will locate singularity. The spin will comprise of two parts where the movement the spin holds will form one part and the next part will be the movement of space going towards the centre. This is exactly the features gravity holds as Kepler's formula indicates in $a^3 = k\ T^2$. The linear k movement shows the direction in which the space a^3 moves as the space compacts and reduces while the rotation T^2 is coming from the object spinning around its centre k^0 or forming its axis by spinning around its singularity.

All hydraulic pumps work on this principle and I show that all stars are hydraulic pumps pumping cosmic liquid into such stars. The planet we call Earth is just another very poorly developed star on its way to become a star in the far, far future. The entire Universe is working on the principle of hydraulic power, which in fact, is the most powerful source of drive thee is. Electricity is just more hydraulic drive that implies cosmic fluid as a liquid source…and that too I am going to prove mathematically as the book progresses.

The space that is drawn and is forever becoming more concentrated and denser as it contracts towards the Earth we named as being the atmosphere. This movement towards a centre by a rotating body drawing space is as much physics as mass is not natural physics used by the cosmos. If any object spins, such an object will influence the surrounding space holding whatever in that space to move towards the centre. Having gravity is forming movement that is inclined to move towards the centre of the Earth. That is called centrifugal force and therefore one could call gravity a centrifugal force. It is movement drawing all the space that is around it towards the centre. The Earth, when spinning has to show centrifugal force because it is part of the spin the Earth shows. All space surrounding the Earth will move towards the Earth and thereby become more compact, or compressed or denser, no matter what name one attach to the process. Mass don't even need to be mentioned in this process!

I am going to explain in this book why everything closer than 2.4674 times the radius of a star will dissolve

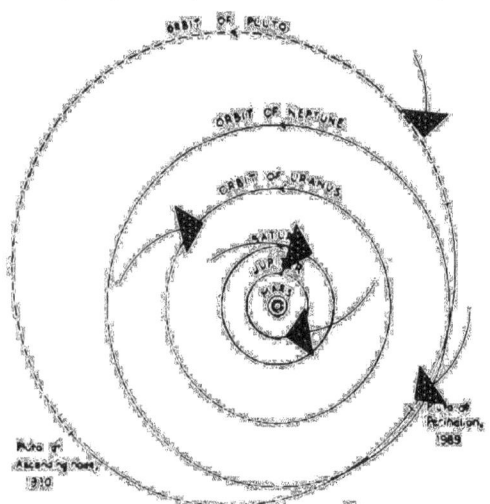

into liquid and become a fluid that a star incorporates into its structure. This is called the Roche limit. As the Earth spins the spinning of the Earth is engulfed with space and the space surrounding of the Earth, the Earth draws from the outside towards the centre. There are so many layers named different names of atmosphere where everyone has a different density and each holds a name. In terms of science naming names to layers the real issue of why it is there in the first place as well as the fact of why the layers form is left to gravity, according to science, which is left to mass, which is left to magic.

The contraction of the space immediately around the Earth becomes dense and hot while the further it expands space towards outer space, the fewer particles the space holds and therefore the colder it gets. I put this incorrect view to the test in other work and show that what we think of as being hot is in fact cold and what we think of as cold is extremely hot. However, that argument I leave for another opportunity because at this point we look at gravity in its most basic form. This comes about because there are two forms of substance forming the Universe. There is a liquid holding a solid and it is movement that makes the solid secure. By spinning within the liquid the solid brings about gravity. Every atom is a centrifugal pump within a liquid forming gravity.

There is one BIG centre pump pumping space-time towards the centre. There are nine smaller pumps, pumping space-time towards each one's individual centre and this is aligning according to the Titius Bode law of positioning the allocated position according to the specific requirements that the Titius Bode law prescribes.

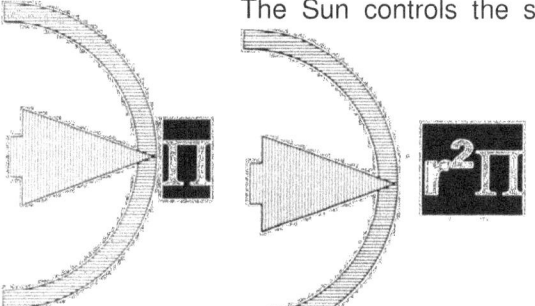

The Sun controls the solar system not by mass but by presenting a **governing singularity** (a term I explain later on) that presents to space a **controlling singularity** (another term I explain later on) and that controlling singularity extends as far as a third of the way to the next star. One can assume it is not a third of the way but is Π by the distance thereof. The Sun produces Π as a controlling mechanism of singularity $Π^0$ to form an area of control as wide as no one could ever have thought. This has no implications on mass because mass in this has no factor to offer.

The main factor required has to be Π since everything so far is relying on the mathematical factor of Π. Everything is spinning and when it spins it involves a circle and a circle can only be a circle by using Π. In that sense when searching for the four phenomena we have to connect the use of Π to every one. I will explain the Π connection in all four of the Phenomena called the **Titius Bode law**, the **Lagrangian positions**, the **Roche limit** and **the Coanda** effect and I will not use any force because forces belongs to witchcraft and in physics that is very absent.

Being a circle requires two factors and both those factors Newton dismissed in his search for gravity. More

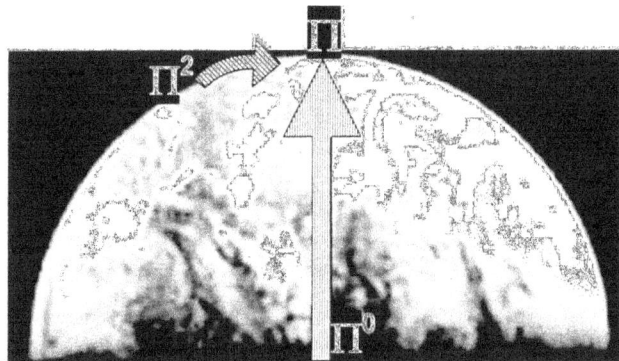

important is the fact that modern science are so well equipped with the skills of mathematics and yet for hundred years after the fact not one in science came to a conclusion about Π having to be involved as well as with the Earth being round therefore the Earth having to have a diameter when dealing with a circle…this requirement fits any circle and the Earth is just another multi dimensional circle.

If r is the diameter, then the position science so feverishly award to mass has a point that actually holds Π as reference in the laws of mathematics. The first cosmic connection is between singularity in infinity and singularity at the edge forming the curving of space. This relation is $Π^0 \Rightarrow Π$. As mentioned there is a line forming within the centre of every rotating object. That I named the **governing singularity** or in numerical term it is either 1^0 or $Π^0$. Throughout all of the more than twenty six books I wrote this far, I show that the Universe and whatever is in it, is governed from and by $Π^0$ or the **governing singularity.**

From $Π^0$ a line forms that is running not only from top to bottom nut also all the way to Π and Π forms the limit of space. In terms of cosmology the line that forms is $Πr^0$ but r^0 coming from $Π^0$ can't be dismissed because it is most critical in determining the next singularity namely the **controlling singularity.** Newtonians wish to give r^0 a measured value so that they can play games by using mathematics ands in the cosmos that doesn't apply as it does on Earth just because r^0 forms $Π^0$ in a completely different manner because r^0 is as much responsible for $Π^0$ that $Π^0$ is responsible for valuing r^0. The Sun has much more gravity that does the Earth have because the Sun holds a much more significant **governing singularity** that we have on Earth.

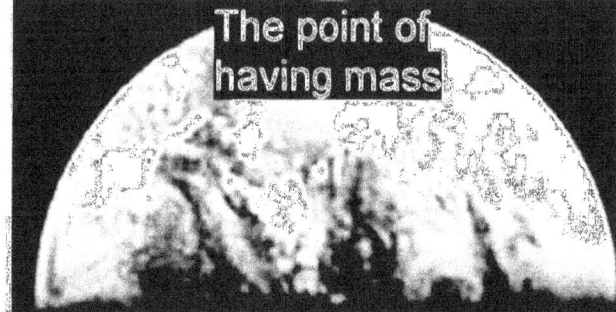

That is because all the atoms within the Sun forms r^0 by measure of every rotating particle contributing to $Π^0$ as every rotating particle holds an individual measure of $Π^0$ and when all these material components put together the individual singularity $Π^0$ the lot form r^0 which then becomes the **controlling singularity.** This however is not as simple as guessing the mass applying and from that then the mathematician can play god by using

$$F = G \frac{M_1 M_2}{r^2}.$$

There is no way anyone on Earth can place a measurable value on r^0 because there is no way in God's Creation that any human can determine the number of atoms comprising Π^0. However it is not $\Pi^0\Pi r^0$ that holds the key to any orbiting object but it is Π^2 that holds the true value gained from $\Pi^0\Pi r^0$. Most important of every concept is the movement Π^2 represents.

Nothing stands still to anything else in the cosmos. When having "mass" the body is standing still in relation to the Earth while the Earth is moving on behalf of the body having "mass". The Earth is taking on all the moving responsibilities and the body "with mass" is taking on the moving speed and density of the moving Earth. Everyone accepts gravity is taking a body "straight" to the centre but it is not. The Earth moves in two ways and this seems to be Π which is the point ending the circle and Π^2 where Π moves = Π^2.

Modern science still supports the Neanderthal idea that a body fall straight towards the Earth as "mass"

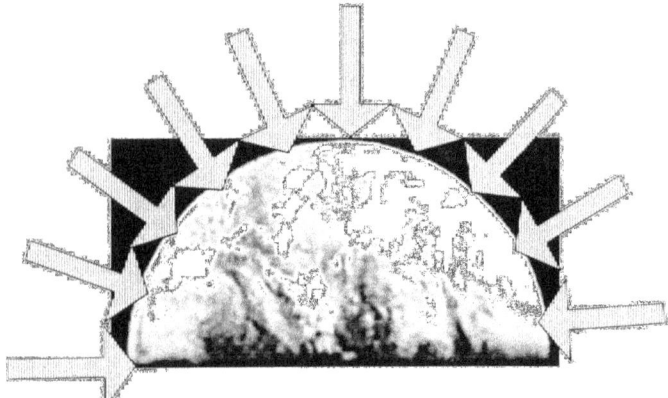

draws the body directly to the centre. This is as outdated as any view science may have on gravity where they hold the informed opinion that mass is having a pulling power that can pull by force where such pulling is aided by magical powers and forces. Any object moves in a straight line as much as the body moves in a circle at the same time. That is what Kepler's formula says if it says anything. A body a^3 moves = $k\ T^2$. That is what Kepler's formula says. Then the formula also says a body a^3 moves = $k\ T^2$ but as it moves it moves straight k as much as it moves in a circle T^2. That is the duplicate value of gravity. The body moves in a circle T^2 that is representing the **governing singularity.** At the same time and

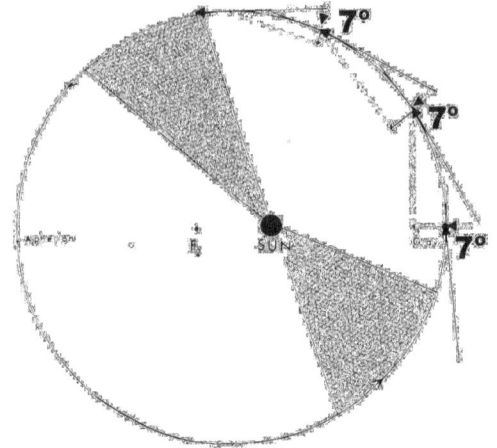

during the same movement it seems to move straight k but this movement will also end in a circle $a^3 \div k = T^2$ and in this movement it

The Earth is going around the Sun by 7°
The Earth is going around its axis by 7° in a cyclic rotation

represents the **controlling singularity.**

There are more positions forming in relevancy that is holding singularity, which I will explain as the information that I provide progresses.

Directional change by 7°

The Earth is diverting cause by 7° whenever it moves. As the Earth spins around the axis the Earth centralise the movement by 7° every time the Earth moves. The directional diverting brings movement not only to the Earth but also to the space surrounding the Earth and in that the Earth forms one huge centrifugal pump that pumps the space towards the centre of the Earth.

In turning towards a centre it compact the space around the Earth and as the space around the Earth compacts by moving towards the centre it compresses the surrounding area of the Earth and in that this compacting space awards a flow of everything within the air and in this process the object is compressed onto the Earth whereby this compacting of space bringing pressure to bear on the object then forms mass where this object presses onto the solid of the Earth. This compressing can only be if the object being compressed is also a solid substance of any sort and when the object forming mass will no longer relinquish space.

Then part of this movement there is anther directional change in movement causing gravity that forms the controlling aspect of gravity and is a directional change also of 7°. This change in direction is the spinning of the Earth around the axis of the Sun and this forms the second part of 7o that forms gravity in the forming of the value of Π.

Although being in a state of having a "mass-attack" such "pulling" on your body moves the body to the centre of the Earth. By the Earth rotating the Earth is moving notwithstanding your body being in the state of having no motion, still moving with the Earth has your body falling by 7° as it circles with the Earth around the axis of the Earth. The Earth falls by 7° when rotating and therefore your body is falling with the Earth being connected to the Earth by 7°.

Not only is the Earth falling by 7° as it revolves around its axis, but also it is circling the centre of the Sun and by doing that the Earth is falling another 7° by rotation. This is pivotal in understanding gravity as a mathematical fact.

That puts the falling of the object completely in relation to the speed that the object holds and that places gravity by falling in direct relation to gravity by orbiting.

When a body falls there is no mass involved because all objects fall equal and this was accepted long before Newton started fantasizing about his mass involvement in gravity applying. That is what Galileo proved eighty years or so prior to Newton. The distance the object orbits measured from the centre of the Earth and the orbit circle holds a direct link to the speed or time in relation to space that the object rotates. If the speed in revolving declines, then the orbit circle declines and this reduces the distance the orbit circle is from the Earth centre and the orbiting diameter reduces. If T^2 reduces then space in which the orbiting object is a^3 declines from the centre k^0 and k the relevancy factor depreciates. It is all in Kepler's formula. The orbit circle T^2 is directly associated with the distance k the orbit takes place a^3 measured from the centre of the Earth k^0 in a ratio of time taken versus space travelled through. This has to do with speed or movement and applies to all objects equally holding no specific relation to size or mass. It is a relation between the orbit circle (circumference) and the distance from the Earth centre (circle radius) and if that is the case, then gravity forms by Π having some sort of involvement and that throws any idea of mass playing a part in forming gravity out of the window where I hope it takes all of Newton's ideas of mass-forming-gravity with when going out the window. In forming gravity the centre line (diameter) holds a specific value to the orbit (circle) and with that being the case then we have to search for the part Π plays in the function gravity has and when doing that we can leave mass out of the frame because big or small, all things fall equal. Galileo was the one that proved that.

So you think that it is much simpler to maintain the argument that gravity is the force created by mass pushing the object onto the Earth only when the object moves at the same pace as the Earth rotates...and mass is always present and not only as I say when the object is on the ground and finds mass or weight!

So you still think that explaining gravity remains as simple as putting gravity in a connotation with a force fed to measure by having mass attracting whatever is attracted and this then allows the simplicity of the Newtonian concept to deal with the confusing part of the entire issue!

This is how I prove mathematically how gravity works. There is no pulling of mass or by mass or even that having mass plays a part in forming gravity. On the contrary, it is the forming of gravity that establishes mass when the space can no longer reduce and the reduced space locks whatever then has mass onto the solid surface of the Earth.

When the object moves while being in space or in contact (in relevance) with the spinning Earth, the object wishes to continue moving straight ahead while the Earth also moves straight ahead by turning 7°.

Therefore, the Earth by spinning is falling away by turning 7°. That clears space or compresses space by the margin of 7° declining (compressing) of air / space. The Earth pins around its axis by 7° and also turns around the Sun by 7°.

The Earth is moving, constantly spinning and in this is contracting space by compression (we call this contracting of space in air the atmosphere) and while the air is getting more compact, it takes whatever is filling with space towards the Earth constantly at a rate of 7°. By the Earth rotating, it is compressing space and with space compressing it is moving objects in the direction of the Earth. That is why objects that is falling, has no mass and only the stupidity of the simple Newtonian mind will force scholars to accept that it is mass that is pulling gravity. The entire idea of gravity is secured in movement of everything in relation to everything else. There is nothing in the Universe that ever could remain still because everything cosmic that is filling the Universe is spinning while it is also at the same time moving in a straight line and by that is following a circle. The Earth is only moving straight ahead because the Sun is spinning and while the Sun is spinning, it is compressing space, which allows the Earth and all other rotating objects to spin around the Sun in a perfect synchronised fashion. This process is going on throughout the entire Universe.

This places Pythagoras in the pivotal role of gravity by forming a calculated value of Π when dissecting Π mathematically. Gravity is Π using the law of Pythagoras.

In any right triangle, the area of the square whose side is the hypotenuse (the side opposite the right angle) is equal to the sum of the areas of the squares whose sides are the two legs (the two sides that meet at a right angle).

The square of the hypotenuse of a right triangle is equal to the sum of the squares on the other two sides.

If we let c be the length of the hypotenuse and a and b be the lengths of the other two sides, the theorem can be expressed as the equation:
$a^2 + b^2 = c^2$ = or, solved for c: $c = \sqrt{a^2 + b^2}$.

From this principle gravity starts by doubling the square of 7 just as I have explained before.

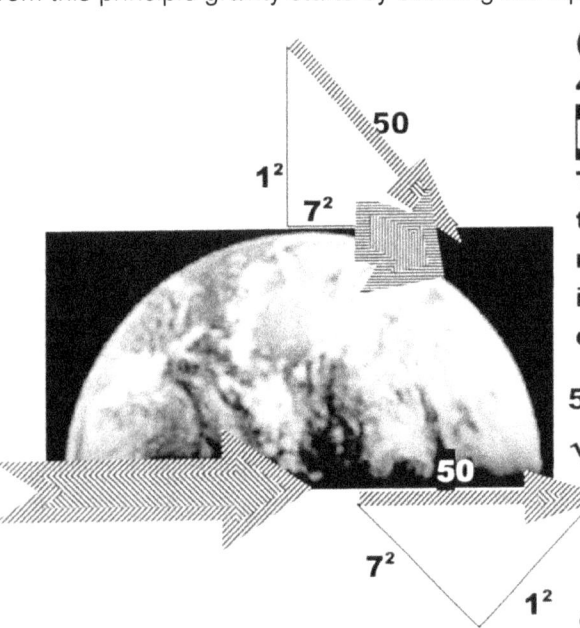

$(7^2 + 1^2) = 50$
$49 + 1 = 50$

Pythagoras
This applies twice in one movement (going in a circle and circling the Sun)

$50 + 50 = 100$
$\sqrt{100} = 10$

Since the movement involves two equal phases that acts as one, therefore the double value of 7 in relation to forming ten becomes what forms Π. We have the movement of seven forming one direction standing in relation to singularity which is the square of 1…According to Pythagoras that will bring about fifty. Since singularity is equal a one and seven is combined with singularity, the equality of singularity brings about the seven uses the same attached 1 in the square making the fifty a combination of another fifty and from putting a double fifty in the square as Pythagoras demands we have ten as a result in relation to seven. In Pythagoras's square the one side, let's say the adjacent side of the triangle forms a square in seven bringing on forty-nine. The other side let's say the opposite side of the triangle uses the square of one (singularity), which also remains one. The sum total of the two forms fifty, which is the measured value of the hypogenous. Since gravity is always applying in the double movement of seven the total of the hypogenous then is fifty plus fifty which is one hundred. The square of one hundred is ten and that brings about the value of one side of Π. When explaining the total worth of Π such explaining in detail requires a lot more information which will claim about as much space that which this article in full would allow. I complete this explaining as well as the explaining of how Π comes about through the forming of gravity much later on in **THE VERACITY OF GRAVITRY**.

That is what gravity is. Gravity is space moving or changing position in time and when an object can retreat no further towards the Earth centre, it only then forms a solid that aligns with the spinning solid material and with that then receives mass… Gravity is the movement of space in regard to any one specific point…and that is also precisely what time is. Nothing is standing still in the entire Universe. There is not one fragment of a sub-atomic particle standing still in relation to any other particle through out the entire Universe that is standing still. Having mass is when one object is standing still in relation to the Earth forming a part of the

Earth while the Earth does all the moving on behalf of the particle having mass as well as the Earth and only happens when through having mass the object becomes part of the rotating Earth.

What does this all of this controversy mean…it means the way **the Brilliant-Master-mind-Newtonian** say the Sun and all the planets formed is total rubbish. The way **the Brilliant-Master-mind-Newtonian** say the Universe came about is hogwash. The age **the Brilliant-Master-mind-Newtonian** gives the Universe is proof of their total incompetence understanding cosmic principles (Newtonians can't even understand or explain any of the four cosmic principles I named the cosmic pillars) and total by lacking such fundamental understanding shows complete ignorance on the side of Newtonian concepts.

The Universe is something **the Brilliant-Master-mind-Newtonian** can't dream to fathom…or begin to understand and then **the Brilliant-Master-mind-Newtonian** wish us to consider their positions they hold in society as the wise experts that can explain it all, while all along they can't even explain gravity. …

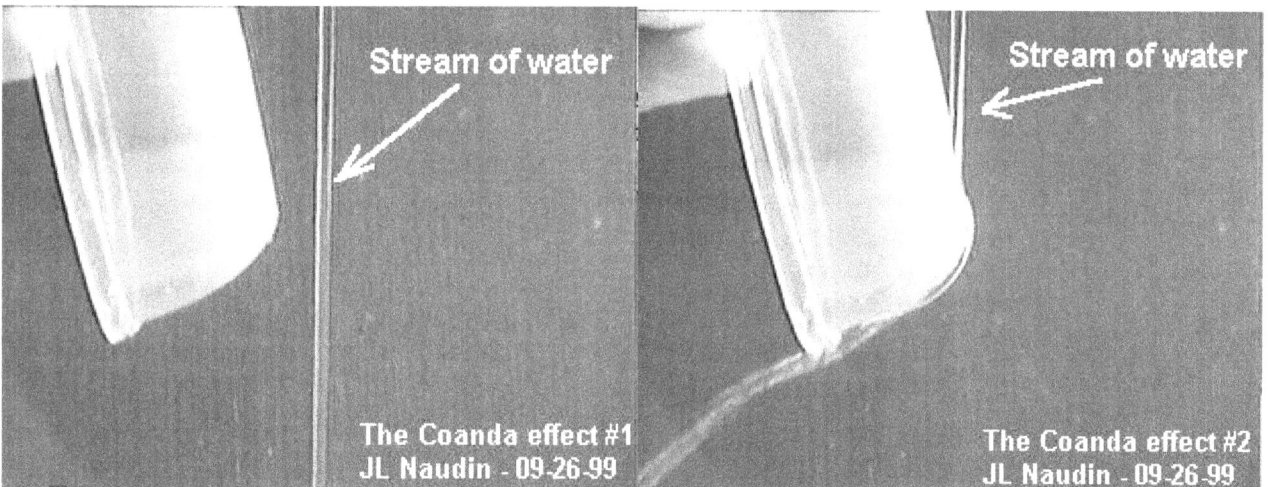

This explaining abandons Newton's idea that gravity is being formed by mass that through some form of magical intervention is pulling on other mass and this is forming gravitational contraction, which is madness. I hope this idea is finally going down the toilet. If one takes the Kepler formula indicating space-time $a^3 = kT^2$ that Kepler introduced, which Kepler received from no less than the cosmos at large, one find the **space a^3** is equal to the movement of the defined space in a **straight line k** as well as a **circle T^2**. In the cosmos no line can go straight without circling as well and no circle can go on without going straight at the same time. That mathematically explains the Coanda principle in detail.

In the books **The Dissertation on Gravity** or the more informing **The Veracity of Gravity** I explain the process in much detail. Gravity is the Coanda effect as the above picture indicates. Should you wish to find more information on **The Dissertation on Gravity** or **The Veracity of Gravity** please visit the web site called www.gravitysveracity.com.

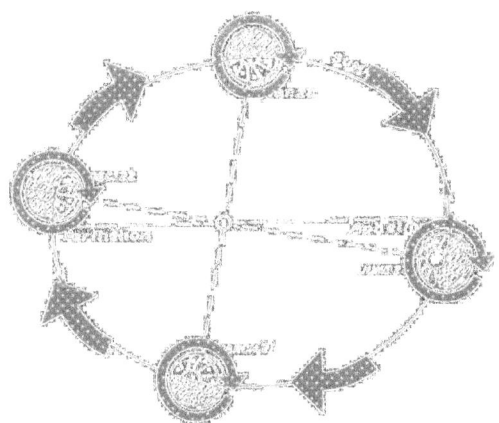

In gravity there is a circle turning. Where there is a circle turning we have Π involved. Also we have a radius involved. The radius or distance from the centre Kepler called **k** and Newton classified this as zero ($\frac{dJ}{dt} = 0$), which is an insult to mathematical principles.

Dividing anything into anything can never become zero $\frac{dJ}{dt} \neq 0$ but at the smallest possible value become singular $\frac{dJ}{dt} = 1^0$. This is a very critical mistake Newton made in his observation of the Universe.

Where a circle formed while spinning in a cube, three factors are involved, namely the radius, the spinning speed and most of all there is Π. Without Π there can be no circle and the Earth is a circle, even Newton should have been aware of this. Therefore there can be no gravity without having Π as a contributing factor to what forms gravity.

Since gravity is tightly interconnected with a circle formed as a sphere and is spinning around an axis, the main issue of research has to start with finding the factor Π. There is no connection with a circle and mass but for sure the circle will find in form an end serving in a measured value as Π. It is common knowledge that in calculating a circle the formula used is $Π2r^2$ or $Πd^2$. Would it not be mathematical plausible then to start looking for mathematics in gravity while leaving Newton's magical mass out of the picture since there might just be some common sense to be found in this. In science and in mathematics we have to see where true mathematics fit and what role every factor has in playing a part. We can't keep on dumping all findings on mass since mass has no part in mathematics.

If we wish to award mass a value as a factor we then must see the role we give mass. Giving mass a value is issuing a body a value as it would have when being part of the Earth. We take a cube of water and we award the water a measured size. One meter by one meter by one meter of water would give a thousand litres of water which would leave a thousand kilograms of weight and that would leave a mass value of one metric ton or a thousand kilograms. The awarding of mass is giving the object a relevancy of being part of the Earth. In Newton's time the Earth was the Universe because people were starting to get used to thinking that the Universe was not spinning around the Earth but the Earth was only a small part of the Solar system and that was a small part of the Milky Way which was a small part of whatever was a small part of another small part of another small part of something getting all-the-while into a bigger picture.

There is also a movement that should go straight but is in fact going in a circle and by never going straight but always circling the Universe becomes eternal on the one end and infinite at the other end. The infinite point I am going to explain in due time. Everything eventually is going in rotation and in that we find the measure of outer space being eternity. In Kepler we find $a^3 = k\ T^2$ where T^2 is a circle but on the other hand **k** is also forming a bigger circle and everything a^3 that is going away is going to return someday.

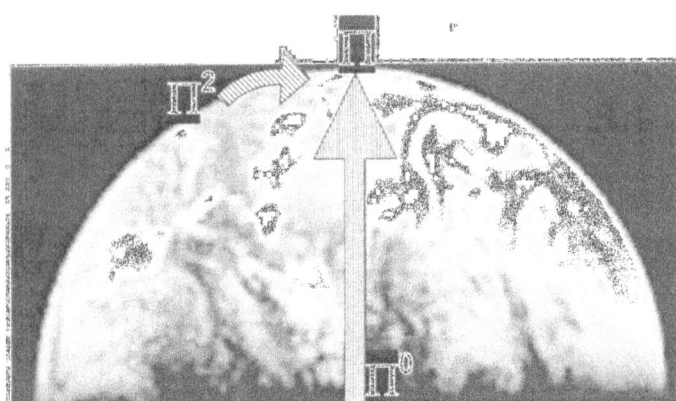

Science always awarded the position an object holds standing on the centre of the Earth with a measurable value of mass. This is not incorrect in normal physics but as far as astrophysics goes there is no mass factor present anywhere. In normally practised physics we are allowed to award mass as a usable tool because on Earth all things adhere to the singularity holding the Earth, however that is as far as the use of mass could go! The value of such a position should be **Π** in the relevancy of the movement that the Earth holds as **Π²**

This is the process whereby gravity forms. Mass has no influence on gravity except for resulting from gravity compressing the Earth.

At this point so far after all my numerous attempts in trying to establish some contact with academics world wide I wrote seven books in a combination I titled **"Matters Time In Space: The Thesis"** covering the entire issue of my work plus the mentioned books wherein I combine all the various letters I wrote to academics through out the eight years of ardent trying to establish some line of communication. The last letter I addressed to academics I include as part of the content of my web page called www.sirnewtonsfraud.com for your insight and which forms part of this and other of my books where I join and elaborate on the letters that I combine to form a unit as a book.

While it probably is the greatest mind to walk the Earth that produced the spectacular used in the idea that mass is responsible for gravity I am not sharing such an opinion notwithstanding my admission that my mind is much less dynamic that those Masters minding physics, yet with my mind being a much more simple mind as those Maters controlling principles accepted by science, I have noticed much more simple aspects of nature that only one with a simple mind as I have could recognise because my mind does not have the capacity for the greatness of the great minds controlling accepted principles applying to physics.

It is said that Singularity is a mathematical point at which certain physical quantities reach infinite values for example, according to the general relativity the curvature of space-time becomes infinite in a black hole. In

The Absolute Relevancy of Singularity — used to explain THE COSMIC CODE

the big bang theory the Universe was born from singularity in which the density and temperature of matter were infinite. If the Universe did start from one single point and time matter and space flowed from that point, then that point must have a relative connecting base that connects everything to all other things because such a point holding singularity must be eternal as space matter and time link eternally. Therefore there must be one point linking the entire Universe when regarding the fact of singularity. Then according to the theory of relativity there has to be one exact point holding time in relevance notwithstanding the fact that time departs from that position and relate differently to all space-time away from such a point.

The cosmos is a line by a multitude of dimensions forming lines in time that criss-crosses to form a Universe filled with space. The Universe is a line by definition because space runs by a line connecting all points to all other points there is. In order to prove my point I wish to ask the reader to define the shortest line there can theoretically be. If he should answer anything but that the shortest line will be at a point where the beginning and is the very same spot such a person will be wrong. The shortest line that can ever be anywhere must have a start and finish holding the exact same spot. The line will be humanly impossible to create but we humans are capable of very little.

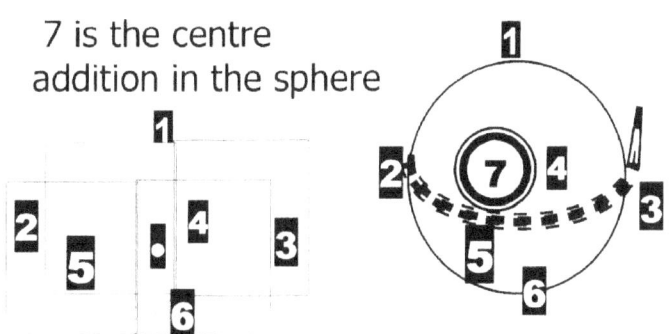

7 is the centre addition in the sphere

Kepler's formula also indicate that a sphere is within a cube that is holding a sphere

When the line has a beginning and an end at the very same spot and it wishes to extend the position as to further the possibility it has, which direction should it favour? Humans in the west would naturally think of extending from left to right while in the east humans may want to go from right to left. It would resemble the direction and the way they write. Some persons will tend to go up or down, but all of the options are about human preference and not mathematical conclusions.

Extending the line in any one direction will favour one direction without a conclusion about not extending in other directions. Such a conclusion has no sound mathematical foundation. The only option about extending will be in all directions equally in order to give a meaningful non-bias flow of mathematical equilibrium. In this idea we find the Cosmic Code developing as the value of Π grows from there into a dimensional mosaic of space.

Wherever the Universe starts, such a starting point has to start with the smallest line growing into space. The shortest line in the realm of possibilities must have a start and finish holding one spot and such a line will also be a dot or a circle. Not favouring one direction puts all directions at equilibrium meaning that any form what ever there may be, can develop from such a spot with the end and the start being the same and still not favouring one specific direction. This reasoning prompted me to look for singularity in such a spot because if the prime spot from which all came was a spot, then the spot must hold the shortest line as the spot became space, but more prominent still, it will hold the smallest form including the smallest circle.

One possibility that the shortest spot can never have is having a starting point on the zero mark. If the mark of zero holds the start it must also hold the end because the end and the beginning have the same position. If the position of zero then is the beginning, the end will also be zero leaving the line without an end as well as without a beginning.

The conclusion from this is that no line can start at zero because that will be a mathematical impossibility. A line or spot starting at zero would therefore be shorter than the shortest line possible. A line growing or extending from zero can never leave zero because of the influence of being zero disqualify any possibility of growth! If the line then had to grow in all directions at the same pace the line must therefore be a circle by multitude. The value of the circle is Π, and that is where creation started.

That gave me the clue where to start looking for singularity. One would find singularity in the value 1 and the value 1 is also equal to Π^0 where it then will connect to Π in order to form a circle. This will then be applicable to all things circling which will apply to all things rotating in a circle. To start my explanation about my cosmic theory I wish to firstly bring some nostalgic and the relevancy will become apparent. Newton used the top to show that the Universe is made up of zero $\dfrac{dJ}{dt} = 0$ while I am about to prove the Universe uses singularity (1^0) as the main ingredient giving the Universe substance.

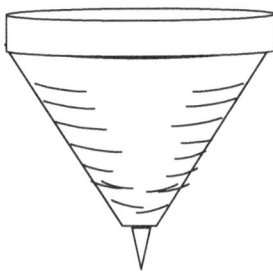 When we were boys we played with a top we called the spinning top. I cannot imagine that there is one boy in the western world that did not hold such a devise in his hand. Tying a string securely around the tapered cone started the operation and then with a jerking or pulling throw the devise is launched in a projectile manner and the big knack to success was getting the nail end firmly on the ground and by the realizing jerk the top was rotating. The champion was always the one boy that could throw his top to spin the fastest and that would create a humming sound. The louder the sound produced the bigger champion. All boys would be very aware of the difference there is between any top spinning and a top laying on the floor while not spinning.

The difference in the top spinning and the top lying on the ground forms the art of top throwing. When a back braking effort produced a throw of enormity, the spinning top would not only produce sound varying in pitch but also create a spin that would seem to have some instability. There are very many limitations about the spin, parameters that determine the slowest and the highest spin rate and spinning is within the parameters of such settings. The question arising is why such parameters are there in the first place?

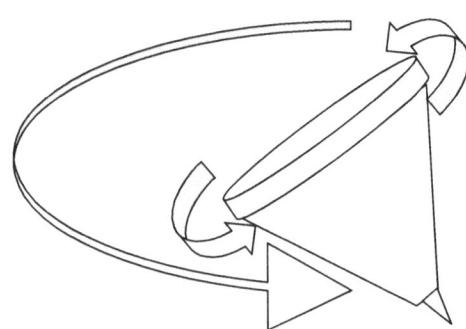 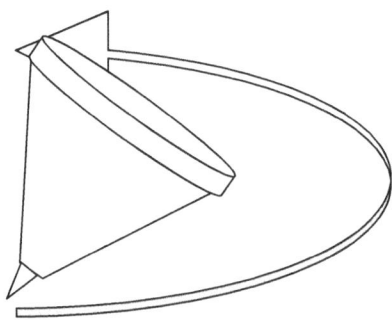

An enormous effort will have the top going oblong while spinning violently and as the pace reduces with time the top will stabilize by coming to an upright position. This period has the top spinning precisely erect. In the upright position it wall then spin for the remainder of the period where it will in the end start tilting to the side and in a last effort throw a few wild oblong turns and fall over. His proves there are certain limits ad borders forming the fastest and the slowest boundaries of spin.

Boys playing games will never realize scientific breakthrough explaining and grown ups do not play with toys. In this little toy played everywhere everyday by almost everyone is the answer most brilliant of human brainpower seek answers about all the cosmic riddles no one seem to understand. In the spin as such one may find two vital boundaries in the motion and the boundaries are marked by a wobble coming about as if the top is fighting some other influence. Spinning too fast pulls the centre of the top to seemingly go off centre and so it also does when spinning too slowly. It is the same influence coming about at both ends of the limitation in the spin. There are influences at work, but force…no; it cannot be forces setting such boundaries. From that I started per cuing what sets such limitations because that limitation must be universal as all matter is spinning in one way or the other.

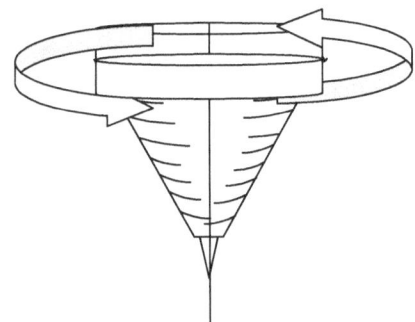 When looking at the cosmos from whichever angle indicates the fact that the cosmos is moving. It is forever spinning and it is going to as much as it is coming from. Everything is on the move and always encircling something of greater importance. A top can spin but the parameters of its spin are limiting the motion it can apply. By not spinning the top is still spinning as the Earth is doing the spinning on its behalf. When spinning too fast the top fights something because the alignment keeping it upright starts to tarnish. The same apply when spinning too slowly but that makes sense. It is the fact that the same affect comes about when spinning too slow that triggers the questions.

It will be recommended to keep in mind that although I use the spinning top as an example that fact is that whatever applies to the spinning top also applies to the atom as much as it applies to the Earth as much as it applies to the largest or smallest star going all the way up to and including the Black Hole. It is well advised to remember that any piece of atom forming material is also a Black Hole since it holds everything a Black Hole holds including that it incorporates light that never afterwards return, just as a Black Hole does. If the top spins it holds all the characteristics we find in atoms and in stars because all of these things spin by virtue of generating a governing singularity that holds a relevant position to a controlling singularity.

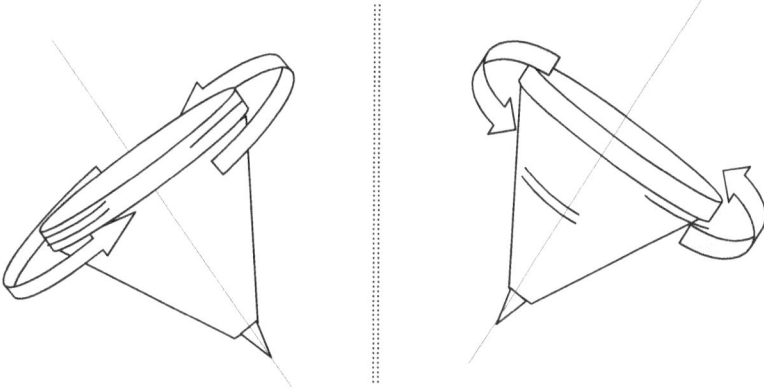

It seems not to be relevant but this I used as a tool in the manner in which I defined gravity; I defined energy, but before that I had to prove the existence of time and time's control over the universe, time's role in the Universe and what time is. This fete of discovering what time is, was up till now not yet been achieved. I had to prove what space is, that time and space is sides of the same coin, with matter forming the separation. The main conclusion that brought about such realisations was my different overall view of science. It's not the explanations science brought to the table that at first that made me question the validity of Newton, but much more is the things Newton cannot explain but is factors in the cosmos nevertheless.

Using logic such as science makes appears foolish. For instance: how can something such as a graviton that is supposedly hidden within an atom produce an influence that can reach as far as Oort's cloud and even beyond. There is just no rational in the time verses events that can explain facts in accordance to the way science puts facts on the table. Take for instance the argument that mass draws on material that has no obvious connection because according to science cosmic objects spin in nothing. Since the time of Newton, the arguments was at first brilliant, but tarnished from being brilliant to being clever to fair too poor and a hundred years ago it went as far down the order as to the point of being stupid. That made me search for new meaning to old news and I started investigating Kepler where I found what Kepler is about and what Kepler's formula is all about. I found what is what Kepler indicated with his formula $a^3 = T^2 k$ is the concept of space-time. The space of an object (a^3) is equal to the time (T^2), which it is in, in every given instant (**k**). If the space becomes smaller ($k^{-1} = T^2 / a^3$), the time duration becomes longer every instant of time's progress.

I found out that Kepler expressed the fact that singularity is a mathematical reality ($k^0 = T^2 k / a^3$). Einstein may be the first to give singularity as a concept a name and Galileo (unwittingly) may have been the first to define it but Kepler was the first to formulate singularity, but in mathematical terms singularity is the most basic principle.

At this point I wish to re-establish a fact that seems lost in all other grandeurs of cosmology and even mathematics. A straight line cannot begin at zero or nil it can only start at infinity. Such a statement will hardly seem appropriate but the relevancy of this fact has no limits.

POINT OF INFINITY

If the line started at zero there was no line to start because zero multiplied by whatever results in zero as the answer. That must also be the cosmic starting point. Einstein introduced such a point and named that point singularity or so I correctly or incorrectly believe. The basic idea of the line is to give rime an indication because time follows a line that produces space and by the line that time provides we can see how time formed space. However, if time and the line started at zero time would not be a line.

The line forming the axis is without space and only holds form, and therefore the line represents a point not having any dimensions while it still is there without ever being there. If ever there is a concept I have to introduce, then it is the concept of how important the axis is and how science up to now missed the biggest issue that is responsible for all movement within the cosmos. The line forming the axis is there but only intelligence will ever form the concept whereby one can realise where the line is without ever seeing the line. Anyone unable to understand this concept can never see the validity of space-time. In the axis line there is a something that is there but only intelligence can bring understanding to the understanding thereof. Only motion of space can resurrect the line coming from the point it holds as a dot. Everything in the cosmos spins and everything that spins has to form a line that doesn't exist but yet the line controls everything that spins around this line that never can hold any space or be part of the Universe. Without having space to fill, the line can never form any viable part of what forms the cosmos, which is space.

The point in reference is the line forming the axis and the axis must be a line that never forms in space because if it did, it would have to rotate in either one of the directions space spins in and by not spinning, it has no space. **That point** albeit hypothetical, is also as much a reality none the less and is placed where that point **must be standing still** because every line **running from that point** in **opposing directions** is also **in opposing directional spin to the other or opposing side.** In considering the spinning motion in the fraction of time in the detailed instant every aspect of rotation will turn in every instant of change in time. Although the points had the same characteristics only one instant before, they oppose the characteristics it had just before and just after the very instant in which they are and to which they relate by similar points also in rotation. Looking at the graph unfold will explain my point about quarterly opposing dimensions and values unfolding.

The circle can reduce one step more when the circle eliminates r completely by returning r to a point of singularity r^0, but the elimination of r as the factor reduces the major factor to the single dimension in Π^0. That will not reduce the cosmos to zero, but it will only eliminate all potential lines r^0 to potential circles $\Pi^0 \Pi r^0$ and from there the circle Πr^0 will come about by manifesting as a line but that manifesting can firstly only establish a circle Πr^2. The only value that singularity can have although the single dimension may host the entire Universe is Π^0. Pick a number and elevate it to the power of zero and in the process one may have established another point holding all points in singularity because that is the value of singularity. Only Π^0 or any other value holding one accompanied by zero as an exponential value can ever be the accurate value of singularity while singularity will then host the rest of all the possibilities in the Universe.

This means that the entire Universe composes of and is made up of singularity... this much I am going to prove. Every point occupied or otherwise constitutes of singularity either under control by movement in a form we call atoms or being passive in a location we call outer space. This position one can derive from Kepler's formula $a^3 = T^2 k$. It is just a question of how to fit this sensibly into Kepler's formula $a^3 = T^2 k$ and find a way that will bring much understanding to cosmology and the way that singularity connects one Universe to form cosmology. The top spinning is what connects space to form the Universe. The top being still on the ground and not spinning holds singularity at a value of the dot forming Π^0 while putting the relevancy on the Earth's roundness by Π. When the top spins the relevancy changes to the line from forming as a dot Π^0 becoming a line Π. The line Π forms as a result of the top forming space Π^3, which is in place as a result of the movement that the top acquires Π^2. It is singularity without space so being a line or a dot makes no difference. The top no longer holds only a dot Π^0 in the centre, but generates the relevance Π by forming $\Pi^0 \Pi r^0$. The top, by moving adjusts Π to form space by movement which is $\Pi = \Pi^3 \div \Pi^2$. All of this is what makes gravity be what it is and all of that Newton missed and Newtonians never saw since all of that is covered by a blanket called mass being responsible for gravity.

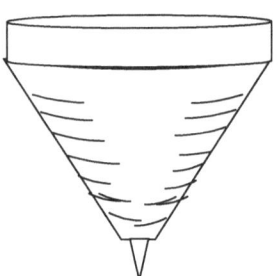

If the line started at zero there was no line to start because zero multiplied by whatever results in zero as the answer. That must also be the cosmic starting point. Einstein introduced such a point and named that point singularity or so I correctly or incorrectly believe. The shortest line there can ever be is in front of every person's eyes everyday and no one ever made any effort to see it. The line running from the centre that holds singularity in the centre of the spinning top must be such a line because smaller than that line can be no line can ever be.

This brings us back to the spinning top I presented at the previous occasion.

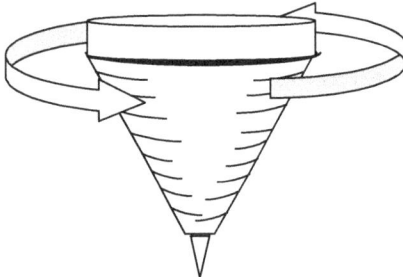

The Absolute Relevancy of Singularity   Page 117   used to explain THE COSMIC CODE

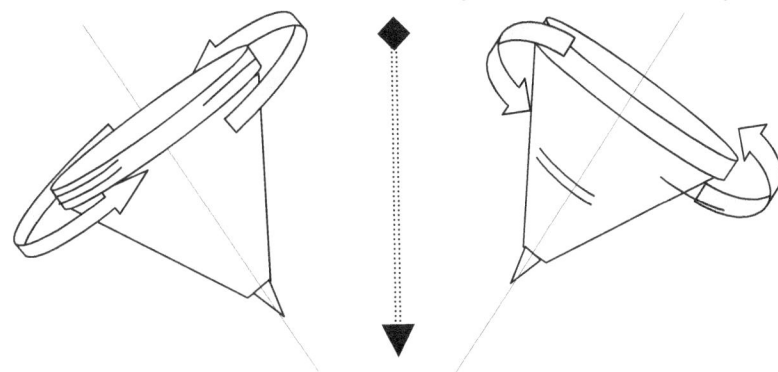

I have asked as many persons as I do not care to remember why the top sinning will remain spinning around one point while turning. The answer I receive from the most educated to the schoolboy is always about momentum bringing about a balance. That is a very simple answer and to say the least a little too simplistic by further analysis. Why would the spinning top go of centre when spinning higher than a specific velocity and lowering the velocity it would stabilize and run square to the Earth only after that it will go circling in an oblong fashion before it then falls over on its side. There is an obvious boundary developing when the top spins to fast and another boundary coming in just before the top collapses onto the ground. These are vital signs that science never cares to mention because science is bent on Newton's idea of "mass".

When the top is spinning it is spinning about its own axis and when it is not spinning it still remains spinning about the Earth's axis therefore when it is spinning it is also spinning about the Earth's axis. That means when the top spins it generates a governing singularity making the Earth's rotation a controlling singularity and when it holds mass by lying still on the Earth, the Earth provides the governing singularity and the top's singularity changes from being a line to become a dot. Therefore the limitations applying can only result as an influence coming from the Earth's axis being in some relevancy with the axis that the top generate. The boundaries that form when the top spins too fast and where the top spins too slowly are a fight for supremacy and for survival of the weakest. The second question now comes screaming across and that is in what manner could the Earths axis ever affect a spinning top since the spin and the spinning top is a gross mismatch to what ever standard the Earth may introduce. It is clear that spinning objects do influence each other in contrast to Newtonian opinion that the top does no work when circling around its axis.

Every round object has a point establishing a very centre, a middle dividing one side from the other. That division determines the space from one side away from the other side. At one point there must be a point that does not fall on either side of the divide which holds no space in the very centre but has a place in the very centre from where this space less place controls the entire spinning of the top. Such a point will still be a circle, because from that side the circle divides into two sectors.

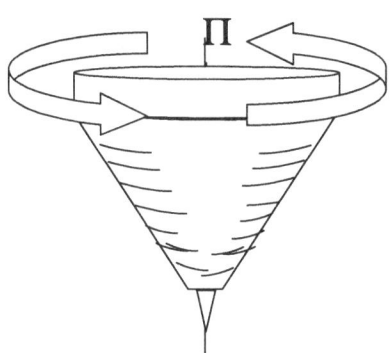

In every spinning object there is a point of infinity, a point that does not turn because it holds the dividing spin. From that point running in all directions the spin is opposing the other side. All spinning activity starts at that point diverting outwards and from that point the spin is either clockwise or anti clockwise in all directions. As I pointed out no line can start at zero because then there is no line and no rotating point can start at zero because then there is no rotation. Crossing that point that holds no space will put the movement on a complete opposite direction from what it had before the crossing took place.

Calculating a square in terms of producing volumetric space involves two aspects that we think of as sides.

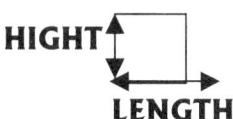

Newtonians think of a square as something that consists of two lines where one holds the one dimension and the other one holds the next dimension. The important factor is establishing consensus about the two

lines. As already agreed no line can start at zero because then there is no line. In cosmology this idea is invalid because in the cosmos a square represents movement from one point to the next point. This is the big mistake that Newton saw in Kepler's cosmology. Newton saw a sphere as $(4\Pi r^3) \div 3$ while Kepler found a Universe becoming three dimensional by movement in $a^3 = T^2 k$. However, for arguments sake let us stick to the Newtonian idea.

By reducing the one line the other line can never reach zero because then there was no such a line to begin with. That makes a straight line also inevitably always a potential square and that makes the straight line half the value of the square being 180°. At a later point I shall continue with this argument, but for the mean while I wish to come back to the circle. This same principal applies to the cube and that means everything there is and ever will be is either a square being part of a cube or a circle becoming multi dimensional and ending up as a sphere. With the straight line forming half the value of a square or a circle we have $360° / 2 = 180°$ in as much as being one line and reserving one line in infinity to eternity. The straight line is just half the value of a square. In that manner the triangle is also half a square and therefore holds the same dimensional value as the straight line being also 180°.

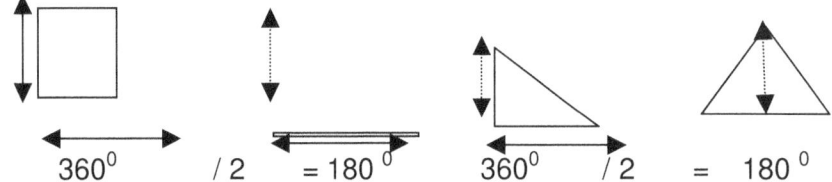

The circle is a square holding a round shape, as the straight line is a square holding one side to infinity. Calculating a circle involves two aspects where the one is either the radius or the diameter that is double the radius. The other is the factor Π

$\Pi \times D^2 / 4$ = circle and $\Pi \times r^2$ = circle

The point of singularity cannot be in space at large because space is not valid where we locate singularity and secondly what ever is there within the limits of singularity, can't spin or then spins too slowly to have a connection with singularity directly. To spin one requires space and that is what singularity does not have.

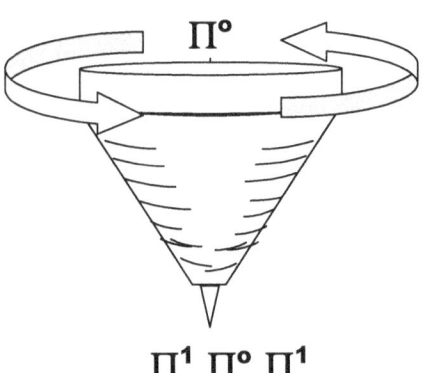

With everything in a cube or a circle or any potential of the two brings about the implication of eternity in a form of singularity or the point of creation. Removing the radius of a circle does not remove the circle, because the circle is there, securing the ring. If the line (or imaginary line if you wish) holding the value of $\Pi^0 = 1$ there has to be a point where the circle is no longer in infinity but claims existing outside the imaginary. Then we will still have Πr^0, which still implicates a circle by the measure of Π. At that point the radius may be equal to slightly more than infinity, but to all calculating purposes it still remain as infinity.

The only way a line can form cosmically is by generating a circle and the only way to form a circle is by generating a line. Space a^3 is defined by the circular motion T^2 that holds the governing singularity k^0 that generates a relevant position k according to the controlling singularity k. That represents Kepler's entire formula as $k^0 = a^3 \div (T^2 k)$. One has to read into Kepler's formula the relevance of singularity being the governing factor T^2 as well as he controlling factor k and that will demarcate the space in relevance. It serves us well to realise that the entire Universe was that small at a point where everything started forming because the spot that developed into the dot is still with every spinning circle...and the Universe is a multitude of spinning circles. It is also very wise to remember that once anything becomes a part of the Universe, it can never leave the Universe since it then has no place to go or no gate to pass through in order to leave the Universe. With the spot becoming a dot, there must have been a time when everything in the entire Universe was that big as the spot is, and that then moved on to form the dot and in that it went on growing in relevance. The point around whichever spins becomes the centre of the Universe by singularity. In establishing such a centre containing singularity we find the reason why bullets travel more straight when they are fired circling and circling is what gives the bullet the accuracy in its trajectory that then established a cartelise singularity that establishes a value forming Π in relation to the centre singularity being 1 or as I named it as singularity Π^0.

The circle can only be with spin applying because with the spin this movement while going on splits infinity from eternity because where the spin does not apply, it has a value of 1^0 and the zero is a factor in the exponent which is another expression for eternity or infinity depending on movement.

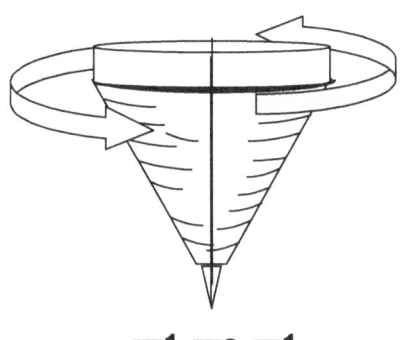

Having edges where Π^0 duplicate to present the edges singularity loses by acquiring space, expanding from the value of Π^0 to the value of $\Pi^1 r^0$ this expanding result in having the same value had singularity had being Π^1 to the one side and Π^1 to the other side. This means going into space from Π^0 must be the point of splitting singularity into two parts of eternity, the eternal value of the first dimension outside eternity.

$\Pi^1 \; \Pi^0 \; \Pi^1$

By receiving space, singularity receives a value outside singularity as Π^0 receives eternal edges formed by Π that never end. Granted the fact that the edges is so small there still is no r to present a circle as r is r^0.

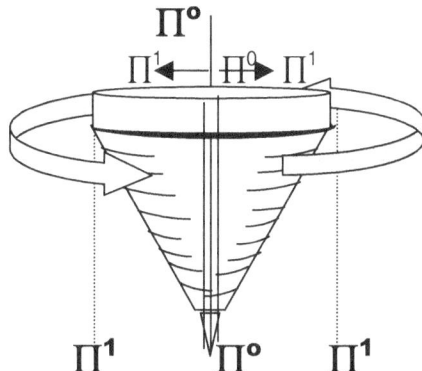

It is the square of Π^1 being Π^{1+1}. That is the first dimension outside singularity Π^0 where singularity has a value of Π^1 in the form of $\Pi^{1+1=2}$. The first claim to space then has a value of Π^2. This applies to both sides of the claim to space outside singularity, and the double proton becomes the dominant factor on matter.

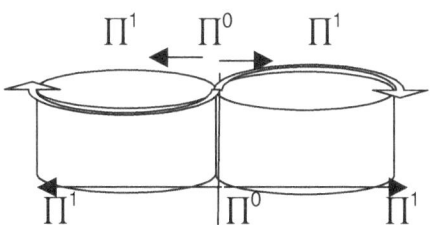

In the spin there are always two directly opposing directions following each other continuously. The two are then sub divided in two more directly opposing directions following each other continuously. Accepting this as a fact concerning gravity is a large step forward in understanding gravity notwithstanding the simplicity involved.

Taken from the point of rotation two sides form two sides where the two sides are in opposition to each other in every aspect that they may contain and with all that they hold.

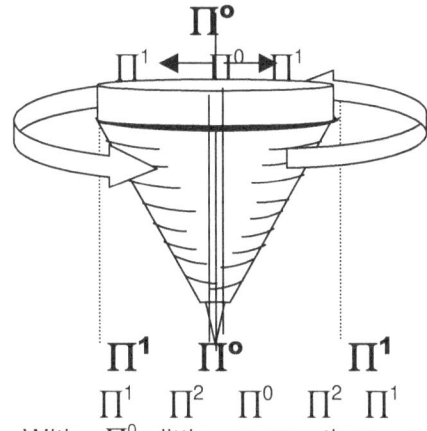

Using the concept that gravity applies Π as the circle factor Π as well as Π^2 replacing r^2 the replacing by Π brings two values as Π and Π^2. Underlying this is another hidden reality where we find singularity Π^0 charging another form of singularity Π and in this hidden Universe the value of the radius r is always presented using r^0 expressed as singularity. However this reality of a cosmos in singularity only comes to be when gravity Π^2 charges a relevancy Π by limiting the space Π^3 in accordance with Kepler's findings that $a^3 = (T^2 k)$, which then embraces $k^0 = a^3 \div (T^2 k)$ forming all structures. In the solid structure I use Π as a value for reasons that will become apparent in due time.

With Π^0 little more than a figment of the imagination there is actually two values of Π^1 facing each other in a relation combining Π^1 to hold the value of $\Pi^{1+1=2} = \Pi^2$ and with two sides being the very same but opposing each other there will therefore also be Π^2 to every side that holds Π^1.

At last I can come to the one part that I disagree with Newtonians, and what I regard as Newton's second biggest infamous or famous blunder. Science, made one enormous blunder, from the following stance. They took the radius of a wheel not to have any influence on the wheel. In doing that, they removed the very fact that keeps the universal attachment together.

$$\frac{dJ}{dt} = 0$$

Dividing anything into anything can never have a value of zero. This disputes mathematics. In mathematics one can't place a factor into another factor and have the outcome form zero. **DJ / dt** can have any number from eternity being 1^{Ω} to infinity 1^0, but in this rage it only excludes one value used by man; it cannot be 0. By placing the one in division of the other, you bring in relevance. You cannot then say there is relevance with no outcome or no relevance. By doing such, you proclaim that one of the factors is non-existent.

$$\frac{dJ}{0} = dt \text{ or } \frac{0}{dt} = dJ$$

In both cases, one of the factors then does not exist. Such a claim is incoherent, because you proclaim that a circle has no radius, or a radius has no circle. When calculating a circle, you multiply either the square of the radius by Π, or the quarter of the diameter at a square by Π.

$\Pi \times r^2$ = CIRCLE

If you remove r it then is $\Pi \times r^2 / r^2$ = CIRCLE.

You cannot then say $r^2/r^2 = 0$ and therefore $\Pi \times 0 = 0$. That is nonsense. $\Pi r^2/r^2$ is also $\Pi \times 1$ so it will always be $\Pi \times 1$, and that is the eternal circle. There can never result zero from $r^2/r^2 \neq 0$. When looking at any rotating object, there has to be a point of no rotation and no rotation means "no rotation", not no existence. No rotation means a factor of 1, not zero.

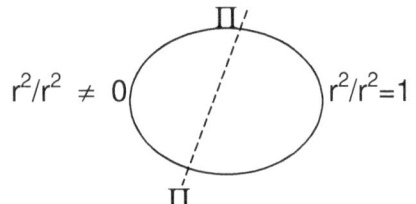

That then is singularity. The eternal Π, the Π that may not have significance but still it is a Π of value. The relativity remains one, eternally one, but it cannot be zero. Therefore, **dJ/dt** cannot be zero.

dJ/dt can become eternal or infinitive or at the worst it can become **dJ/dt = 1** and with that I full heartedly agree.

When explaining this to any child, they can immediately see that the principle Newton apply is totally out of bound with mathematical law. Explain this inconsistency to any Newtonian High Priest and they are as blind as only those that don't want to see cold bee. I cannot find one Newtonian, large or small, important or ordinary, clever or not that bright to accept that this is a clear mistake on the pert of Newton. They just would not read my work because I show Newton's obvious mistake. Every solar structure is spinning around an individual axis while the whole lot is spinning around a mutual axis the Sun provides. This shows a double singularity applying and that forms gravity by doubling the square of 7. The spinning leads to years and years become time. The spin that shows on the different planets is the most crucial aspect of their orbiting the Sun.

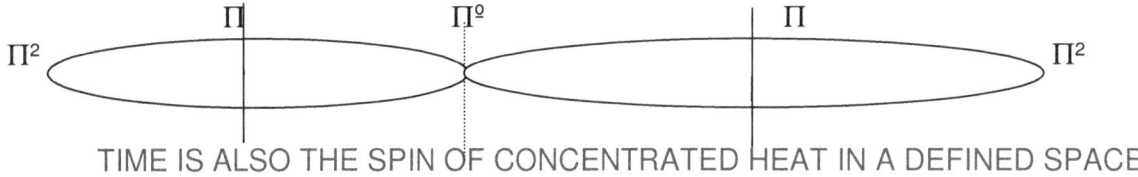

TIME IS ALSO THE SPIN OF CONCENTRATED HEAT IN A DEFINED SPACE

All spinning matter has the point where the spin is in that line forming the axis is not there because the radius is too small to measure by any human means ($\Pi \times r^0 = \Pi$). That point Π^0 is standing still in relation to the rest of the spin Π^2. In relation to that logic I do not accept Newtonian science holding the radius of a spinning object invalid in work done by the spin, because from the spin applying the Universe is fixed in place. To tell the truth, so far I found it has been above my ability to convey this to any person with a decent physics education where as indicating this to any one equally as poorly educated as I am, they see the point I try to show immediately.

The Absolute Relevancy of Singularity　　　　　Page 121　　　　used to explain THE COSMIC CODE

It is not in the calculation that I disagree, because that calculation is totally accurate. It is in the outcome, the resolve that I strongly object.

Applying Newton's second law **F=ma** One arrive at the formula

$GMm / r^2 = m(\omega^2 r)$

By replacing ($\omega^2 r$) with $2\Pi / T$ we obtain Kepler's third law

This law predicts that $T^2 = a^3$ (This is where I disagree because throughout Kepler is $a^3 = T^2 k$.)

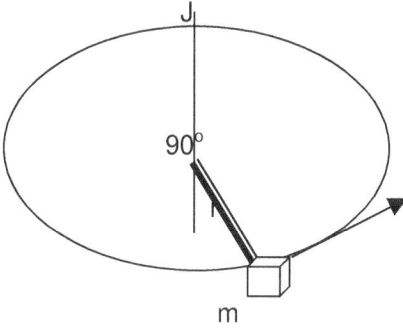

p = m.v

The mass (m) multiplying the speed (v) forms a new value J AND THEREFORE j CONTINUOUS TO IMPLY $J = I\omega$.

$J = r \times p$ where $p = (v = r \times \omega)$

$J = r.m.v = m.r^2.\omega = I.\omega$ and becomes interpreted as $J = I\omega$

This establishes that $r = dJ/dt$

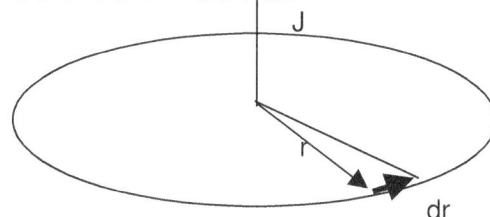

$r = dJ/dt$ In the case of planets in orbit around the Sun r forms a value of zero because $dJ/dt = 0$.

What this statement implies is that r does not exist. When anything has a value of zero it is for all purposes non-existent. Only when an object is following s straight line can the radius be non-existent because the radius alters value through time development.

Taking the argument back to Kepler's law, we have to see the formula in terms of how Kepler intended it to be read where $a^3 = T^2 k$ is placing the space a^3 equal = to the relative **k** movement T^2.

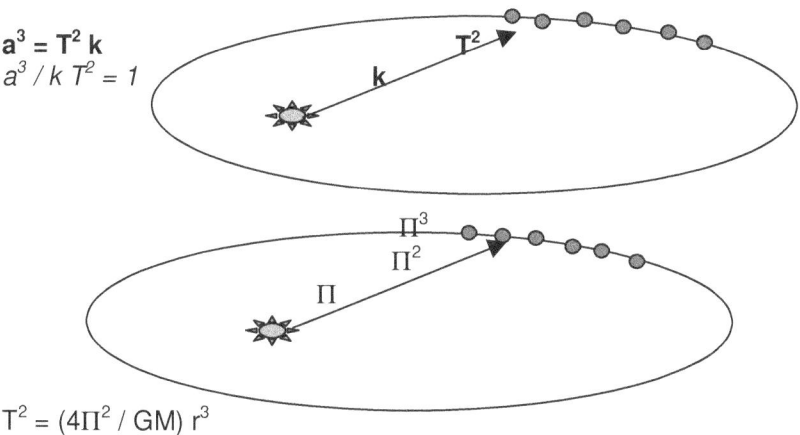

$a^3 = T^2 k$
$a^3 / k T^2 = 1$

$T^2 = (4\Pi^2 / GM) r^3$

The spinning or not spinning is not part of the issue because at the point of absolute singularity the object never spins. Therefore spinning or not spinning does not apply to the point of singularity because singularity never spins in any event. The point that is not spinning is governing all that which is spinning and from that point the next point holding singularity (the axis of the Sun) becomes the controlling singularity and without these points the cosmos will come apart. The singularity maintaining the spin around the axis of the Earth as well as the axis around the Sun and the axis around the Milky Way places the governing control of all that spins within the Milky Way and that puts everything even beyond the Milky Way in a Universal perfect order.

Since Newton became an institution the fashionable trend to follow was to ignore the logic in favour of complete stupidity. One of the most basic and fundamental rules in mathematics is that when something is dividing into something else such dividing will never result in zero because by placing anything in a relevancy it can at best bring equality but never remove one aspect or factor from the equation.

$r = dJ /$

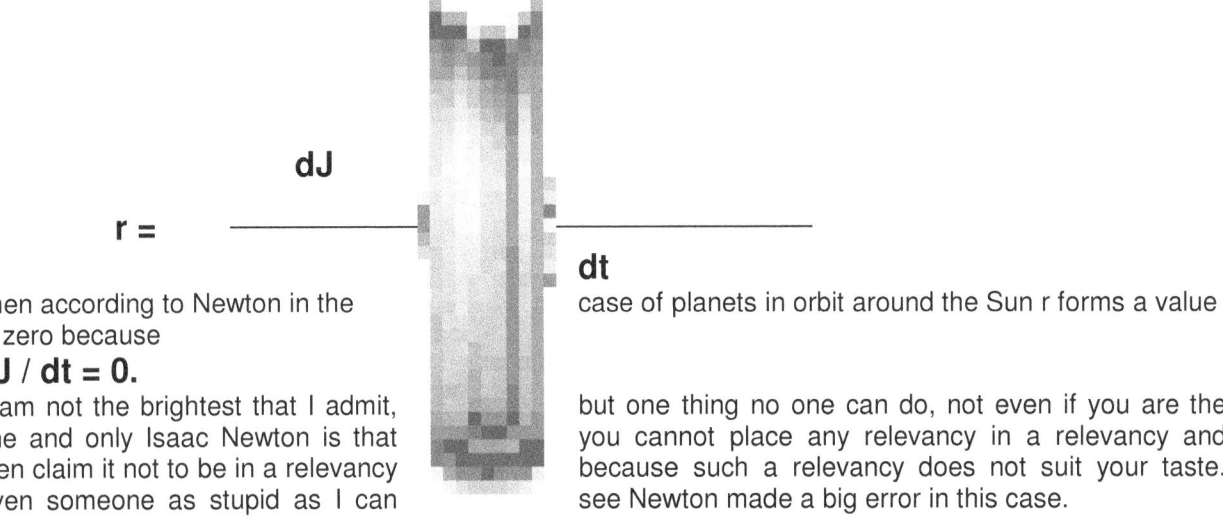

$$r = \frac{dJ}{dt}$$

Then according to Newton in the case of planets in orbit around the Sun r forms a value of zero because

$dJ / dt = 0.$

I am not the brightest that I admit, but one thing no one can do, not even if you are the one and only Isaac Newton is that you cannot place any relevancy in a relevancy and then claim it not to be in a relevancy because such a relevancy does not suit your taste. Even someone as stupid as I can see Newton made a big error in this case.

You cannot put something in relation to another object and then decide there is no relation because you find no relevancy in the relevancy.

$r = dJ /$

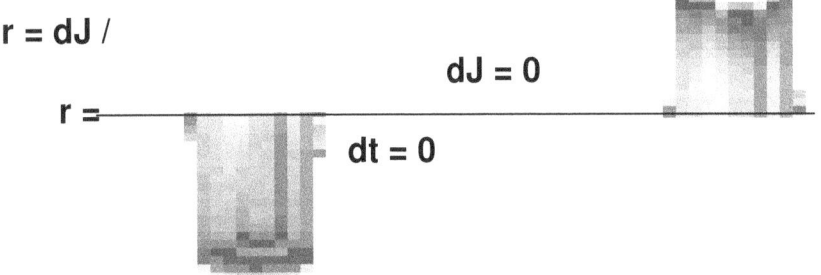

$$r = \frac{dJ = 0}{dt = 0}$$

$dJ / dt \neq 0.$ If

If $dJ / dt = 0$ then either $dJ = 0$ or then $dt = 0$. That is a mathematical principle, much larger than even Newton

One cannot claim there is a wheel and then remove the spokes on one half because according to your taste, you do not like the spokes. The wheel will not spin. The one half must perfectly match the other half, or the mismatch will not spin.

The only way to cheat yourself out of the situation is to remove the wheel and spokes altogether, and you are left with what you say there is: then you sit with NOTHING, which is precisely what Newtonians so inaccurately try to fit into outer space. That simply does not apply in cosmology. The object rotates the centre structure and therefore there has to be a radius holding the circling orbited in relation to the centre structure. By not having a wheel that rotates, the wheel becomes the factor of one, and the rotation becomes zero. The wheel does not disappear. In the cosmos, everything is rotating because nothing ever stands still. Therefore the mean equilibrium, the common factor there is to share, has to be one, Infinity holding eternity captured, the infinite Π^0 governing the eternal Π, because all rotating Π^2 of objects has $\Pi^0\Pi$ in singularity, and sharing singularity, gives every object in space a relation with all other objects in space. After trying for many years to bring those Newtonians I tried to contact the candle, I concluded that Newtonians are incapable of realizing that mathematical principle as reality. Newtonians whish to live in a fantasy of magical make believe where mass can produce gravity only by the say so of Newton.

Every one must realise that if $GMm / r^2 = m (\omega^2 r)$ did apply, the comet must hit the Sun in total destruction. That does not happen! The comet rotates the sun, and the Sun by itself has a point of singularity where Π

remains without r. The comet, holding the orbit, also has a point of singularity, but since there is space separating the two objects, they cannot share a mean point of singularity, the very point of existing. Since singularity means just that, being single, there cannot be two. The comet and the Sun have a mean point of singularity but the space they occupy divides their common singularity. That is why they orbit in an oval path, a path where the one structure holds on to more space from its point of singularity towards the space it claims. Since they do not claim equal space, BY THE DENSITY they hold, the space will not be in proportion.

They do share in the common fact of singularity and singularity cannot be two, because then it then will be "dualarity" or duplicity (in case there is such a word) where both find the space they occupy, with the space they hold, will be their individual eccentricity from singularity. The two objects are holding eccentric space around their individual but common singularity. That point of singularity is Π the circle without the radius because the singularity removes all forms or values of r, leaving Π to be singularity.

The very inner centre may or may not spin and the fact that it does or does not spin is all the same because that centre part never spins in any case. Therefore the boundaries set by the spinning motion do not only depend on the spinning motion of the object in question but has to stand related to another body bringing about a larger spin influence. Granted is the fact that the influence the Earth has on the top may be that of gravity but if that is the case then surely the Sun has also influence on the Earth and other rotating objects through gravity. It needs more investigation because it may bring about evidence we are not aware of.

When a rocket is fired and the spin is not present there will be no stable trajectory. The only way to secure the stability of the trajectory is to allow spin (Π^2) that enables a point holding (Π) as this will locate and establish singularity (Π^0). Establishing singularity is the most fundamental principle about gravity we can ever find. This is the one part that is most important when we go in search of gravity secured by singularity that forms the absolute relevance of everything filling Universe we have. Everything is a rotating object that holds any point allocated in Universe to form the centre of the Universe because everything in the entire Universe spins around any given point and that then forms the centre of the Universe.

Every centre of every atom forms the centre of the Universe by spin! Again I indicate the precise location of such a point. What is in the Universe, is spinning and therefore what I am referring to, applies to everything holding a place in the Universe and therefore this which I mention directly links everything holding any space whatsoever in the entire Universe to one single point around which all spin. In the **precise middle** of all **objects in rotation** is a precise centre dividing the object in sectors that will **start the spinning initiation** from that centre point. Thus, the spinning object **will have a middle point**, a very specific **centre point that does not spin** and only holds Π as a specific value because no radius can apply. But also the one value such a line **cannot have is zero** because the line **is there and holds contact** with the rest of the material bringing about that **zero does not start any** line and therefore the **value of the line must be infinite**, just as described in **accordance** and by **the definition of singularity.**

Everyone calls this line that forms the axis. Everyone knows about the axis and yet through so many thousands of years of using an axis, no person ever thought to scrutinise the principle behind the axis. Yet in all the millennia everyone was aware of the line that forms called the axis, no one took time to see it holds singularity at Π^0 presenting Π. The only conclusive value singularity can have is 1 or Π^0. The axis controls all particles spinning around the line being the axis while the axis in itself forming the line represents no particles because the axis represents no space. If there was space within the axis, the space had to spin in some or other direction. Having no space would mean occupying no space which means forming no part of the Universe filled with space and yet it controls all the space as wide as the mind can imagine. Without space it does not form a part of the cosmos, but forms the cosmos as wide and as deep as the cosmos goes. The axis could not be seen but with applying intelligence the axis could be witnessed. Having no part in the cosmos in space, the axis could only be understood and never be seen. The axis could be proven but never be shown. The axis is what controls the Universe from end to end because when there is no end there the axis provides one end to what never can have another end and the axis governs whatever spins in relation to such a line. Again I wish to press this issue to form clarity. The line forming the axis is without space and only holds form, and therefore the line represents a point not having any dimensions while it still is there without ever being there. If ever there is a concept I have to introduce, then it is the concept of how important the axis is and how science up to now missed the biggest issue that is responsible for all movement within the cosmos. The line forming the axis is there but only intelligence will ever form the concept whereby one can realise where the line is without ever seeing the line. Anyone unable to understand this concept can never see the validity of space-time. In the axis line there is a something that is there but only intelligence can bring understanding to the understanding thereof. Only motion of space can resurrect the line coming from the point it holds as a dot. Everything in the cosmos spins and everything that spins has to form a line that doesn't exist but yet the line controls everything that spins around this line that never can hold any space or be part of the Universe. Without having space to fill, the line can never form any viable part of what forms the cosmos, which is space.

As I am introducing a very new idea, I wish to explain in better detail what I try to convey. While the toy top is spinning one will find singularity by moving the rotating line or radius progressively to the middle by reducing the length the line has from the edge to the middle. At one point all further reducing must end but the ending cannot include zero or nothing because the rest of the line is still attached to the rest of the top. As the rotating direction moves inwards, the rings will become smaller and smaller. Then we reach a point everyone thinks of as being the axis around which everything rotates. The line only forms when everything around the line spins by establishing a circle to the value of Π.

Boys playing games will never realize scientific breakthrough explaining and grown ups do not play with toys. In this little toy played everywhere everyday by almost every one is the answer most brilliant of human Brainpower seek answers about all the cosmic riddles no one seem to understand.

An enormous effort will have the top going oblong while spinning violently and as the pace reduced the top will stabilize by coming to an upright position. In the upright position it wall then spin for the remainder of the period where it will in the end start tilting to the side and in a last effort throw a few wild oblong turns and fall over.

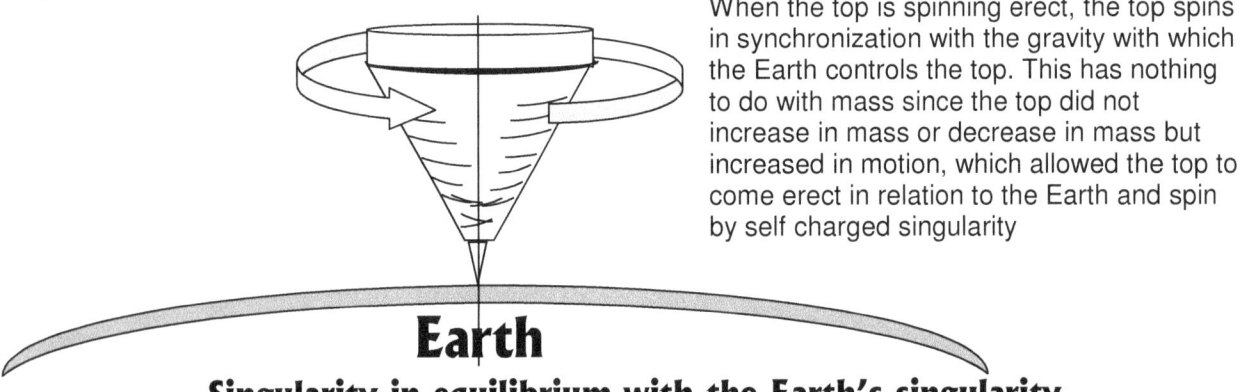

When the top is spinning erect, the top spins in synchronization with the gravity with which the Earth controls the top. This has nothing to do with mass since the top did not increase in mass or decrease in mass but increased in motion, which allowed the top to come erect in relation to the Earth and spin by self charged singularity

Singularity in equilibrium with the Earth's singularity

At both the slowest end of the spinning scale and the fastest end of the scale it is movement that sets the upper and the lower limits of spin. Mass as a factor is nowhere to be found so mass has no input.

The top uses its governing singularity to fight to try and get the upper hand and escape from the Earth controlling gravity. This is a fight for control.

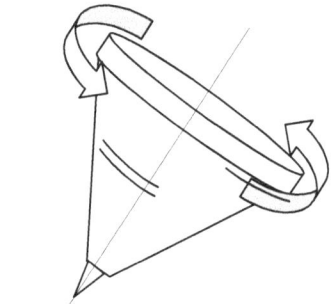

Singularity of the top exceeding the Earth's singularity

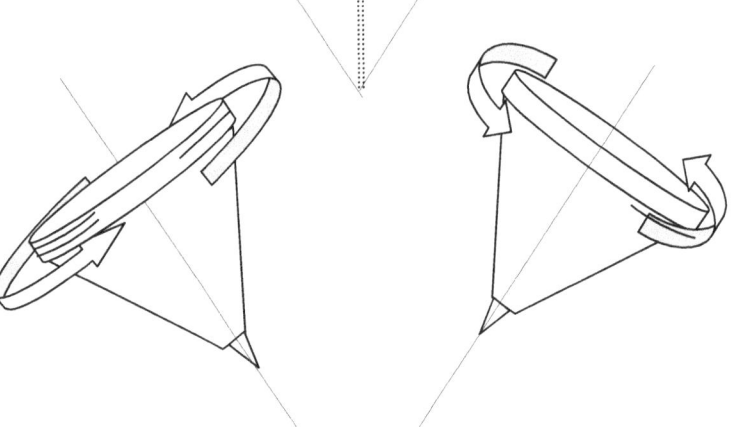

The top's governing singularity is in a fight for survival with the Earth's controlling singularity to survive as an independent cosmic entity. This is a fight for surviving and again shows no input by mass.

The Earth's singularity dominating and exceeding the singularity top produce through spin as the top collapses and fall.

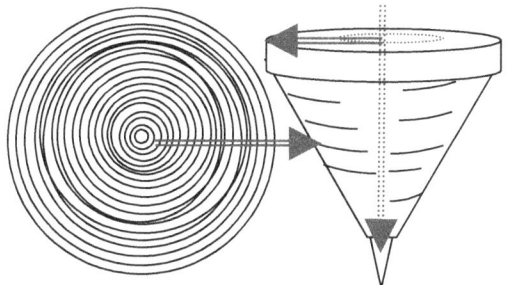

This observation places a much bigger question mark on the statement of Newton where Newton proclaims that there is no influence on two rotating cosmic structures. Newton said when completing as circle no work is done. When hearing this statement in my first years of studying physics my paths with Newton split and I can recollect the day it happened vividly as clear as I can remember anything.

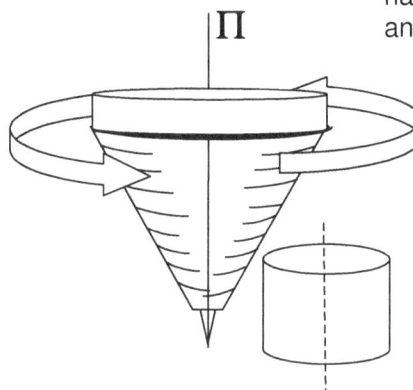

We may proceed to the wider picture that the cosmos hold. What is it the Newtonians fail to see? They fail to see singularity because they are insanely bent backwards on Newton's non-existing mass factor.

If an electron is orbiting around an atom, the inside of the atom must be a circle. If the atom were not a circle, it would not spin. If there is a circle because the electron spins, there then has to be a cube in which the circle spins. The electron cannot rotate around a cube; therefore, the inside of the atom is a circle. This means that the atom then has a governing center holding the Earth as a controlling axis.

In a circle, there is a radius that initiates the circle. The calculation of such a circle is $\Pi \times r^2$.

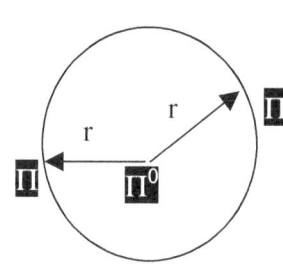

The radius r runs from the circle outwards, from a circle centre point towards Π, the value of the circle. In the centre of the circle, there is a point where the radius starts and the radius forms a line. It runs outwards from that point where the line starts in all directions towards the circle Π. That is where the value of Π as a factor has a role to play. Technically, there then has to be a point where r is singularity, an absolute exponential zero. However, the circle therefore remains Π. The circle does not disappear; it remains there for all to see. It is only the radius that removes. Then there has to be a point where Π is singularity, an absolute exponential zero. However, the circle therefore remains Π^0.

$$\frac{\Pi r^2}{r^2} = \Pi$$

If one removes the radius from the circle, the circle remains, only holding the value of Π. By removing the value of r, to replace the value with r^0 then Π becomes singularity but with no place to be. Singularity is the place where there is no space to be in place. However, Π remains because once r receives the slightest of space Π will find space. Then the circle will grow to Πr^2 and r would determine the space. Without space, there is no r but there is a circle with the value of Π. Singularity is in every single rotating object, be it the proton or the Universe.

Every person blessed with eyesight can observe that that is not the case.

SPACE IS AS RELATIVE AS TIME. $\quad a^3 / T^2 k = \text{one} \quad \text{AND} \quad a^3 = \frac{1}{T^2 k}$.

I am not disputing Newton as far as general physics go for there Newton was the Master every one claim he is; I am disputing the relevance of Newton's scientific breakthrough when he ventured into Kepler's cosmology. Kepler formulation of space time $a^3 / T^2 k$ deals with singularity holding positional relevance relating to space formed by time and this concept was as far into the future in Newton's era as space flight was at the time. When Newton saw his apple fall from the tree it was no experience of cosmic proportions that took place! It was not two objects of cosmic proportions, colliding in a show of spectacular. It was, after all, only an apple falling from a tree with no cosmic implications of any sorts. The truth is that life placed the apple in the tree by growing the fruit and it is life that let go of the apple and in contradiction to all the atheistic cheer about nothing trying to show the opposite, the only applying truth is that life is only a factor on Earth. Therefore having life let go of an apple is something that can only happen on Earth. One can hardly connect any cosmic implications to such an event!

There, on that occasion Newton, and science, made one enormous blunder, from this stance. When Newton tried to mathematically force a proof using physics as an example Newton tried to show that by spinning anything, such spinning nullifies all other factors. That was to try and bring proof to the suggestion that it is mass that pulls gravity. They took the radius of a wheel not to have any influence on the wheel. In doing that, they removed the very fact that keeps the universal attachment together.

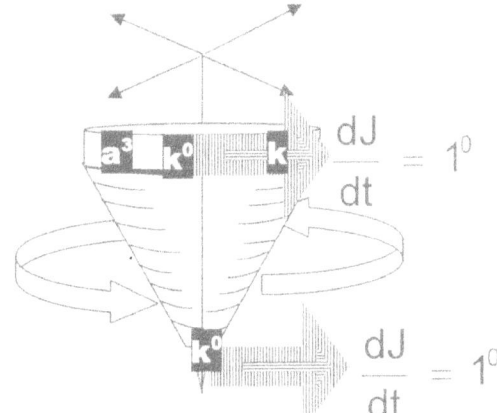

$\frac{dJ}{dt} = 0$ Claiming this disputes mathematical principle at the very core. DJ / dt can have any number from eternity to infinity, only excluding one; it cannot be 0. By placing the one in division of the other, you bring in relevance. You cannot then say there is no relevance. By doing such, you proclaim that one of the factors is non-existent.

$\frac{dJ}{0} = dt$ or $\frac{0}{dt} = dJ$ In both cases, one of the factors then does not exist. Such a claim is incoherent, because you proclaim that a circle has no radius, or a radius has no circle. When calculating a circle, you multiply either the square of the radius by Π, or the quarter of the diameter at a square by Π. Newton's claim suggests that a wheel in rotation will return to the same spot it had previously, as it does not affect the spin.

What should be in place is a continuous un broken line connecting every point serving singularity to every other point serving singularity. It should be (1+1+1+1+1+1.....) going as far as light can travel and still be (1x1=1). That explains the cosmos from every angle we may question the cosmos.

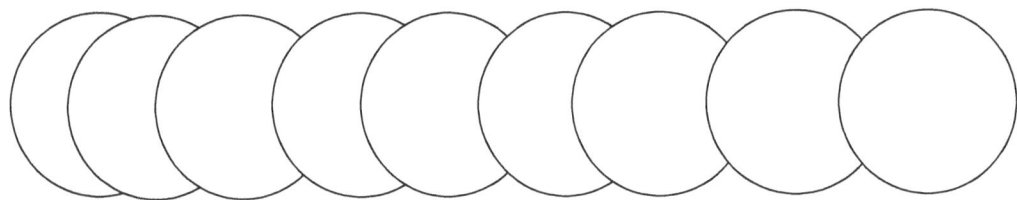

That is impossible since rotation brings about motion changing the principles of the location.

One do seem to get the impression that little changes as the rotation will bring some forward motion and some returning to the original position.

Even by using one half of a wheel as a marking spot for controlling the spin, it would still bring considerable confusion but one can clearly see that Newton's presumption does not quite match reality

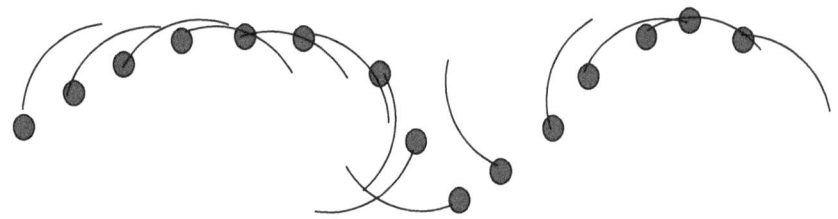

Shortening the arch changes the complexity considerably as one can then see a changing of the arch does not nearly bring the return of the dot to the previous spot.

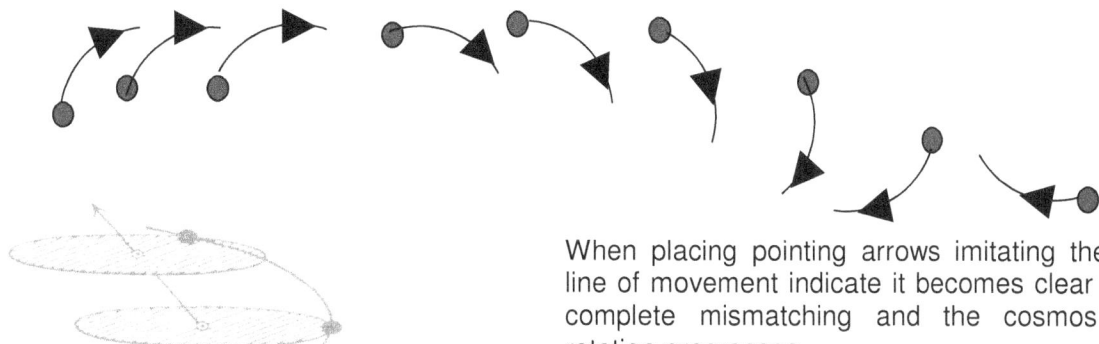

When placing pointing arrows imitating the direction the line of movement indicate it becomes clear that there is a complete mismatching and the cosmos changes as rotation progresses.

Even by not supporting the claim that structures influence space, there is acceptance that the gravity tugging between cosmic structures does take place. When mentioning that I must immediately state that this very fact I am about to dispute in the forth coming few pages. However let us leave it at that for the time being. Let us just say there are admitting that structures influence one another. t is moreover the individual singularity in maintaining the major singularity, which sustains the governing singularity providing equilibrium in space-time.

Not only does atomic individual singularity maintain self-preservation, but in doing that it also sustains a governing singularity holding structural composition and forms within a cluster of matter for example a star. As there is between stars so there are in the same manner a mutual or bonding singularity between atoms in stars, which we see as fusion.

The sectors provide individual singularity as a means in sustaining governing singularity by which provision comes through maintaining governing singularity the required spin in maintaining cooling. If this process did not apply, there would be no connecting individual singularity to major singularity.

Every quarter provide a distinct value that indicates the progress of the flow of time from the one point Π to the next point Π.

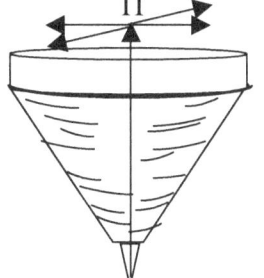

Any changers occurring in Π will lead to a un-unequal triangle providing two different values to r and will alternate the link between r and Π^2 bringing about different form (Π) and time (Π^2). When singularity forming the lines of the triangle is not in equilibrium the triangle will destroy the matching of half circle.

The sectors each provide an equal factor of individual singularity that gives a means whereby the sustaining of the governing singularity could be managed and by which the spin acts as a provision that comes through maintaining the governing singularity to the required spin which then is maintaining cooling or if better understood in those terms, then we call it the balance. If this process did not apply, there would be no connecting individual singularity to major singularity

In every sector the directional flow will provide a distinct meeting of Π linking r to Π^2 and this allow the time component in the rotation.

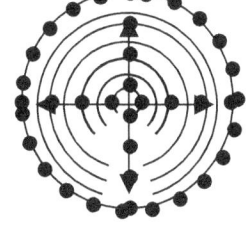

As the meeting of r points to a very distinct different value in direction of movement r will therefore change in direction where such a point of meeting opposes the other points in meeting by direction and if not equal at all times, that will lead to destruction of the form Π. Where in the event of any value changes by Π such changing will completely change Π^2 and r.

However, to think as Newton did and as Newtonians do that the space a^3 is then only equal to the circle T^2 where the rotation $\frac{dJ}{dt} = 0$ nullifies the relevance **k** or in fact the radius r is obscene to say the least. If this was true it will have a galactica completely decompose after the first rotation and have galactica abandon the structural bonding as the galactica will loosen the attachment.

The Sun is more or less somewhere close to on the outskirt of the Milky Way and the Sun is in an oval circle orbit around the Milky Way. The law of orbit is in principle that all orbiting structures follow an oval path because the circle has to compensate for growth in relation to the straight line or the governing singularity has to allow for the controlling singularity, which leads to the circle expanding..

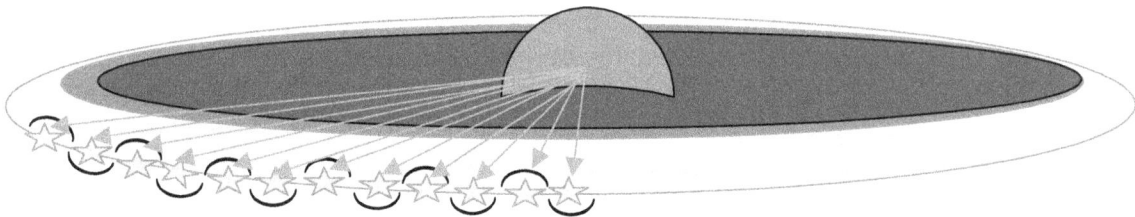

Exaggerated to a large extend the influence the Milky Way has to have on the Earth orbit comes to focus when a pattern comes in pace as the Earth follow not a circle but a wave around the Sun while the Sun sets its motion around the Milky Way. The fact that the planets orbit the Sun and the fact that the Sun orbits the Milky Way indicate an influence undeniable. The fact that the Sun is heading farther away from the influence should then lead to a variation in the planets orbiting wave. The Earth never, not once land on the exact same spot by the completion of one more year cycle. It is therefore not possible that the

$\Pi \times r^2$ = CIRCLE

If you remove r it then is $\Pi \times r^2 / r^2$ = CIRCLE.

You cannot then say $r^2/r^2 = 0$ and therefore $\Pi \times 0 = 0$. That is nonsense. $\Pi r^2/r^2$ will always be $\Pi \times 1$, and that is the eternal circle.

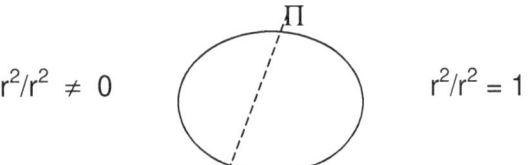

When looking at any rotating object, there has to be a point of no rotation and no rotation means "no rotation", not no existence. No rotation means a factor of 1, not zero.

That then is singularity. The eternal Π, the Π that may not have significance but still it is a Π of value. The relativity remains one, eternally one, but it cannot be zero. Therefore, dJ/dt cannot be zero.

    dJ/dt can become eternal or infinitive or at the worst it can become
    dJ/dt = 1

Taking this case t the wheel by not having a wheel rotate, the wheel becomes the factor of one, and the rotation becomes zero. The wheel does not disappear.

In the cosmos, everything is rotating because nothing ever stands still. Therefore the mean equilibrium, the common factor there is to share, has to be the dimension from where all dimensions develop, the dimension of a circle becoming a straight line – a straight line becoming a square – a square becoming a cube – a cube allowing a circle to spin within and in that the concept is holding eternity and infinity to a point of one, eternity, the eternal Π, because all rotating objects has Π coming from singularity, and sharing singularity by rotation, which gives every object in space a relation with all other objects in space. After trying for many years to bring them the light I see, I concluded that Newtonians are incapable of realizing this mathematical principle as forming the only comic reality.

With Π^0 little more than a figment of the intellectual imagination while being the most critical reality, there is actually two values of Π^1 facing each other in a relation combining Π^1 to hold the value of $\Pi^{1+1=2} = \Pi^2$ and with two sides being the very same but opposing each other there will therefore also be Π^2 to every side that holds Π^1. This is a fact coming from the fact that the value of $\Pi^0\Pi$ can only realise by generating Π^2

From the above I can conclude that gravity is not 9,81 Nm/s, it is Π^2 = 9,8696.

MY THEORY

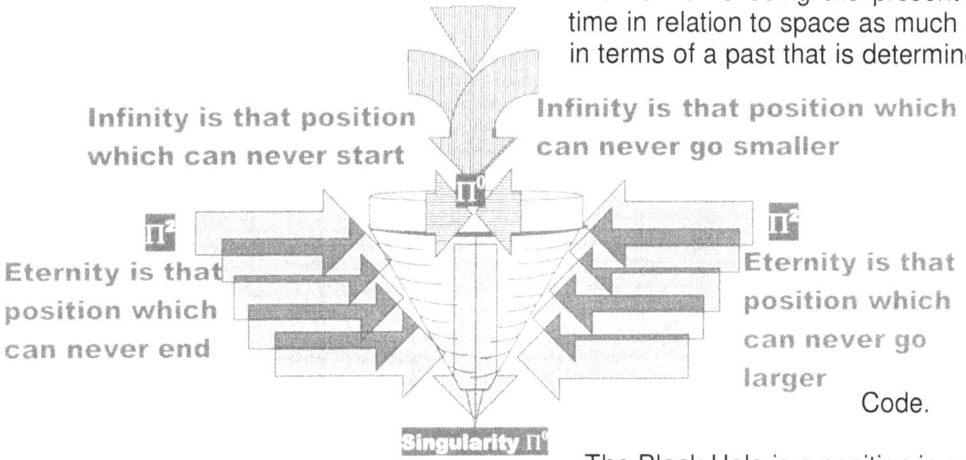

Infinity is that position which can never start

Infinity is that position which can never go smaller

Eternity is that position which can never end

Eternity is that position which can never go larger

Singularity Π^4

The flow time being the present forms space by moving time in relation to space as much as relocating the present in terms of a past that is determined by the movement that secures the future. From this we can deduct that the Universe in a three-dimensional form starts at $7/10(\Pi^6) \div 6$ = 112, which is a value forming the start of the element table and that I explain in the Cosmic Code.

The Black Hole is a position in space where space became so small that eternity rejoins infinity in the absence of movement of space leaving only time to move as it reunites time. This is where the Universe ends because infinity meets eternity in unification of the two factors forming time.

Take a look at everything depicting gravity throughout the Universe and you see a picture of Π forming a circle of whatever discipline. Gravity is Π in all facets in nature.

If we wish to get behind the forming of gravity we better start by investigating Π. At the beginning of the book I said gravity is the diverting of a straight line by forming a circle in turning 7°. The reason why this is the case I explain in many books but that will be extending the purpose of this book by delving much too deep into unfamiliar territory as far as the purpose of this introduction of my work goes

Around axis by 7°

Around the Sun turning by 7°

The first question that one can ask is why would there be the value of (Π) forming between orbiting structures limits and positioning the centre holding singularity in infinity.

When the object moves while being in space or in contact (in relevance) with the spinning Earth, the object wishes to continue moving straight ahead while the Earth also moves straight ahead by turning 7°.

Therefore, the Earth by spinning is falling away by turning 7°. That clears space or compresses space by the margin of 7° declining (compressing) of air / space. The Earth pins around its axis by 7° and also turns around the Sun by 7°.

The Earth is moving, constantly spinning and in this is contracting space by compression (we call this contracting of space in air the atmosphere) and while the air is getting more compact, it takes whatever is filling with space towards the Earth constantly at a rate of 7°. By the Earth rotating, it is compressing space and with space compressing it is moving objects in the direction of the Earth. That is why objects that is falling, has no mass and only the stupidity of the simple Newtonian mind will force scholars to accept that it is mass that is pulling gravity.

The entire idea of gravity is secured in movement of everything in relation to everything else. There is nothing in the Universe that ever could remain still because everything cosmic that is filling the Universe is spinning while it is also at the same time moving in a straight line and by that is following a circle. The Earth is only moving straight ahead because the Sun is spinning and while the Sun is spinning, it is compressing space, which allows the Earth and all other rotating objects to spin around the Sun in a perfect synchronised fashion. This process is going on throughout the entire Universe.

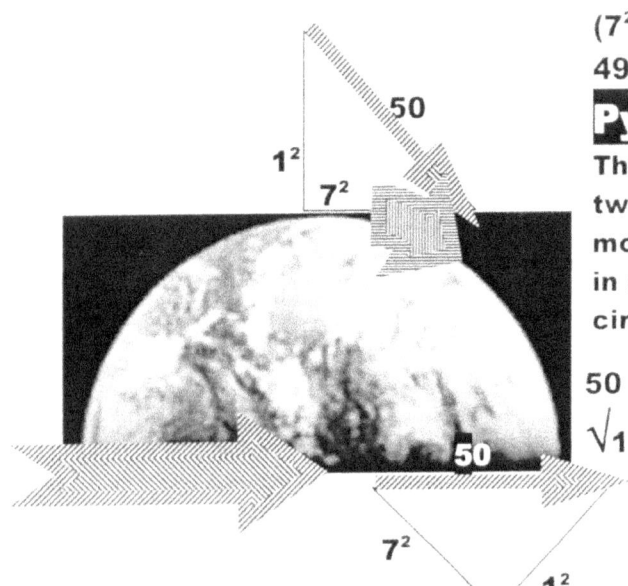

$(7^2 + 1^2) = 50$

$49 + 1 = 50$

Pythagoras

This applies twice in one movement (going in a circle and circling the Sun)

$50 + 50 = 100$

$\sqrt{100} = 10$

This places Pythagoras in the pivotal role of gravity by forming a calculated value of Π when dissecting Π mathematically. Gravity is Π using the law of Pythagoras.

There are three movement involving Π² from which gravity takes effect. At this point however, I will only introduce the two of the three ways in which the value of gravity Π² takes charge of displacing space-time in a process of dismissing space-time by diminishing space through the flow of time. It might sounds the same thing but it is processes putting the Universe in dimensional positioning.

Since the movement involves two equal phases that acts as one, therefore the double value of 7 in relation to forming ten becomes what forms Π. We have the movement of seven forming one direction standing in relation to singularity which is the square of 1…According to Pythagoras that will bring about fifty. Since singularity is equal a one and seven is combined with singularity, the equality of singularity brings about the seven uses the same attached 1 in the square making the fifty a combination of another fifty and from putting a double fifty in the square as Pythagoras demands we have ten as a result in relation to seven. In Pythagoras's square the one side, let's say the adjacent side of the triangle forms a square in seven bringing on forty-nine. The other side let's say the opposite side of the triangle uses the square of one (singularity), which also remains one. The sum total of the two forms fifty, which is the measured value of the hypogenous. Since gravity is always applying in the double movement of seven the total of the hypogenous then is fifty plus fifty which is one hundred. The square of one hundred is ten and that brings about the value of one side of Π. When explaining the total worth of Π such explaining in detail requires a lot more information which will claim about as much space that which this article in full would allow. I complete this explaining as well as the explaining of how Π comes about through the forming of gravity much later on in **THE VERACITY OF GRAVITRY.**

$3Π^2 / 5 (10^2)$

$7^2 = 49;\ 1^2 = 1$
$49+1=50$

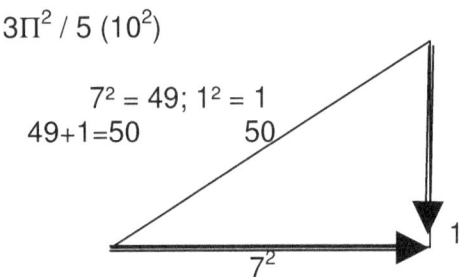

$50 = 5 (10^2)$ where the complete Pythagoras is $2(5)(10^2) = 100$

$\sqrt{100} = 10$ the value of space.

The fact of this comes as 49 plus one becomes 50 and that is in the three dimensions of space $7^2 \times 1^2$ where 7 holds the relation to one and 21.9921 /7 again where 7 relates to one or then to the diminishing of space. At this point it is most important to remember that Pythagoras works on the application of the sum of the square of the two sides. When seven has a direction in the fourth dimension applied to it, the opposing dimension will be one and this applies in time relevancy, therefore the interchanging in time between infinity will place movement of matter at $7^2 \times 1$ relating to circular and $7^2 + 1$ with $7^2 \times 1 = 49$ plus one (singularity) always being a factor of one. Space in time however, never can be a cube, because the flat Universe depicting a flat surface hold Π°Π and from that we will always be a square provided by movement of space in space with one side pointing the direction of time from time to the past (1) to time to the present (1) to time to the future (1). This means with space in singularity in relation to time in singularity there will always be $Π^0 = Π^3 / ΠΠ^2 = 1$. if we see a square the square is in the movement in space of space.

In singularity this holds the answer. The Earth is circling around its axis forming 50. Forming 50 also includes the square of 1, which is singularity. We find not two forms of singularity, which I have introduced before, but there are three forms of singularity. However, the third one only become worth mentioning at this stage because beforehand it played no part in explaining. The influence singularity has on the cosmic is

The Absolute Relevancy of Singularity — used to explain THE COSMIC CODE

portrayed in the significance we find in the four cosmic pillars. All four pillars function exclusively by the manner that singularity forms a governing singularity holding a lesser dominant but absolutely controlling singularity (hence the name I gave to the role the singularity has) and then finally the principle singularity. This is very well but in Π there is 10 + 10 + 1.991 and that is the repeat of 10 and not of 7 as the example I show will indicate.

The definition of singularity is as follows:
Singularity: a mathematical point at which certain physical quantities reach infinite values for example, according to the general relativity the curvature of space-time becomes infinite in a black hole. What this means is that singularity has no space and holds all value equal to each other as much as equal to 1^0 or 1. This we will not be able to understand because the value of this is definably measured on the "other side" of the Universe, the side we have no contact with but where we will land when we die. On the other side we will see why 1^{500} that should represent the Sun's governing singularity will be equal to every single singularity (1^0 1^1 1^2 1^3 1^4 4^0 5^0 6^0) while the Sun's governing singularity is also equal to the compliment of all the atoms centres within the Sun with each forming singularity. To us witnessing that singularity is 1 = 1 wherever it is, singularity makes no sense because $1^0 = 1^1 = 1^2 = 1^3 = 1^4 = 4^0 = 5^0 = 6^0$ but on the other side of the Universe where space makes no sense, this mathematical difference all being equal on this side will have sense. The proof of singularity interacting we find with the four cosmic pillars and the way the pillars show how singularity interacts. The evidence of singularity claiming status and taking charge is undeniably evidential.

Would it not be more scientific to use a graph having such a perfect formula to work with as in the case of $a^3 = T^2 \, k$? From the graph any one can read different seasons applying to a small planet in relation to such a big sun that should be able to shine on the earth from any point all across the poles.

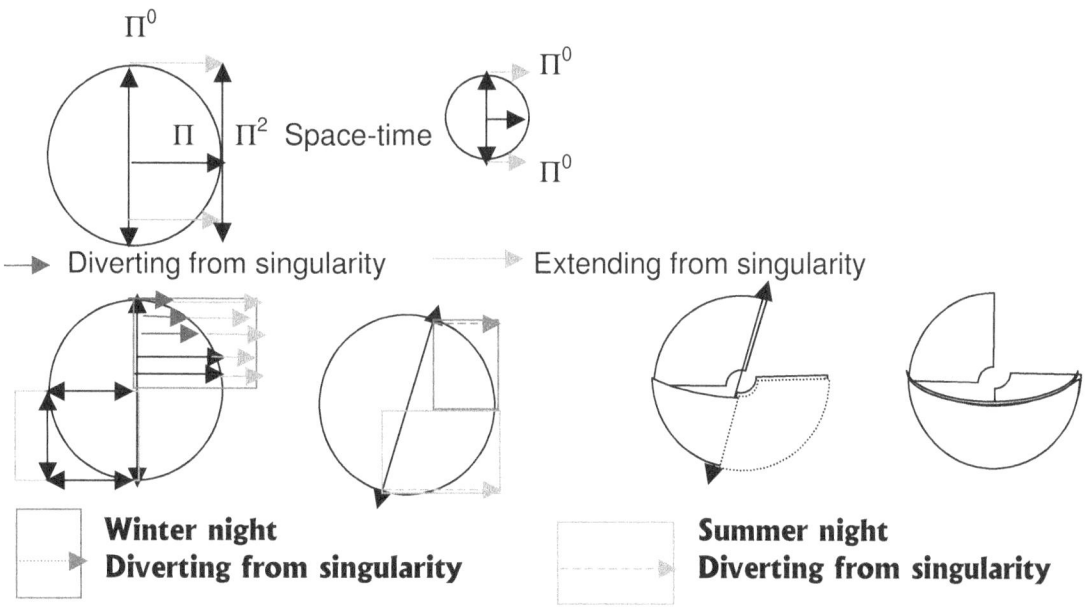

As singularity is eternally cold, it also has to affect the standing between matters, as matter diverts from singularity. The more occupied space there are between singularity and space ends, the more it will be to the advantage of the effect we consider heat, and the less space allowing time the more it will advantage the statement we refer to as cold. But hot and cold are human relevancies not acknowledged by nature. Science does acknowledge a flow from singularity towards space by reducing heat; therefore admitting singularity must hold temperatures indefinite.

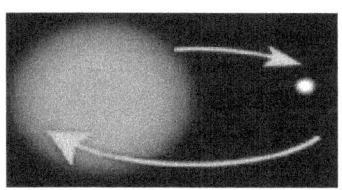

The Roche limit in the practical sense.

The Roche limit is about governing singularity taking charge of the mutual singularity within the atoms that spins in the gravitational range of $\Pi^2 \div 4$ from where $\Pi\Pi^2$ the forms a rotational influence hence $\Pi^2 \div 4$ a the radius of the star or planet.

The Absolute Relevancy of Singularity   Page 132   used to explain THE COSMIC CODE

The Roche limit is:

The region surrounding each star in a binary system, within which any material is gravitationally bound to that particular star. The boundary of the Roche lobes is an equipotential surface, and the lobes touch at the inner Lagrangian point, L_1, through which mass transfer may occur if one of the components expands to fill its lobe. It names after the French mathematician Edouard Albert Roche (1820-83).

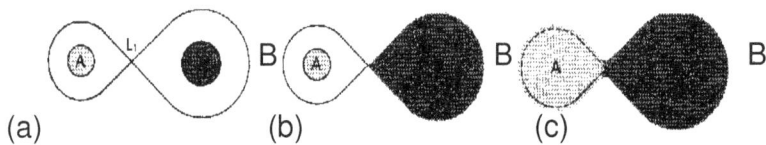

(a)   (b)   (c)

THE ROCHE LOBE: In a binary system, the Roche lobes of components A and B meet at the L_1 Lagrangian point. (a) In a detached system, neither star fills its Roche lobe. (b) In a semidetached system, one massive component, B, fills its Roche lobe. (c) In a contact binary, both components overfill their Roche lobes and share a common envelope.

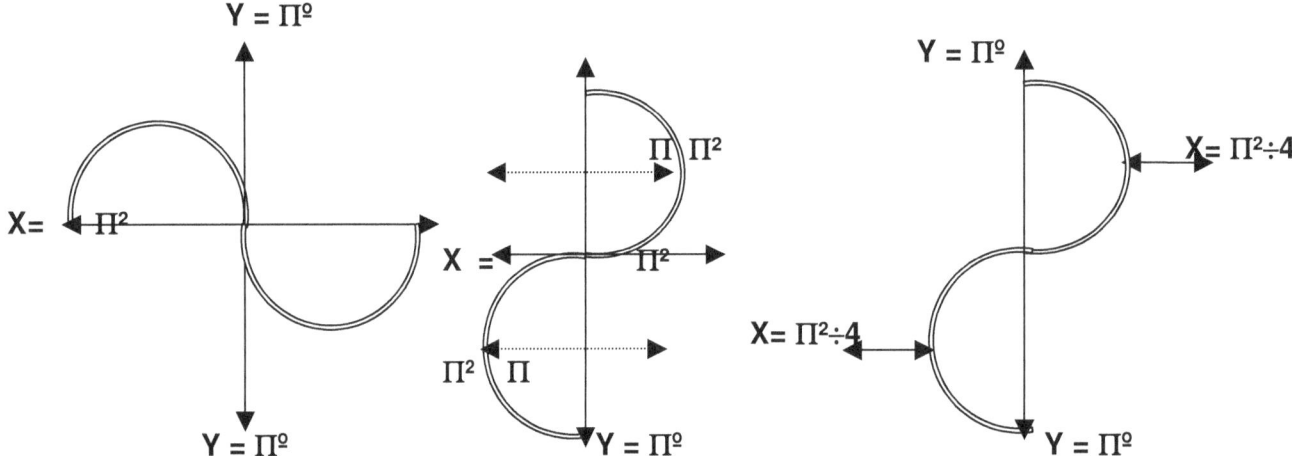

The Roche limit proves that there is interaction between the **governing singularity** situated within the centre of the star and all **mutual singularity** forming the centre within the atoms forming the body of the star. The governing singularity can take charge of points holding mutual singularity within the atom centre and where there is mismatched singularity the governing singularity will liquefy those atoms not complying with the value applying between the atomic mutual singularity and the governing singularity where the atoms forms **individual singularity**. The governing singularity works hand in hand with the compliment of the mutual singularity and the mutual singularity forms a compliment of value that forms the governing singularity where this is the totality of the **individual singularity** compliment. The controlling singularity has the ability to liquefy the individual singularity within the second star's atom structure where the atoms does not perform in accordance with the governing singularity setting the standards for the mutual singularity.

This is of course also applying to the centre of a galactica where all the points serving singularity within every atom forming a structural part of the galactica (**Individual singularity**) forms a measured compliment of what then becomes the galactica governing singularity. Proof of this we have in the Hawkins Black Hole that forms within the centre of some super large galactica. The spin potential generated by the individual points holding singularity that forms a **mutual singularity** taking charge of the galactica governing singularity holds so much influence that it charges a Black Hole within the centre of the particular galactica.

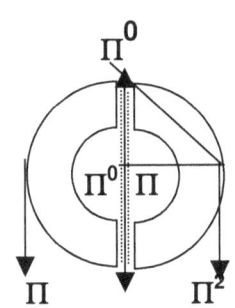

The reason why the "governing" singularity Π^0 stands still in relation to the controlling singularity which is with what the electron movement binds the atom to form the Universal unit we call the atom, is that the governing singularity forms part of infinity and the mutual singularity combines to form the controlling singularity which then together forms the part which is eternity. This evidently goes as far as forming a bond within the entire galactica and as far down as forming a bond within all atoms within the planet structures. The entire unit is bonded by what serves as singularity and by forming singularity it connects to the governing singularity in a similar way as how the Sun holds connection by singularity relevance to the mutual singularity and in that all the planets hold equal speed in their movement going in orbit around the Sun, notwithstanding the mass differences or the orbit length.

The lot only adhere to the Titius Bode law and mass plays no part. The Titius Bode law as much as all the other phenomena works on the principal applying singularity and not mathematical volumetric space. It is because Π^0 sets the standard, forms the basis, applies all rules and cannot be zero, it therefore has to hold the relevancy applying as one forming the eternal one, from where the universe with all content must apply.

I know the following is confusing with all the names I had to give to singularity holding different positions and as much as I hate naming things I found that by not naming these positions holding relevance in singularity the lot becomes even more confusing, but be assured that even with the names given still too me the names make the lot as much confusing as when not having names.

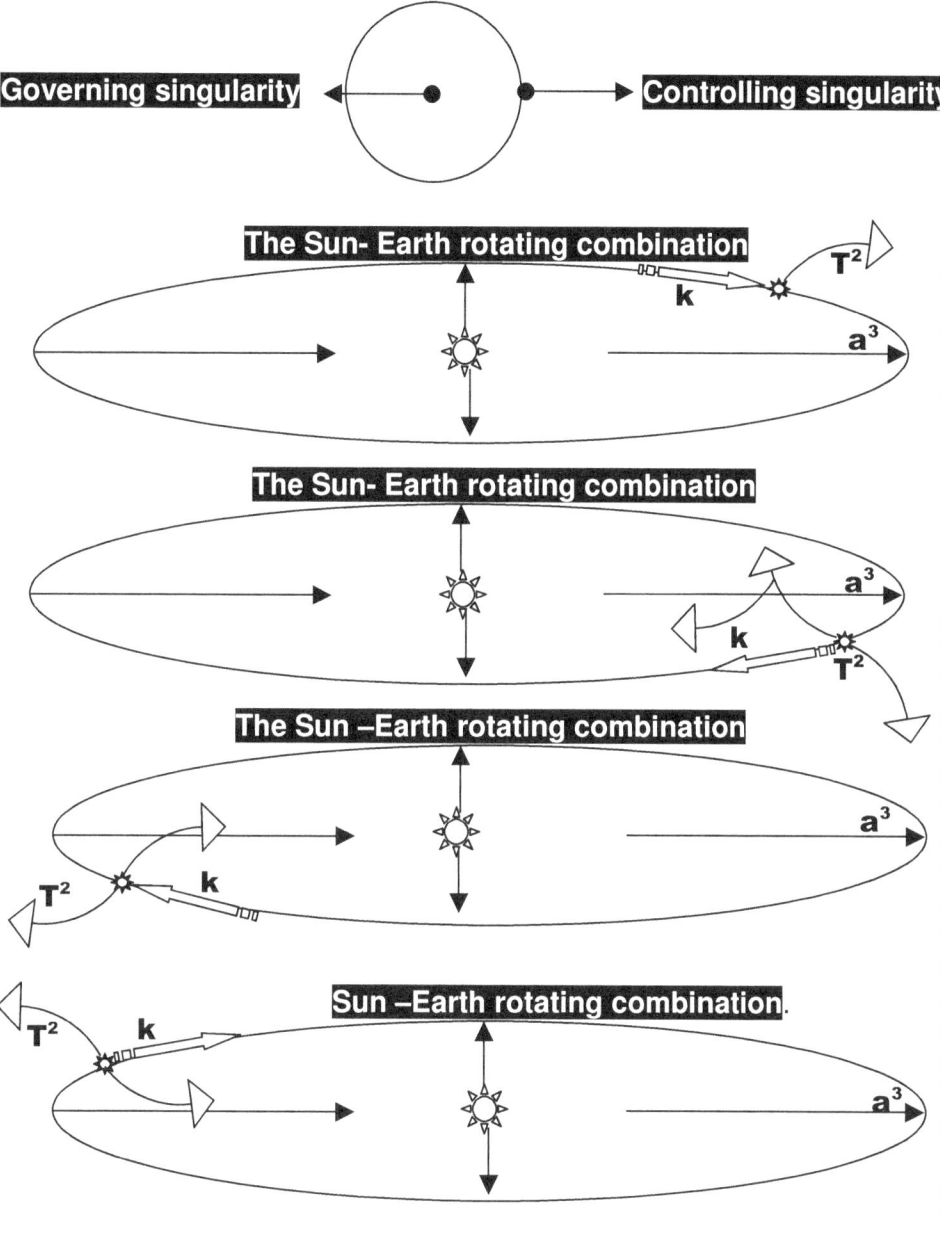

The Moon has mutual singularity in relation to the Earth having the governing singularity that is performing as individual singularity with the Sun's range then holding the controlling singularity. The Moon only spins around the Earth's axis and therefore forms part of producing mutual singularity in relation to the solar system having the principle singularity in relation to then the Sun having the governing singularity and the Earth holding individual singularity. All the debris not spinning around their individual axis holds a position of mutual singularity in terms of planets offering individual singularity in the principle singularity controlled by the Sun's governing singularity. The debris holding mutual singularity spins around other objects that hold an axis giving individual singularity.

Whatever name applies the difference in grouping does not form the factor since it is not in the name but the movement that forms the relevancy. The Universe moves as big as it is where in this movement we have is a compliment formed as three dimensions by movement in a circle relating to movement in a line relating to what is not moving and then the relevancy changes where the relevancy goes into the single dimension of singularity and everything is in relevance standing still while everything is moving that was before not moving. This works on the positioning of space swapping where the relevancy applies to the Earth and then applies to the solar system or to the atoms because everything is in conjunction with all the other material filling space but only in direct relevancy according to the positioning of **k**. In the one instance the Sun will fill the role as forming the controlling singularity with the Earth forming the governing singularity and with the Sun then being the governing singularity the entire solar system becomes the mutual singularity within which the planets then forms the individual singularity. In brief I wish to explain how galactica and stars develop by initiating progress when applying singularity performing in relevance.

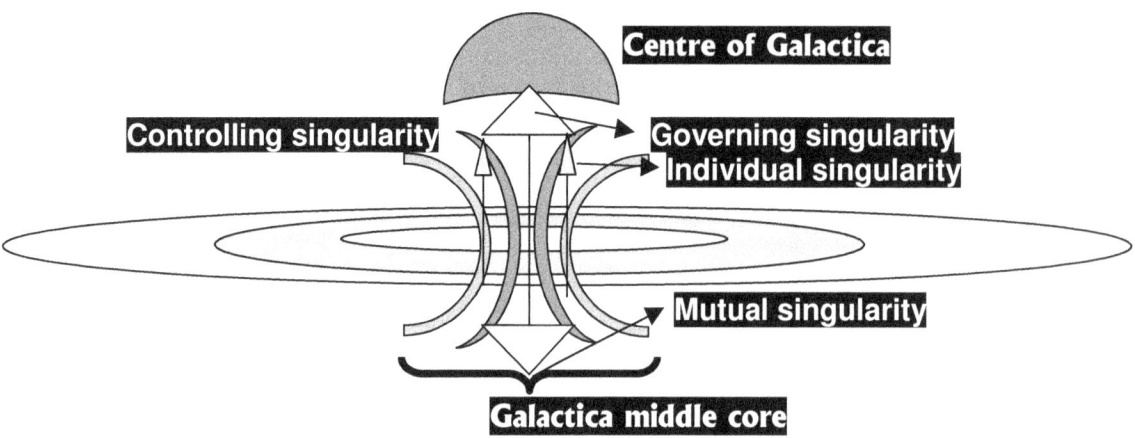

The Earth for instance is the equivalent (more or less) of all the atoms spinning within the Earth. The atoms hold a value of individual singularity forming the mutual worth of the singularity factor but the total worth is the compliment of spinning differentiation that the atoms form between the governing singularity standing still and the mutual singularity allowing the Earth to spin. This goes as far down as where material is forming space goes. In the atom there are several points holding singularity of which some perform as being part of the governing singularity with the sub-atomic particles spinning around as part of the individual singularity where all these points form the factor acting as the mutual singularity. Then this entire lot forms a mutual singularity made up with the combined effort of all the factors forming the individual singularity spinning around the atomic governing singularity.

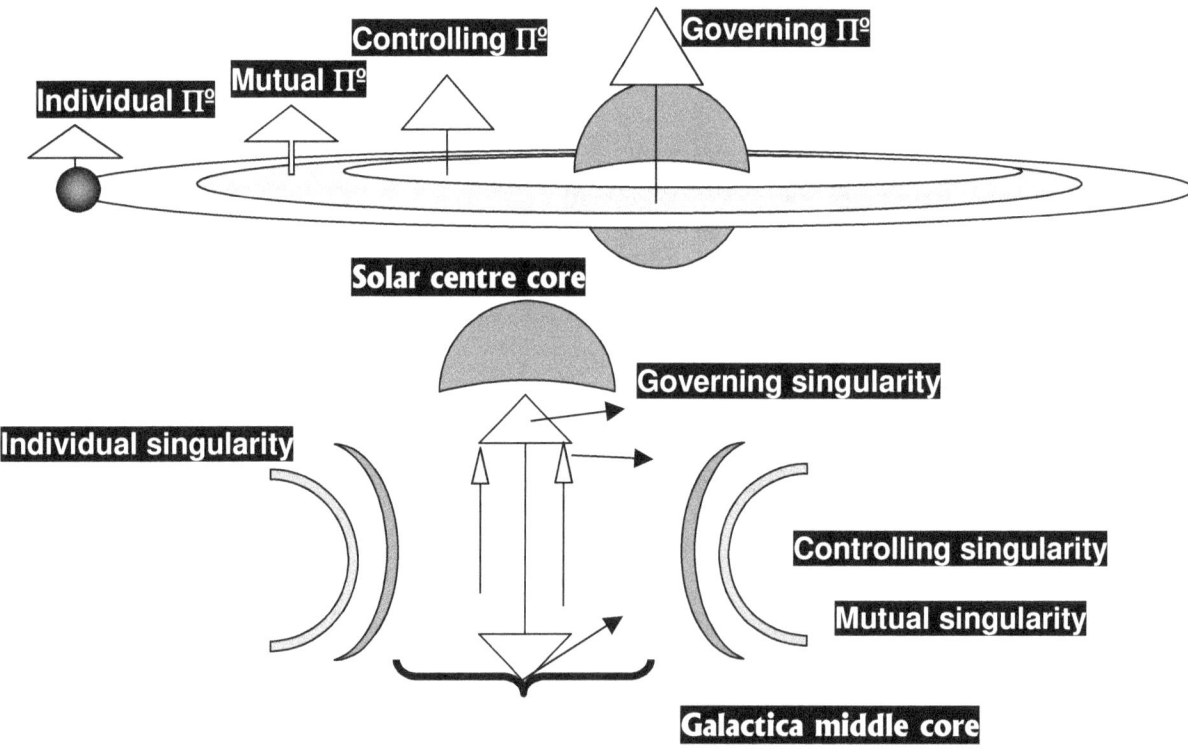

The galactica holds a centre formed by liquid with stars being solid granules in the centre. The young pre-developed star is as hot as the frozen liquid is holding the liquid-centre. The movement of the stars on the outside of such a centre cools the inner core of the galactica to such a degree that the young stars inside freezes in liquid.

Initially the sun was part of the liquid inner centre of the Milky Way in the very centre. There was a mutual singularity forming the entire galactica that was holding the Sun amongst many other particles as a particle of the space-time of the mutual singularity. To the outside were other stars benefited from the growing relevancy that provided an increase in span blessing the outside with a larger developed gravity and those on the outside of the galactica had more matter surrounding their individual singularity as a result of a wider circle allowing a bigger relevancy to take place.

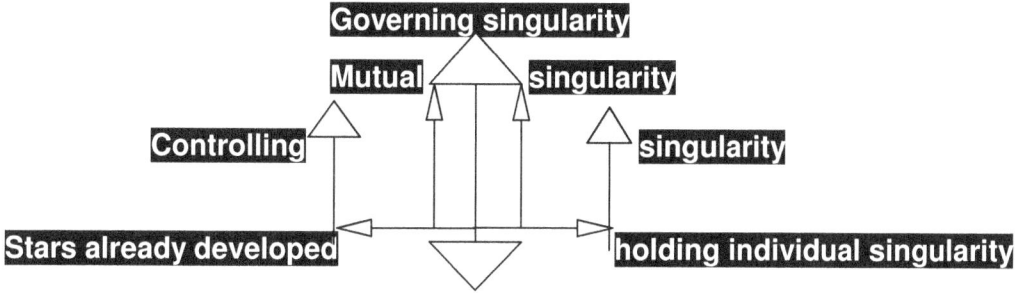

The Sun was totally dependent on the Milky Way centre to provide the governing singularity while the yet underdeveloped Sun formed the mutual singularity as individual singularity not yet participating individually. As the outside circle of the galactica centre drew flat in terms of the centre still forming a sphere, the Sun grow from the position it had when it was only mutual singularity without spinning around an individual axis to where it now is forming one coherent structural part of the individual singularity its spinning develops as it spins around a governing singularity and by spin it establishes in reference to its controlling singularity. No, it is not different from what it was but time positioned the relevance to such a way that it is the relevance that changed and not the Sun. For the purpose of informing I will elaborate on this statement in due time.

As time progressed by forming space the Sun's individual singularity held more space and as it applied more individual status it also claimed more space and with progress of time the Sun moved away from the governing singularity in the centre of the Milky Way as the Sun's individual singularity became stronger in the overall context forming a part of the mutual singularity within the context of the principle singularity.

By forming a strong governing singularity the Sun stared moving and that enhanced the individual singularity position. This then allowed the Sun to form not only part of the mutual singularity as before but having a controlling singularity within the Milky Way producing the principle singularity, the Sun started to perform as an individual (spinning) singularity in terms of establishing the governing singularity as this differentiation in movement is the result of what the gravity forms. The solar system as a unit holds the principle singularity and then the Sun holds the controlling singularity. When the Earth spins it has the governing singularity in terms of the solar system offering the mutual singularity. When the Earth spins around the Sun, the Sun holds the governing singularity and the Earth holds the individual singularity and while the Moon forms part of the spinning Earth, the Moon holds mutual singularity with the Earth as well as the rest of all particles within the solar system and while the Earth spins around an individual axis the Earth holds individual singularity while forming part of the mutual singularity that relates to the Sun holding the controlling singularity in ratio with the solar system forming the principle singularity and the Sun performing as the governing singularity. In all of this one form the liquid eternity, while the other forms the solid infinity.

Spinning around

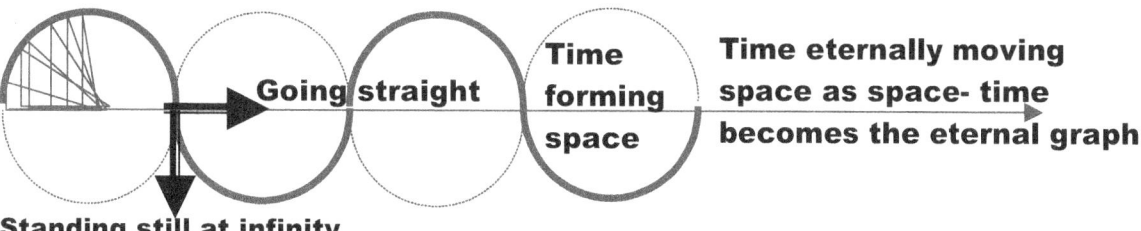

Standing still at infinity

There is always something spinning in relation to something else standing still. This forms the principle of singularity. The Sun could form the governing singularity while the solar system is the principle singularity but any individual singularity (any planets) could be the governing singularity while the solar system forms the mutual singularity. It all depends on something standing still in relation to something else spinning while the third relevance is moving linear. This gives nothing the opportunity or ability to leave the solar system or the Milky Way because movement is always connecting something standing still to something going straight in terms of something else turning.

This relation singularity holds in positions with each other forming a controlling singularity in relation to a governing singularity kills the Newtonian myth of inter galactica contact with any forms of extra terrestrial beings visiting us from "nearby" (in Newtonian mentality this could be anything from about 70 to 120 or more light years away) which is another part of the Newtonian myth of how science wishes to play childish games with magic to hide their misunderstanding and not understanding of science. All of what I now have mentioned Newtonian science pushes under a blanket of mass forming gravity without showing any proof that this is the case. All the planets align without mass proving to have any part in such an alignment. The question science should ask is what keeps individual structures in orbit while forming part of the solar system that is part of the Milky Way. To find an answer cannot be so difficult and in that answer the information will kill off any notion of inter-galactica space travel and show it to be a myth born in stupidity and conceived with the inability to think (or in other words to be Newtonian).

Should anything wish to leave the solar system that object has to be stronger that all the gravity within the centre of the Sun and should anything wish to leave the Milky Way it should develop more gravity that does all the objects do spinning within the Milky way having gravity and that gravity is all put together. The departing object must generate more linear movement than the all the spinning movement of what the entire governing singularity controls within the range of the controlling singularity performed by the mutual singularity within the principle singularity. If anything wishes to leave the Milky Way it has to be stronger than the totality of the mutual singularity performing the linear part of movement in context to the governing singularity controlling spin which matches the spin performed by all the individual singularity and when it gained movement equal to match such spinning gravity can become inter galactica space flight.

The simplicity in which science tries to dodge the reality is mystifying while the question in need of answering is mystifyingly simple. What stops all the planets to go straight and leave the solar system in disarray while it is very clear that there is perfect order in God's Creation? If it was simple as just leaving at free will or on mass bringing discrepancy by producing non matching gravity the solar system would have dissolved into chaos just after it collected dust to form planets.

The question in everyone's mind would be where my proof is about the presumptions I make and what evidence I deliver to substantiate my claims. It is written in the four pillars and the way the pillars interact.
In this we have tow forms of Π forming space-time that results in gravity. On the side forming space we have 10 being the contribution space makes forming a double as the one planet is responsible for forming 10 and the other planet is also responsible for forming 10 as part of their contribution to space while the third structure stands still (Π°) as to give time validity to form while time shows growth coming from 0.991 to 1 as the factor forming Π. This is space that establishes Π. This then stands in relation to the 7° of spin that forms the Earth's (or any other planet for that matter) gravity. The singularity forming 7 is 1 and singularity is not similar but is the same. That is because by duplicating one we reproduce the same point serving singularity 1 X 1 = 1 so therefore 1 has to double the same one by using the same one twice.

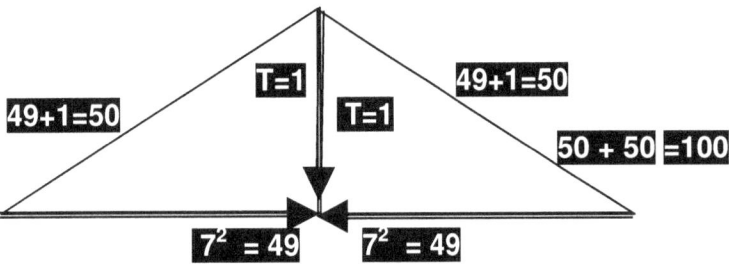

The movement we see forming gravity is not in the one but is in seven duplicating by the square to form 49. The one can never move because the one is singularity and even if singularity goes square (move) it remains 1. In that we have 7 interacting with the same on twice because although the two points we see are space apart, the two points are representing time since time is 1 in singularity. The two points spinning around singularity is using the same instant time forms although the points are way apart.

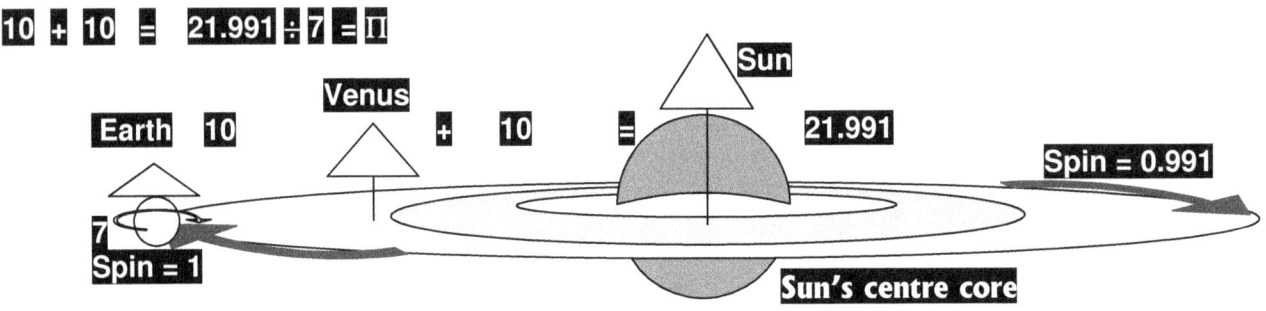

Therefore the 7 turning is turning about the same point and in that we have 10 duplicating while 7^2 is adding to time 1^2. The space must represent the two positions 10 holds while the singularity duplicates 7 and not doubles 7 by the square of movement. In that Π is 3.14156 which is 21.991 ÷ 7. From that we can see how relative space-time is relating to singularity. In the space frame we have the relevancy of singularity spinning about an axis that holds governing singularity (1.991) in relation to mutual singularity (10) as well as controlling singularity (10). This then confirms the space part of space-time forming the Titius Bode law and it is the Titius Bode law applying that forms the gravity that forms the relevancy of time that is expanding space by forming space. It is the way singularity controls the Universe in time forming space that the phenomena apply.

Then on the material side we have three (two planets and the Sun) all spinning which is redirecting movement to the tune of 7° and 7+7+7+0.991= 21.991. This again and once more stands in relation to the 7° of spin that forms the Earth's (or any other planet for that matter) gravity in relation to the atoms within the Earth. By having space forming Π and material spinning forming $\Pi°\Pi$ the interaction of space-time then will become gravity Π^2.

That is the one way that the relevancy of space-time takes place forming 3 dimensions. It must again be said that time $\Pi°\Pi$ goes flat while space-moving $\Pi\Pi^2$ remains 3 dimensional.

In this explanation we have the individual singularity (7) standing relevant to the mutual singularity of the other factors spinning in the solar system (7) being relevant in spin to the Sun spinning by implicating the controlling singularity (7) in terms of the governing singularity applying the movement or growth (0.991). In this instance the triple 7 places the movement of spin that forms gravity in relevance to the singularity bonding each other in the bondage of Π forming and this becomes the other half of what forms Π^2.

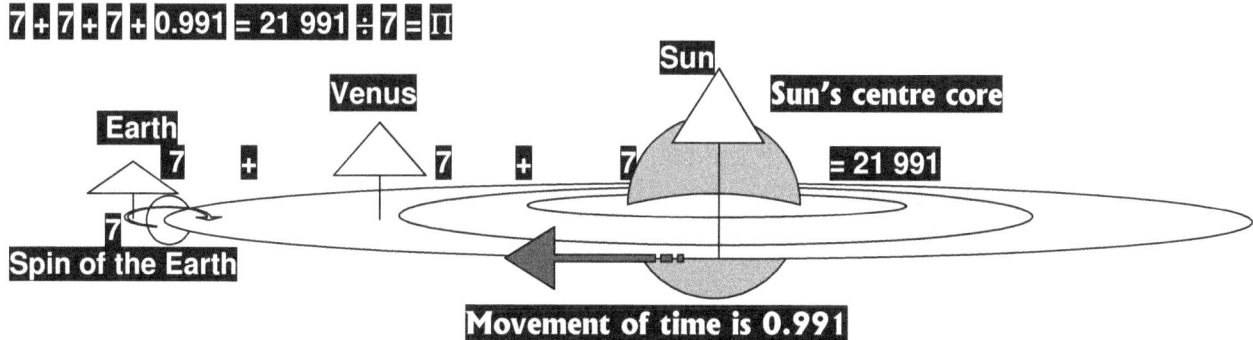

With every planet spinning in synchronised order the three forming the Titius Bode law will each contribute 7° of spin that adds to Π forming. In relation to that we have on the side forming space a double 10 forming in relation to 1 standing still and the movement coming about contributes the growth time contributes that is 0.991. With material forming Π (7+7+7+0.991=21.991) and space contributing (10+10+1+0.991=Π) we fins time forming Πx $\Pi=\Pi^2$. The movement we see forming gravity is not in the one but is in seven duplicating by the square to form 49. The one can never move because the one is singularity and even if singularity goes square (move) it remains 1.

Singularity controls the Roche limit by committing borders forming relevancies.

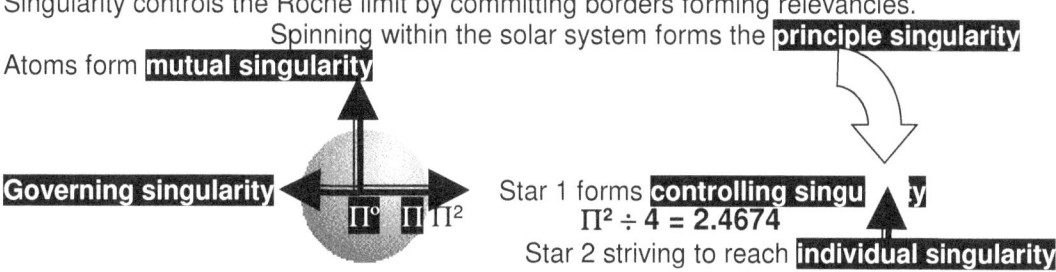

Looking at the overall picture we may find that 7 spins around 1 but since 7 spins around the same one it is spinning the same 7 that then creates a dual in 10 where ten serves as a value that forms space. The mathematical implication in the Titius Bode law works on the principle that we find the number seven doubling as every planet holds a governing singularity that works in tandem with a controlling singularity and the direction changes of the two factors always work in tandem.

This is most impotent. One can never see one singularity bring motion about without the other principle also influencing the outcome. From every planet the value of the 7 applying will change and in every case another double seven will be in place since every planet is a Universe apart from the rest of the Universe. In the picture forming space-time by explaining the Titius Bode planet layout we have 7 by the double as well as the planet holding its own position according to the Sun, which is 10.

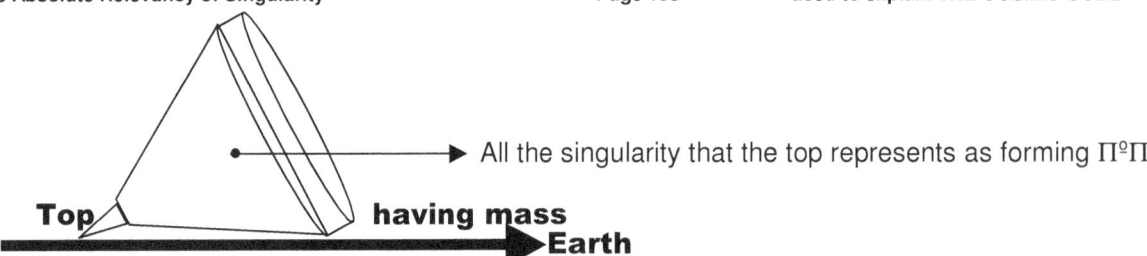

When one observes the cosmos one observes the night sky as one big black cloud that is forming a darkness that absorbs all light seen from our perspective. The darkness of the night sky allows light to disappear into an abundance of darkness that absorbs all the visible light from our view. From what we observe the night sky also fills with tiny specks of light seen here and there and in between, and when we use better the lenses we find more lights are visible here and there and filling the night sky as it is filling our view everywhere. But in all the blackness and all the vastness and all the sparingly filled spaces there are three relevancies, which result in gravity. Let us acknowledge what we see from the controlling centre. What we see as the blackness is singularity forming as eternity. When there is eternity then that would link to infinity where all of eternity will spin around any one point forming infinity.

Looking at a scene we are more familiar with we find the top not spinning lies flat on the Earth. The top not spinning has mass. The top being in a position where it is having mass gives the top a distinct position as to form the edge of the spinning Earth Π and therefore the singularity in the top Π^0 connects to the Earth's governing singularity Π^0 that positions the top to have part of the controlling singularity Π and by having mass while moving with the Earth and not moving in own perspective the top forms part of the Earth's governing singularity that is forming the gravity as moving in relation to singularity $\Pi\Pi^2$.

The top rotates when holding a position of mass but this applies only while it is holding a position being part of the unit spinning or part of the Earth unit and that as such is placing the top standing still as to form the edge of the Earth and then the top is a small part of the Earth's gravity as being $\Pi^0\Pi$. This position I have sketched places the top in relation to singularity without forming space. With the top having mass while standing still on the Earth it means the top is on own accord standing motionless as the top only moves with the Earth that moves while the Earth moves on behalf of the top as well as the Earth and that gives only the Earth singularity while the top is forming the edge of gravity having mutual singularity since the top with mass holds Π in the $\Pi^0\Pi$ relation it has with the Earth. As soon as the top starts to move on own accord, which means the top is spinning in an erect position, the top then gains individual singularity as a status in terms of the mutual singularity it still holds by the spin being able to establishing a governing singularity in the centre of the top bringing the Earth's role in terms of the Earth forming the controlling singularity.

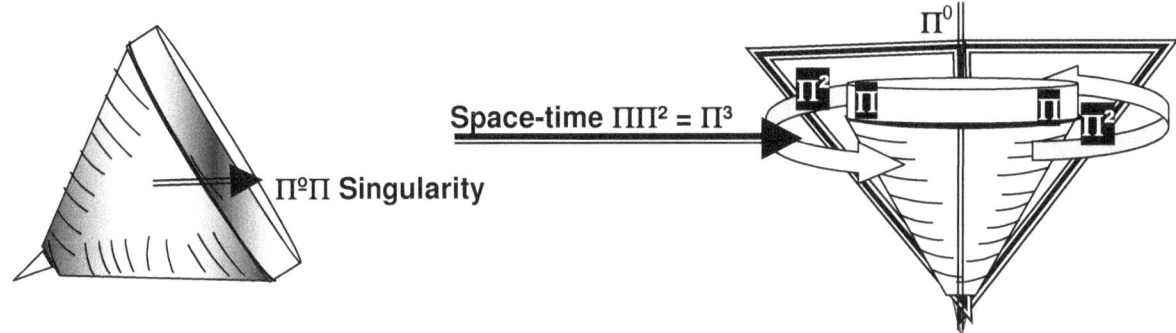

The Universe holds two forms where one consists of space in a three dimensional form and the other consist of time in a space less form of singularity forming relevancies. Three forms singularity at $\Pi^0\Pi$ where singularity holds the Universe in terms of Π. This part has no space and that is what goes singular. Then another part also forms in tandem with the first one mentioned and that holds space by movement in relation to what forms space which is $\Pi\Pi^2 = \Pi^3$. This shows clearly that there is a side formed by space where time becomes space, but then there is a side that sands only in regard to singularity forming relevancies and that id the accurate part of the Universe. Where Π^0 singularity is broken by Π forming and that interrupts singularity Π^0 continuing because Newton stated that $\frac{dJ}{dt} = 1^0$ we find that in that there is the reality cosmology is in search of.

We find any Galactica or the solar system forms relevancies where in these relevancies it is singularity in forming space that becomes the true value.

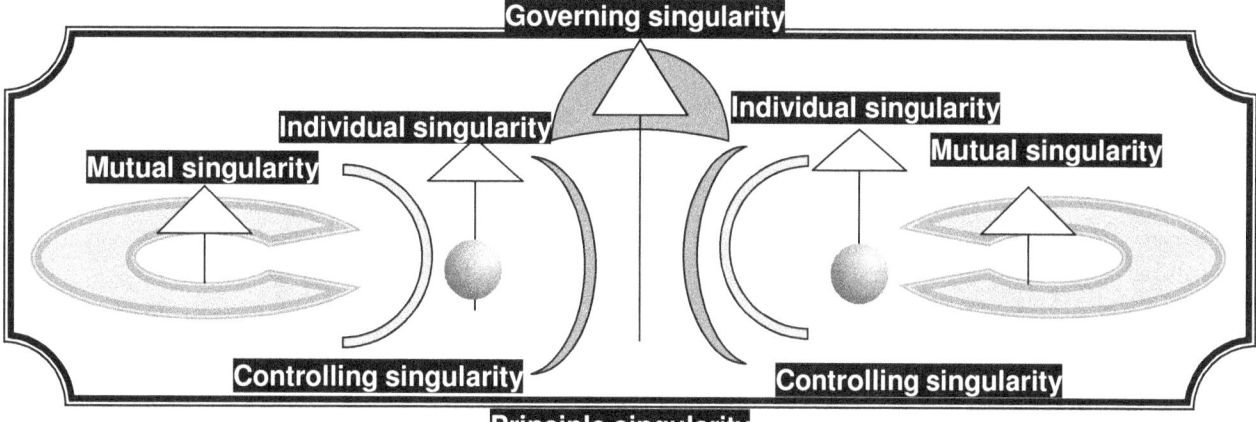

From a human perspective on singularity I have given each value a name as to put order and understanding to the concept I explain. In cosmic terms this is meaningless because in cosmic terms they have the value of $1^0 = 2^0 = 3^0 = 4^0 = 5^0 = 1$.

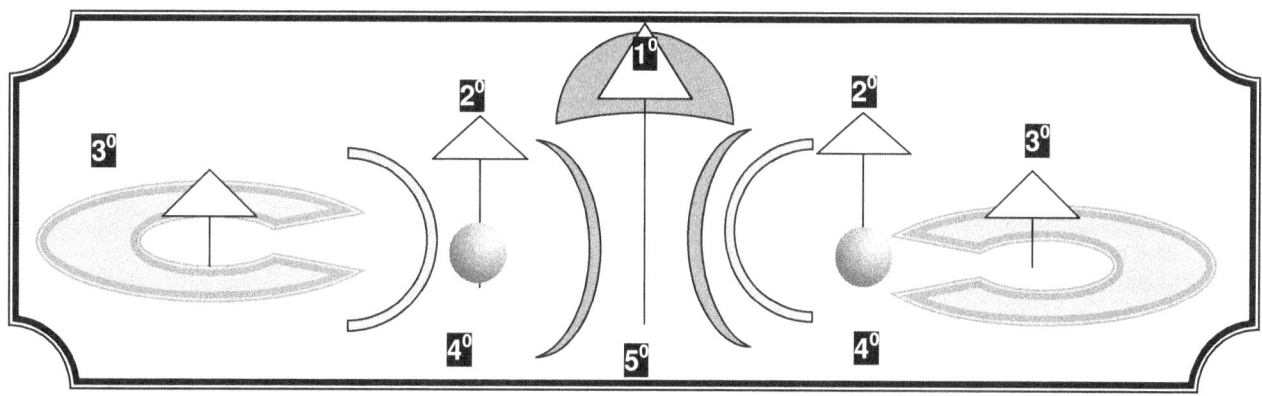

When putting the real worth into the equation we find the true value of what goes flat because in this it is singularity that takes prominence in the relevancy and singularity taking a value is flat all the time. Only the singularity holds prominence because the space is void of reality. The space is there because the space moves and just because the space moves, therefore space in time finds credibility. It is $k^0 = a^3 \div T^2 k$. But in that only singularity holds a prominent meaning because the rest refers to singularity holding a meaning.

This is exactly the Newtonian way of thinking. They gave the physics and mathematics a name as classical physics and by giving it a name they then exempted Newton's mathematical thinking from mathematical reality. By naming Newton's mathematical thinking which is the purist form of mathematical shit as far as mathematical reality goes science has exempted Newtonian's thinking of mass pulling mass from reality by giving that maths a name. Mainstream science calls it classical mathematics and classical physics. There is no way in hell or heaven or anywhere else in between that mass and pull mass to give credence to gravity forming, yet by giving it a name mainstream science has validated the concept and now the validated concept applies. I have written books on the matter and I am giving one away for free. Just go to www.sirnewtonsfraud.com and download a free copy from LULU.com...it is for free...then all you have to do is to read the book which is for free and prove me wrong about any detail I say!

The validity of the concept is not in the naming thereof...Just as much is the validity of space not in the viewing thereof. We have space forming a constant that is not reality in the overall picture of reality. We have shifting of singularity by the measure of relevancy where space is $k^0 = a^3 \div T^2 k$ and that puts a relevancy in place of $a^3 = T^2 k$.

Returning to some small part of Newtonian bogus fiction Newton said that that factor of space $a^3 = T^2$ is equal to the factor of time on the grounds of his invalid conception that the spinning neutralises the space $\frac{dJ}{dt} = 0$. As I already explained this should read that $\frac{dJ}{dt} = 1^0$ and that is where the proof is that the Universe only holds one reality above all other and that is that the cosmos is singularity forming relevancies which space categorises.

PLANET	PERIOD (Years) (T)	MOVEMENT (T^2)	DISTANCE k	SPACE (a^3)	RATIO
Mercury	0.241	0.058	0.39	0.059	0.983
Venus	0.615	0.378	0.728	0.381	0.992
Earth	1.000	1.000	1.000	1.000	1.000
Mars	1.881	3.54	1.524	3.54	1.000
Jupiter	11.86	140.66	5.20	140.6	1.000
Saturn	29.46	867.9	9.54	868.25	0.999
Uranus	84.008	7069	19.19	7067	1.000
Neptune	164.8	27159	30.07	27189	0.999
Pluto	248.4	61703	39.46	61443	1.004

I agree with Newton's concept but with mathematical tables being in place one can hardly ignore or denounce the existence of **k** as a factor and therefore one cannot put $a^3 = T^2$. The table is there and the numbers are not to be disregarded. Therefore one must search for the true meaning of **k**. I have already showed that space in time in the history thereof.

Space is time in forming a past. Space becomes what time was at the point where time formed the particular space I relation to Π. The fact of Π not only refers to form but validates the Universe. At this point I can introduce my theory on the ***Absolute Relevancy of Singularity***. The ***Absolute Relevancy of Singularity*** is in the fact that everything that is formed is formed by using singularity. Whether one values singularity as 1^0 or 2^0 or 3^0 or 4^0 or whatever number to the power of zero, the lot is still worth the same throughout. That puts the Universe equal at any give spot.

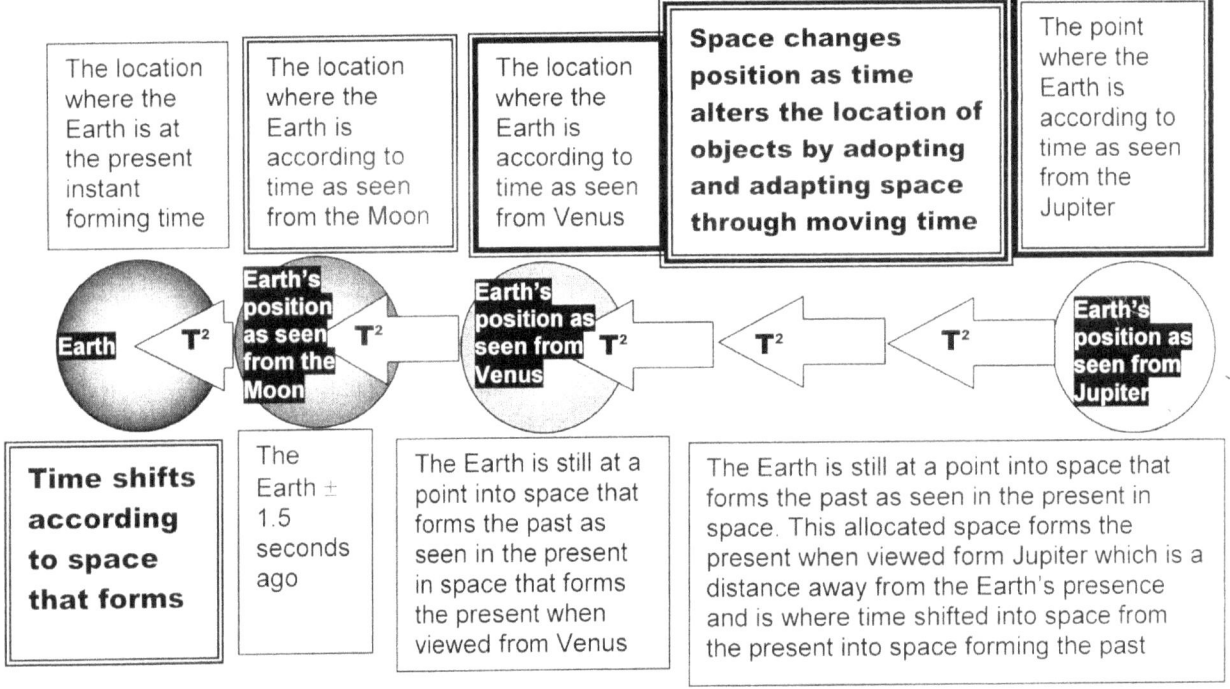

The position I now have standing on the Earth forming part of the Earth gives me a relevancy in relation to the Earth centre in a time position that holds me tied to a position placing me as holding my mutual singularity in terms of the governing singularity. I am Π relating to Π°. However, that is not the only Earth there is. From Venus there is another Earth at that very instant slightly behind the Earth that I hold a position on. Venus is Π relating to Π°. Then from say Pluto we have another Earth just behind the Earth that is visible from Mars which is just behind the Earth that is visible from the Moon and that is 1.5 seconds behind the earth I am standing on. In every instance it is Π relating to Π°. It might be the movement ΠΠ² that brings the significance to the concept but the concept as such remains that in every instance it is Π relating to Π° that forms the relevancy from where ΠΠ² will claim validity.

That means the Universe is the same at any given or collected number of points and big or small strong or weak luminous or dark, all of these human conditions have no meaning in the cosmos and all of these differentiating factors we apply are only developing stages coming and going through the movement of time. The entire Universe is connected by time, which is 1^0 or 2^0 or 3^0 or 4^0 being one particular spot that is everywhere, and anywhere and that is standing related to everything else forming the entirety we call the Universe that moves. In this that moves in relation to one point being anywhere that doesn't move time connects further by space lingering as the history of time. The Universe would be one spot where infinity is holding eternity captured through all eternity. Infinity still holds eternity captured but time moves away from the single spot by moving into space as time forms space and in that space diverts eternity from infinity. First of all we have to look at the Hubble "expanding" of the Universe. This Hubble "expanding" of the Universe is still a normal product of moving into space.

If anything in the Universe moves in a circle, albeit the Earth or the Solar system or the Milky Way or whatever does moves will move according to a circle but also at the same point move in a bigger circle as that thing that moves in a circle also has to move further away in a straight line because space is the movement thereof according to Kepler's ($a^3 = T^2k$) which is space a^3 forming a circle T^2 which is the forming of space that is going straight **k** which is also the forming of space. Everything in the Universe except the being of life is an illusion. Because everything you see is formed by something with no worth according to cosmic principles, that which we witness has no credence. Having everything made of singularity and with singularity unable to form space but only able to represent space by time shifting space, every point in the Universe is an illusion with only having any worth on the other side of the Universe which is the side we do not see and experienced from this point where we are we do not share the other point where singularity makes sense. From where we are, we can realise with intelligence there has to be 1^0 however that point in reality can't be there so 1^0 has to form a value of dominance on another side where 1^0 connects stronger to $1^0 + 1^0 + 1^0 + 1^0 + 1^0 + 1^0 + 1^0 + 1^0 \; \Pi\Pi^2$ than it does to $1^0 + 1^0 + 1^0 \; \Pi\Pi^2$. From the side we view it doesn't make sense that 1^0 is more than 100^0 because everything we see is written in singularity and uses singularity except when shifting singularity in relation to singularity.

At the point in the centre of the circle line must start. In the beginning when I explained the way I figured how the line start I said a lot of dots has to continue in order to form a line. It would be 1 + 1 + 1 etc. because the line must form by holding singularity after that point does mathematics begin but in the line that forms all factors holds 1. The lie can only form when all the points forming the line have the value of 1 being 1^0. In that conclusion one realises something must separate singularity from all other factors because singularity hosts all other factors but is by own initiative Π. Only when singularity meets the end value can the end value have Π where the final ring of the spinning circle forms Π.

From a point always in the centre of whatever spins singularity runs in the straight line as $\Pi^{\underline{o}}$ extends the circle that forms outwards and away from the centre holding singularity $\Pi^{\underline{o}}$ forming the next point $\Pi^{\underline{o}}$ that acts as Π because of the movement involved. From where $\Pi^{\underline{o}}$ centres the rotation a next point will develop as time flows that has a value of $\Pi r^{\underline{o}}$ because $\Pi^{\underline{o}}$ has to extend mathematicly by $\Pi r^{\underline{o}}$ The value of Π **that brings about the straight line** connects in singularity and this affects the relation between the two points forming a dome of structures in the movement **$(\Pi/2)^2$ by means of the triangle forming Pythagoras.**

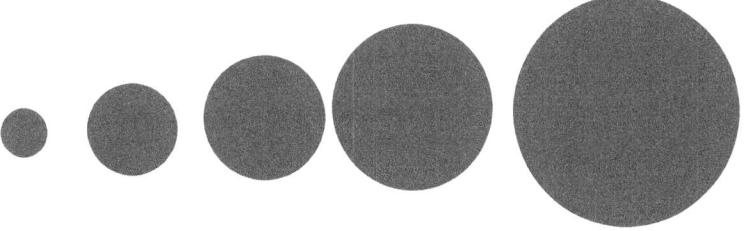 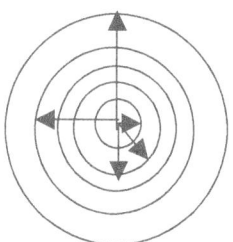

Looking at the affect of gravity it shows the precise quality of no distinctive point as gravity never seems to end at a point but flows all over affecting all that holds a position in its sphere of influence. The gravity coming from China meets the gravity coming from America at no particular spot but intermingles without distinction. That is because the radius we look for forms a connecting circle $\Pi^{\underline{o}}$ validating Π but never r.

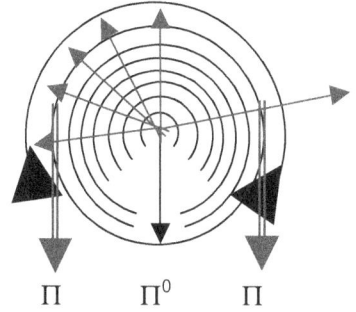

The following concept is what the entire Universe as far as creation is riding on. It is what describes the forming of the Universe and it is what consistency delivers creation in all its splendour. The pinpoint positioning of singularity Π^0 with Π positioning space to either side forming the border set by singularity. Every time that time $\Pi^{\underline{o}}$ moves on it deposits space to the value of $\Pi r^{\underline{o}}$ and this forming we cal space –time. It has no real radius or distance but only has time connecting space by depositing singularity in relation to singularity Π by the positioning of singularity $\Pi^{\underline{o}}$

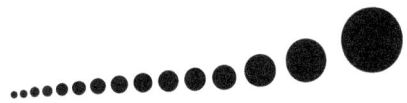

The new direction pointing to a new location in relation to the previous point will oppose the previous point it had in relation to direction considering the centre point.

The line that forms connects by singularity Π^0 not forming a connection in radius but only in singularity Π

In the accompanying sketches the circle would come about from a straight line but in physics it would never validate r or radius as the growing influence is by the appreciation of Π, but to influence Π would lead to a breakdown in r as Π and r are different entities. The dark circles shows a continuous growth by extending Π every time and since Π is the same part as the previous Π, only extending that immeasurable bit of a millimetre each time, the circle will be truly continuous without any signs of a break.

That will be the spot of origin forming the relevance in Π. That will hold the eternal spot…the smallest spot ever because all spots that ever can be were secured in a position in the centre of that spot that must continue as a line that forms. Because of the progress singularity follows from the single dimension singularity only allow mathematics a start at Π^0 progressing further too onto Π^0 and from there the line is born as $\Pi^0\Pi^0\Pi^0$ and to $\Pi^0\Pi^0\Pi^0\ \Pi^0$ etc. where Π^0 then may form the concept and value of r. But the line starts at $\Pi^0 = r^0$. This forms because cosmology is singularity based and the value is $\Pi\Pi^0$. This line $\Pi^0\Pi^0\Pi^0$ of singularity can only continue because every spinning atom preserves Π^0 in the very centre and since $\Pi^0 = \Pi^0 = \Pi^0$ the line is the same without finding conclusion except at the end where it forms mass at Π. At the point where Π forms, the movement Π^2 of the circle defines the space Π^3 of the circle and it confirms the centre Π^0 of the circle through the rotation. Let's call this the solid forming or if you wish, let's call it Kepler's singularity. After that singularity forms a line $\Pi^0 = \Pi^0 = \Pi^0$ where this forms another line again as Newton stipulated it by $\dfrac{dJ}{dt} = 1^0$. Let's call that the liquid singularity or Newton's singularity and the relevance of singularity having a solid base compared to the singularity holding a liquid base comes about by the movement of gravity.

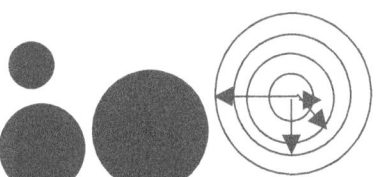

Looking at the affect of gravity it shows the precise quality of no distinctive point, as gravity never seems to end at a point but flows all over affecting all that holds a position in its sphere of influence. The gravity coming from China meets the gravity coming from America at no particular spot but intermingles without distinction.

Using the concept that gravity applies Π as the circle factor Π as well as Π replacing r^2 the replacing of Π brings two values as Π and Π^2. I found that is the case with gravity and will be apparent when explaining the sound barrier as well as the Roche lobe.

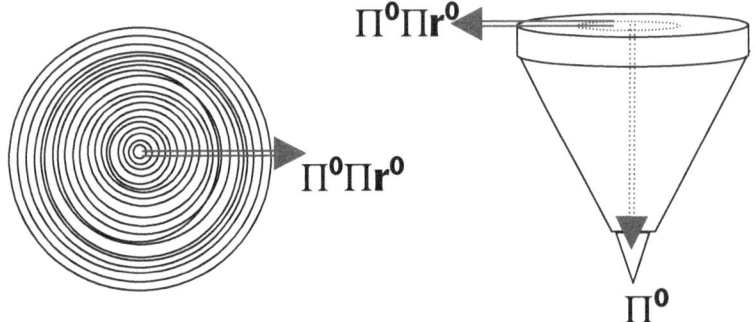

From these conclusions I prove that gravity is the result of four cosmic phenomena interacting to form the value of Π which by movement becomes the value of gravity Π^2 and gravity is equal to cosmic time applying. In order to understand the development of the cosmos and moreover the start of the cosmos and the progress in the cosmos as the cosmos formed one has to understand the measure of Π. One has to see that Π is not merely 22 over 7 or that Π is a ratio that no one ever bothered to clarify, but Π is the key that unlocks every lock that hides a secret in the Universe. One has to microscopically dissect the measure of Π to find the cosmos in measure.

One has to understand where 7 fits in Π. The fact that Π is 7 at the bottom and that 7 relates to a double value of 10 is a key issue. Further is it very important to see why Πtwo by adding 1.991 on the top part of the equation is 10 times. In this measured value is what holds the building blocks of the entirety we call the Universe. It is behind Π that we will find the four phenomena, which I named the four pillars performing as gravity as they form gravity. It is by the actions of Π that the Universe develops.

The Hubble expanding goes by implementing gravity as Π in the square through the four pillars on which gravity and time rests. It is behind Π we discover the meaning of singularity and how singularity forms the absolute and only building block as a form that forms the Universe. It is in Π we find the Cosmic Code unlocking the meaning of the Universe.

Space is time in that confirmed its presence in the cosmos by moving from the present into space and onto the past. By forming a present time then has to move on and it leaves a legacy being space. Time is movement of everything that relocates everything by moving from the present onto the past and as it confirms the past, time forms space by going into a past. Space becomes what time was at the point where time formed the particular space in relation to Π. As time became the present time had to move on and as time moved on it left space that represents that instant in time in relation to other space that was in some position at a specific location at such a point in time wherever that point in relevancy might be. The fact of Π not only refers to form but validates the Universe by splitting infinity from eternity.

By forming space time using Π is in the process of relocating positions Π^2 and in this it forms a network Π^0 consisting of space Π^3 in relation Π to infinity Π^0 that always stays motionless. If not for movement the Universe would be one line holding time by repeating singularity Π^0 uninterrupted and it is in the diverting of eternity to a position away from infinity that the Universe comes about. This is what happens in a Black Hole where no movement within the Black Hole places eternity that always moves in a standing position to infinity that never moves. Without movement the entire Universe will fall back into and onto one point and everything we thought is real and solid will disappear into that one point holding infinity onto eternity where infinity and eternity the reunites. The Universe is an unreal concept with nothing being a reality but for the movement whereby Π confirms everything in a location in relevancy to all other things in a specific time slot or space.

When I, as a person that forms a part of the Earth by the virtue of having mass that connects me to the Earth, stands on the Earth, my position in relation to the Earth gives me a specific relation to time and the Earth. That gives the Moon a future of say one point five seconds and that gives the Earth a past in reference to the Moon of one point five seconds. Where I am at any specific point in the present, that point I am holding is that which secures my present point in time. The Sun is eight and a half minutes into my past with all the space being in between the Earth and the Sun and by my view of the Sun I have a present time slot, as it also gives me a past of eight and a half minutes in relation to the Sun since the light travelled eight and a half minutes through space to confirm my past during that present instant. That secures my past by eight and a half minutes at the point of giving me a present location in time. However, that also secures my future I have from the point I now have in the present by the margin of eight and a half minutes because that establishes a flow of light that would last another eight and a half minutes of filling a presence worth eight and a half minutes while travelling through space by moving with time and every spot filled on the way would secure a position that I will have in a future presence for the next eight and a half minutes, which then becomes my future as it fills my past.

Looking from Alfa Centauri the position Alfa Centauri holds the Earth in gives the Earth a past of say four point six years while securing the present and having that present secures the Earth a future of say four point six years. By securing movement it forms time in having a past in relation to the present that by the same margin also secures a future in relation to a definite past.

Take this in relation to Kepler's formula we then find the Earth (**a^3**) is in relation as viewed from Alfa Centauri (**k**) four point six years (**T^2**). That secures the three dimensional status the Earth has within the space (**a^3**) forming the Universe in terms of a present (**k^0**) that depends on a location (**k**) secured by a future (**T^2**) that will come by movement where the future also doubles as a past (**k = a^3 ÷ T^2** and **k^{-1}=T^2÷a^3**). That is time and that is how time forms space and that is how space-time forms the Universe. That then forms time in relation to space where time that moves forms space by holding time secured in positions in relevance to where every point was in time gone by.

If we put this in terms of singularity we find that the Earth (Π^3) is in relation as viewed from Alfa Centauri (Π) four point six years (Π^2). That secures the three dimensional status the Earth has (Π^3) in terms of a present (Π^0) that depends on a location (Π) secured by a future (Π^2) that will come by movement where the future ($\Pi = \Pi^3 \div \Pi^2$) also doubles as a past ($\Pi^1 = \Pi^2 \div \Pi^3$).

That is time and that is how time forms space and that is how space-time forms the Universe. The relevance (Π) that forms in relation to the present (Π^0) will relate to movement (Π^2) and the movement is circular which ensures that the relevancy forming is circular (Π) by securing that the movement is circular (Π^2) in terms of one specific point (Π^0) in infinity which then secures a roundness (Π^3) that forms an everlasting eternity ($\Pi\Pi^2$) which validates an never ending circle.

The **governing singularity** (Π^0) holds a **positional validity** (Π^3) of three dimensions in terms of any **relevance** (Π) formed by the **controlling singularity** (Π^2) thus mathematically it equates to $\Pi^0 = \Pi^3 \div (\Pi\Pi^2)$.

If a **relevance** (Π) did not validate a **positional validity** (Π^3) a **governing singularity** (Π^0) in terms of movement formed by **the gravity** that produces the **controlling singularity** (Π^2) a three-dimensional status (Π^3) would not be obtained and thereby the Universe would not be in place.

Time is the movement of space in relation to any one point securing such movement and everything in the Universe moves in relation to any other single point that forms in any location that then has to stand still to form the centre of the Universe wherefrom that point must be motionlessness to allow everything else movement. In that manner the Universe is constructed and there is no valid solid Universe because the Universe is constructed from singularity (Π^0) that holds no valid space (Π^3) other than being in position (Π) while having gravity (Π^2) that forms the time (Π^2), which is also the movement (Π^2) of space (Π^3).

Should the Roche find proof in singularity by applying Pythagoras, and then so should the atomic sub particles find proof. Once again one should consider the law of Pythagoras to determine the sub atomic values as singularity provide dimensions sustaining form.

Singularity forms the line placing the divide of the cosmos in the square. One the one side will be the value of the triangle at $\Pi\Pi\Pi$ and on the other side will be Π^3.

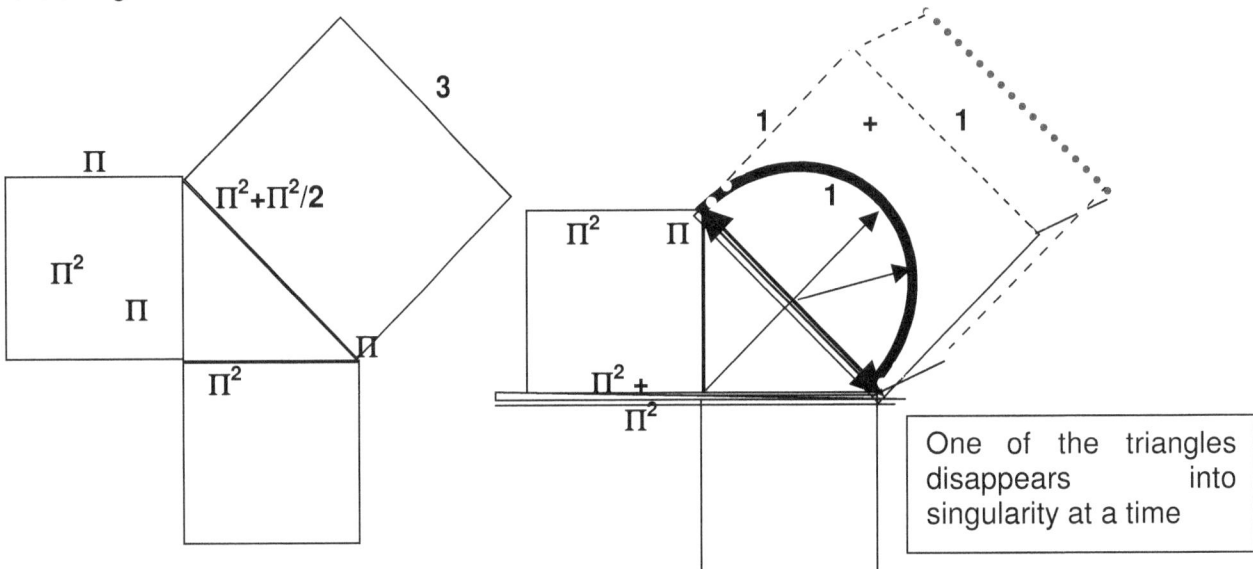

One of the triangles disappears into singularity at a time

From Pythagoras the triangle arrives and by substituting r with Π the square on both triangles become Π^2 being time as space moves from position to position between Π. When the triangle draws flat disappearing into singularity the triangle becomes a square with one side as close to infinity as can be forming Π^2 and the other remaining on the straight line holding the value of Π^2. The triangle duplicating the straight line becomes $\Pi^2 + \Pi^2$. With the neutron position holding the half circle on the triangle the base of the half circle will then be Π^2 with Π extending to form the half circle.

Therefore from Pythagoras come the proton value being $\Pi^2 + \Pi^2$ and subsequently arriving from the proton holding the half circle to a position in accordance with the triangle and the law of Pythagoras the neutron is $\Pi^2 \Pi$ and with the positional dimension of the electron standing in for the cube will hold three side facing Π becoming 3

From the value Pythagoras lends to the atom the dual atomic relevancy comes in place.
$\$=\{\Pi^3=3\}(\Pi^2 + \Pi^2)(\Pi^2\ \Pi) = 1836$ or on the "outside"

$\$=(\Pi^2 + \Pi^2)(\Pi^2\ \Pi)\ 3 = 1836$. This is the relevancy that applies to the atom in relation to the electron serving the Neutron serving the proton and the "mass" is merely progress in density by dismissing of space-time towards infinity. No atom can collapse on itself by the mass being too much.

Mass is a factor that gravity brings about and is caused by gravity and is not the factor bringing about gravity. Science knows there has to be a difference because in space an object might be weightless, although it retains its mass, and no one can say the difference, except to put it down to "gravity". Us, the tax paying public, is letting these Master Minded Academics get off the hook so easily, because every one is too scared to ask "why and how". In the pages above, I pointed to the most basic mistakes about the "gravity" which science ignores, because the answers they have they do not wish know. This statement includes Nobel Prize winning work that shows elements of total and blatantly misguiding. I challenge any person to prove how an atom can collapse on itself, by force, by weight, by pressure or any other means.

.Looking at the top spinning we could see it form a centre I named the governing singularity. Then there has to be a controlling singularity relating to another and more dominant governing singularity produced by an individual singularity (a spinning object creating singularity within the influencing range of the spinning object.).

Establishing Governing Singularity

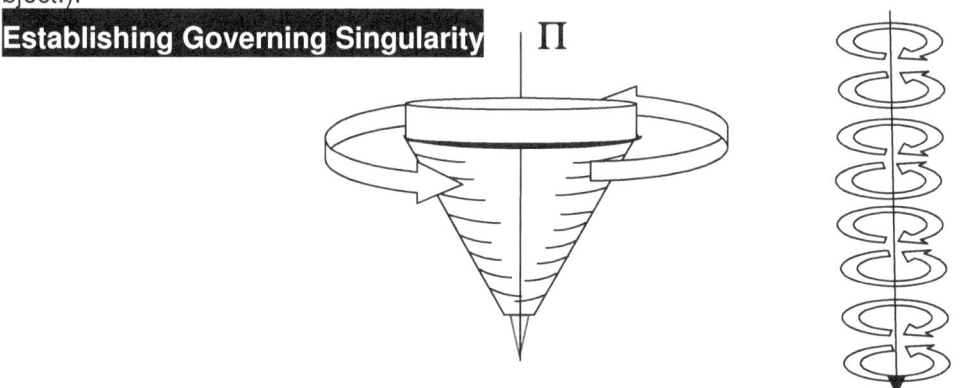

However since it is clear that the governing singularity could never move, by relocating its position in terms of the controlling singularity it stands to reason that the controlling singularity has the ability to relocate the immovable governing singularity to a new stance at a new position.

Relocate Governing Singularity

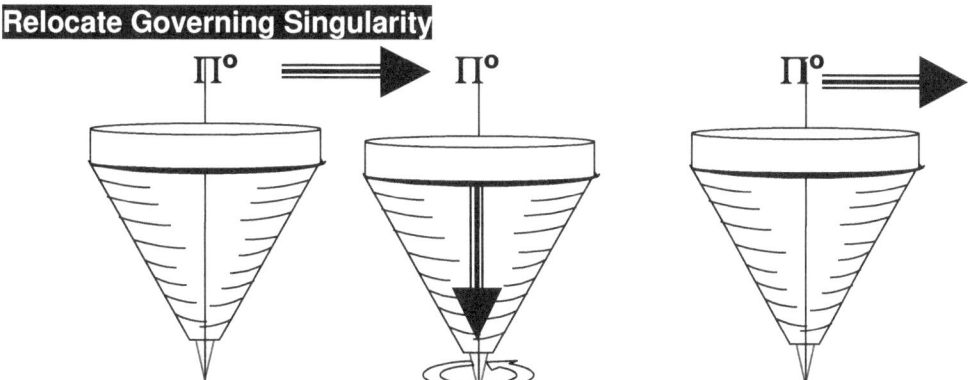

This is a fact that is there. It is present and no person Newtonian or otherwise can dispute such a fact. Then it would be the most logical thing to take this characteristic movement that portrays singularity relevancy behaviour and see how movement started from that. What we now are about to discover predates the Big Bang by many an eternity ion the trod. It makes the Big Bang as recent as yesterday's news coming on TV today. However, before going back we first have to see where we are. We have to have a re-look at singularity. Singularity is where the centre of the spinning circle is.

In the **precise middle** of all **objects in rotation** is a precise centre dividing the object in sectors that will **start the spinning initiation** from that centre point. Thus, the spinning object **will have a middle point**, a very specific **centre point that does not spin** and only holds Π as a specific value. One value such a line **cannot have is zero** because **zero does not start any** line and therefore the **value of the line must be infinite**, just as described in **accordance** and by **the definition of singularity. That point** albeit hypothetical, is also as much a reality none the less and is placed where that point **must be standing still** because every line **running from that point** in **opposing directions** are also **in opposing directional**

spin the other or opposing side. Everything is on the move and always encircling something of greater importance. A top can spin but the parameters of its spin are limiting the motion it can apply. By not spinning the top is still spinning as the Earth is doing the spinning on its behalf.

In the circle using r²Π the r has to have distinctive qualities placing it as a factor apart from Π. The factor r is a line and at the end of the line comes a circle about that forms by the value of Π. Where the growth shows no separate distinction but a continuous flow from the precise centre to the precise edge the flow would become in relation with Π depicting the circle and Π . If we consider replacing r with the true line that forms we find r has to form by many dots following one another as reference to any point on the circle from the centre to the edge. By using r as a distinction in the circle division is possible but by using Π there is no distinction possible making it a solid flow where singularity holding a value of Π° forms dots that forms the radius that ends with Π. But that forces us to return to the centre holding the prominent prime value of singularity Π°.

The radius has to form by Π° connecting and that leaves a huge problem to overcome because we have in the centre no space Π° and without space the dots can't form a line and without a line no radius could apply and without any radius being valid there can be no space formed by a validation of r using Π° as a measure and without space there can be no Universe any Newtonian could relate to! That throws mathematics out the window and replaces mathematical formulas with much better common sense and intellectual logic! I mention this because where we delve into cosmology; at such a point the cleverness of mathematics has no valid ness because this point we now enter in cosmology predates mathematics. It is wise to remember that the Universe created mathematics and mathematics is the result of the Universe. Mathematicians always try to convince everyone that mathematics formed a Universe whereas it is a Universe that came about that established mathematics as a value.

From this centre line that is only theoretical definable, but is still there all the same, an opposing value always form that becomes real and distinct when rotating, but even more distinct when not rotating because then the line grows so much it covers all the matter, to a securing spin value of zero, the most original value it had. When not rotating, it is as thick as the material will go. When rotation begins, the line shrinks back to a hypothetical position claiming zero spin that is not less distinct but more distinct because from that point every rotating piece of what ever is then spinning will clearly carry the singularity value of Π implicating rotation. To become a dot Π from a spot Π° was an increase in relevancy that to this day can never again repeat in relevancy of what applied at the time.

> When the cosmos came to motion, motion was not yet defined. When the cosmos brought about motion, the first motion was relevancies. Cold parted from hot. Eternity parted from infinity. Motion parted from motion absence. Infinity broke the laboriousness of eternity for the duration of infinity. The spot became and grew into the dot.

I am not going into this argument very explicitly as to how the cosmos came about because that is extremely complex and since it is very difficult and involved, I leave that for other more suitable books. I wish to elaborate on movement and no more but this will give a look into the start of the cosmos when applying the four cosmic pillars.

The spot •1 The spot growing from 0.991 to one as we can see from the top spinning

The spot •1 2• becoming a dot by forming the circle, which is one of the three

The spot becoming a dot by forming the a triangle which is one of the three

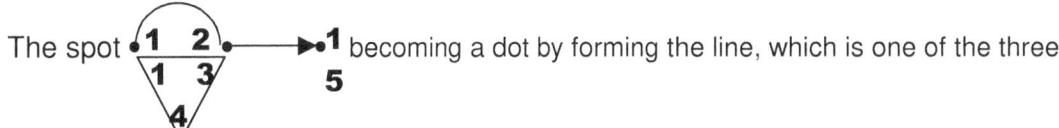

becoming a dot by forming the line, which is one of the three

In that formation as to duplicate one by becoming five we get the Lagrangian point system reflecting on the reason for the mathematical equality there is between the half circle, the triangle and the straight line. We know about the inverse square law forming the four quadrants and we can see in using the Lagrangian points phenomenon that a fifth point forms where the first point has to duplicate. If point 5 comes in place in the space of the spin Π^2 by the four quadrants forming ($\Pi^2 \div 4$) the Roche limit comes into place and that then liquefies the point that is forming too close to the original point.

Then four half circles with triangles becoming lines that forms five points jointly to unite in what becomes Π. I show this to point out how far down relevancies go and that no matter how one looks at singularity it validates Π. That is when Π becomes relevant.

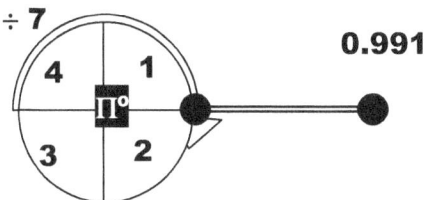

The five dots as a whole and not one of the five dots that formed holds any space and therefore the lot as in counting as many as there could be holds no space and therefore the five dots do not form any cohesion with the Universe as far as we with eyesight can observe. The five dots are not with space and not in the cosmos and with that being true then the five dots are not part of the space we acknowledge. The only way that the five dots that formed in relevancy are there is because the one validates the other and if one disappear the lot will disappear into one dot. The only fact that place all the dots in sequence is that all the dots will spin around any one dot and it is the movement that puts all the dots there ever could be in place with one another but this forms space and the movement is the only factor forming space.

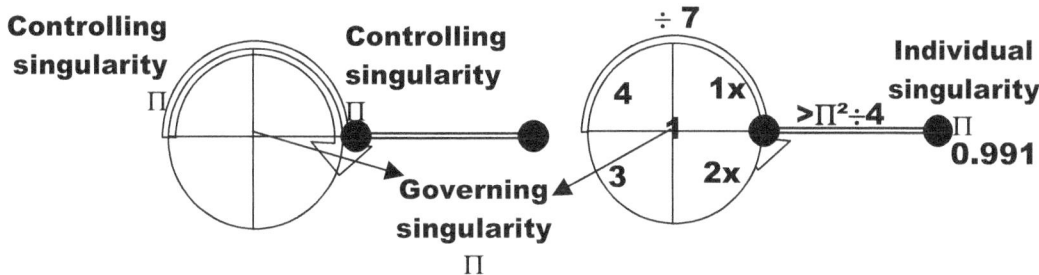

With the spin there is grouping or pairs required that oppose to match the spin that establishes Π^2.

Einstein was absolutely correct when he thought the entire Universe draws flat and then goes back to being 3 dimensional…and yet Einstein is absolutely wrong in his mathematical surmise that the Universe goes flat and then expands back into 3 dimensions. When we go to this level where we see the Universe flat or in dimensions we go into the position where the Universe forms space or don't yet form space. However, the space we think is not the space we visualise because what we see as space is not space but time formed by light recollecting the history of time and should never be confused with space which forms by singularity confirming space. This is the point where light has not started yet or where light has just begun. It is definitely not where we see a Universe as such while that Universe we see being in place according to us is not being a part of reality but is a historical replica of what was when time was in place at that point as time went on. It is critical to realise that every atom is a Black Hole which holds the beginning and the end of the Universe at one point and is as much then a Black Hole in the making as much as every atom is a Black Hole into where light disappears from which it is never to be released again. The space we see and singularity we don't see is all vested in the atom and not confined to the location we think of as outer space.

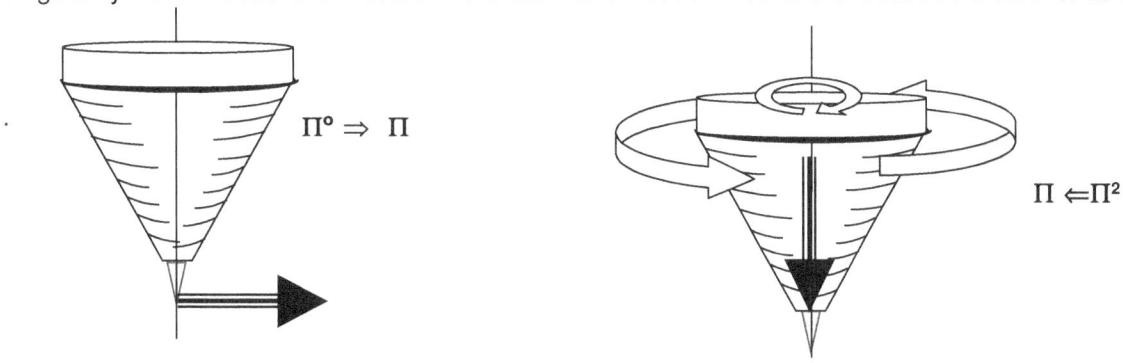

There is a point where the spin of the atom is the spin that re-affirms singularity in its various compositions of singularity such as placing governing singularity in relation to the controlling singularity which is then in terms of the individual singularity placed as mutual singularity in the principle singularity. This is the Universe but if this is the Universe we are unaware of and that is the Universe that is flat or void of dimensions. The space in outer space we think of in terms of being the Universe and is that which we are aware of where the spin is putting a position in relation to the past by securing light in compositions or layers placing relevancies in space in relations to events at the moment it happened layer after layer which is then that what we call space while that is not space but time in replica.

This is space representing time as events happened in relation to that instant in recollection of time preserved by space as time went on to become historical. This is how singularity holds space in the form of light and then there is a point where singularity puts relevancy to task to re-affirm singularity in contact through out the invisible Universe. There is that space that is holding layers made up of light putting in place time in space holding positions in recollection of what took place the moment that which happened took place as a relevancy to one another.

That is not the Universe because that is fiction that is using time to form layers of light is in relation to how things were of what was back when. It is layers formed by time as positions were and is showing how it is no longer but for light positioning the location in terms of when it happened. This form of events concluding the history of time everyone think of in terms of space but that is space-time with the exclamation on time because it is space formed by time in various instances in time in history of time. This history of time we live in as reality is not space but is a Universe in make believe and as void of reality as the atheist's concept of reality. This is time in history allocating light in locations as it was in various stages of periods that was once upon a time but certainly is no more.

Then there is the now and the present, the location the atom holds in reference to all other atoms being in relevancy to the position all atoms are holding at that perfect instant formed by the present in terms of the past moving onto time formed as the future. That is singularity applying in relevancy to gravity. That is the movement of material confirming singularity or $\Pi \Leftarrow \Pi^2$. Then there is a stage where that which can't move does move because it is not there where it was but has moved onto the future at a new location and in that it has to move because it is no longer there where it was but has shifted to another location where it is there at the new location. That which can't move because it has no space moved into the space of the future from a point in the past. That is $\Pi^o \Rightarrow \Pi$.

The following that I present should not be regarded as history in terms of outer space presenting fictional space formed as outer space but it should rather be seen as the instant forming the present that came form the past as much as it is going to the future. The scenario revolves on $\Pi^o \Rightarrow \Pi$ confirming $\Pi \Leftarrow \Pi^2$ while at the same time $\Pi \Leftarrow \Pi^2$ re-affirms $\Pi^o \Rightarrow \Pi$.

When the spot Π^0 became functional and established all relevancies possible, heat parts from cold as eternity $\Pi \Leftarrow \Pi^2$ parts from infinity $\Pi^o \Rightarrow \Pi$. The expansion is not clear motion but more a parting of relevancies where a centre formed a relevancy because the centre could not provide motion. The expansion that comes about is when that which can never have space move into the realms of that which will always form space. Without being capable of motion the centre established four points, which also served singularity. From the inverse square law we know that the centre doubled by producing the four points holding singularity. That which is eternally cold overheats and grows into that which is eternally hot and by growing it expands and by expanding it cools and by cooling it contracts. Instigating temperature as heat expands and cooling contracts can only achieve movement. This is the only way movement could come about. Movement is only vested within the atoms that then transfer it to the star.

I wish to explain the so-called "gravitational pull" at this point. The fact that when this is detected it indicates planets is as big a load of rubbish as the idea that mass forms gravity. Mass plays no part in gravity forming so size of "planets" seen by gravitational pull goes out the window.

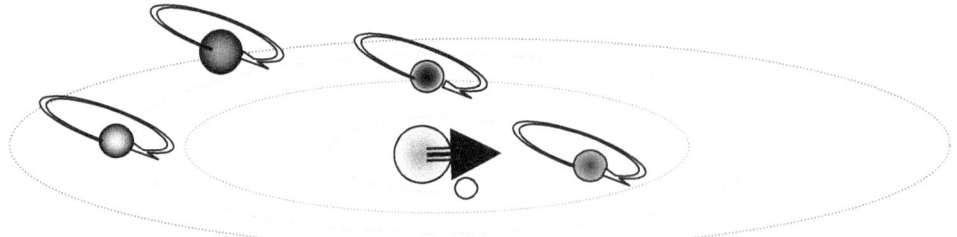

Would it not be more scientific to use a graph having such a perfect formula to work with as in the case of $a^3 = T^2 k$? From the graph any one can read different seasons applying to a small planet in relation to such a big sun that should be able to shine on the earth from any point all across the poles. The space we see is formed by singularity supporting each other as relevancies move in two factors where one shows the lateral movement replacing the governing singularity as the controlling singularity gets confirmed $k = a^3 \div T^2$.

Then the space we see finds re-formation as the governing singularity finds confirmation by movement securing the governing singularity as the controlling singularity relocates. There is no "drag" or "pull" or "push" of any sorts but is simply confirmed by Kepler's view on the cosmos $T^2 = a^3 \div k$. This is where the Universe goes flat because singularity interchanges relevancies while established space holding material remains solid. It is essential to remember that 1^2 and also is equal 2^0 and this again is equal to 1^3. Therefore in singularity being equal at all times the Universe can go flat and remain three dimensional without ever changing.

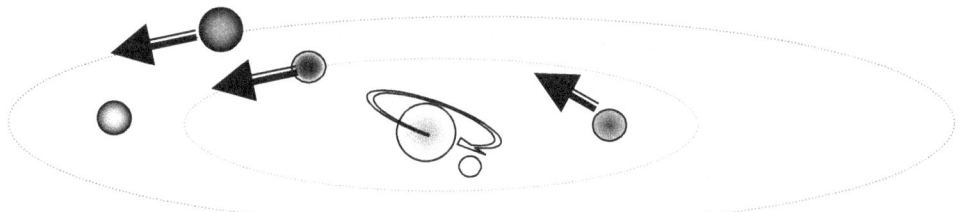

By allowing movement to excite the centre spot, the centre spot comes to be confirmed by the circle that forms in the confirmation of the centre spot. This is evidence we see just buy looking at the spinning top. The movement can only come about in accordance with physics because of the heat that forms in relevancy as heat parted from the cold bringing about the division that followed and that was the motion that formed. When singularity does no move then singularity overheats and the movement is contained by the expansion. When singularity moves, then singularity cools and the movement is produced by the contracting. By expanding the space that is claimed becomes more and therefore the heat had to move but being singularity it could not get singularity to move. In an attempt to establish growth, singularity activated six spots of which four was having motion drawn into relevance by negotiating positioning locations to the four spots that was providing what was to be motion and three that was to be securing the position the centre holds.

There are four directions circling around singularity a forming a ring with two forming in locations we will refer to as above and as below or north and south. The three in line was in singularity not being able to move but the four was also in singularity and just as incapable of moving.

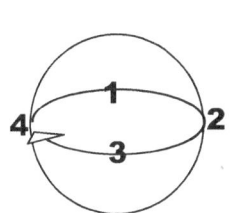 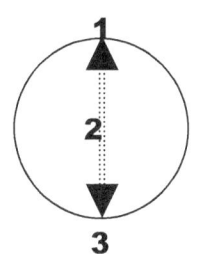 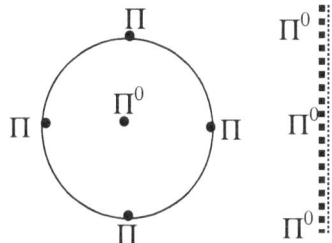

Three points forms a line covering singularity where the centre singularity recovered heat to grow and two points served as an axis to allow the rotation and to assist the duplication. There is one centre connecting the duplication of three as well as the recovery of one (the fourth one) that is applying the tie aspect. Therefore motion consists of three positions in relation to a centre, which forms as space in relevancy to the motion and the space receives a controlling centre.

The space formed by movement is what is known as material and is not what science call outer space because outer space is layers of light forming illusions acting as space which in fact is space that is deposited by time holding no substance being only singularity connecting time gone by being without true value. The space formed by spinning is the material spinning in a way of replacing or relocating material in new positions that is interrupted by activating singularity.

Einstein is correct when he said that the Universe is going flat and it disappears in flatness but this does not mean it disappears altogether. When it "goes flat" it remains in pale but the relevancy in one cycle apply three dimensional and then as the material relocates into new positions the relevancy apply to singularity while space still remains occupied by material. It is the relevancy applying that alternates and from this oscillation we find material filling space applying the one instant and the next instant it is material relocating to new locations while singularity applies.

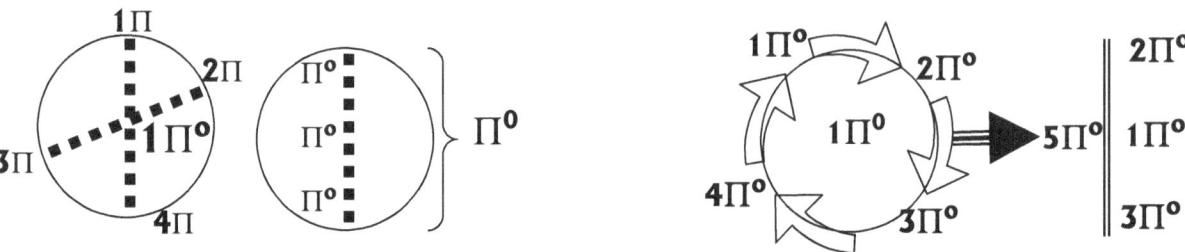

That's why the Universe is ΠΠ² filling space...and...That's why the Universe is Π°Π within singularity

The Universe does not disappear into a state of flatness but alternates between a single dimension of Π°Π and a three-dimensional state of Π²Π = Π³. Te Universe alternates between singularity shifting with time and time forming a three dimensional interpretation of singularity forming dimensional relevancies. Again I have to stress that this does not concern outer space because outer space is special deposits of the history of time that doesn't really exist unless it is interpreted as such by showing intellectual understanding. The only space that applies is the impenetrable space of material that concentrates space to a density predating the density of light. The Universe oscillates between space-time Π³ = ΠΠ² and singularity Π° Π

If not for this mentioned interaction between singularity Π° Π and space Π³ = ΠΠ² there would be no movement to generate gravity in the space we now use. The interaction was used before the Big Bang to initiate singularity duplicating but since light came into place this interaction is used to re-unite infinity with eternity and abolish the space that came into being at the moment of the Big Bang cosmic birth. The interaction brings about repositioning particles in the time-driven movement of forming space in relation to reposition points serving singularity. This is the method in which movement takes place.

Before the Big Bang and before space brought about relevancies applying, all the points came as relevancies applying between the governing singularity forming Π° and the controlling singularity forming Π. This means the most basic component of materials formed. At that time the cosmos was in the process of singularity establishing relevancies and what formed were forming more of what was reproducing or duplicating what was in relation to the next of what was to come but only the four committed to time were expected to move. The four points in rotation came as a result of discrepancies that became time that produced form and that established the relation with the one but had to perform the motion by expanding was as much incapable of motion as the centre was that charged the four with motion in the first place.

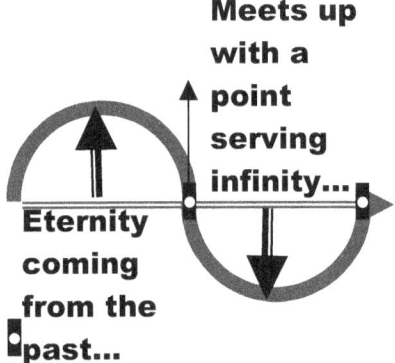

Meets up with a point serving infinity...

Eternity coming from the past...

...And by changing relevancies, which became spin eternity proved movement ability as it alternated time to the future that represented time from the past but only in reverse order since the one formed the past while the next position in movement formed the future of the past while time in the perfect was being in the instant allocated by infinity.

Before time produced space-time that later on (very, very later on) produced material without filling space hence the Universe being the size of a neutron when the Big Bang commenced.

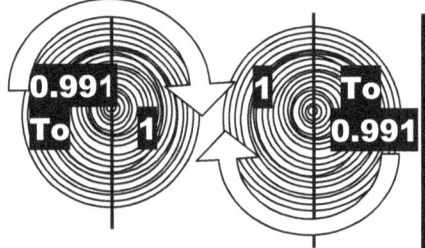

Singularity will always have to cross the divide where the one side of the dived will be Π - 3 x 7 and the other side will be 1. Π forms time in infinity or material and 3 forms time in motion or time in outer space where 7 is the diverting thereof. Singularity will always grow because time is the growth of singularity forming not space but relevance in growth and growing into and breaking the divide is cosmic growth, which also is gravity and is "the Universe expanding" which is what the Universe can't do...it can't expand!

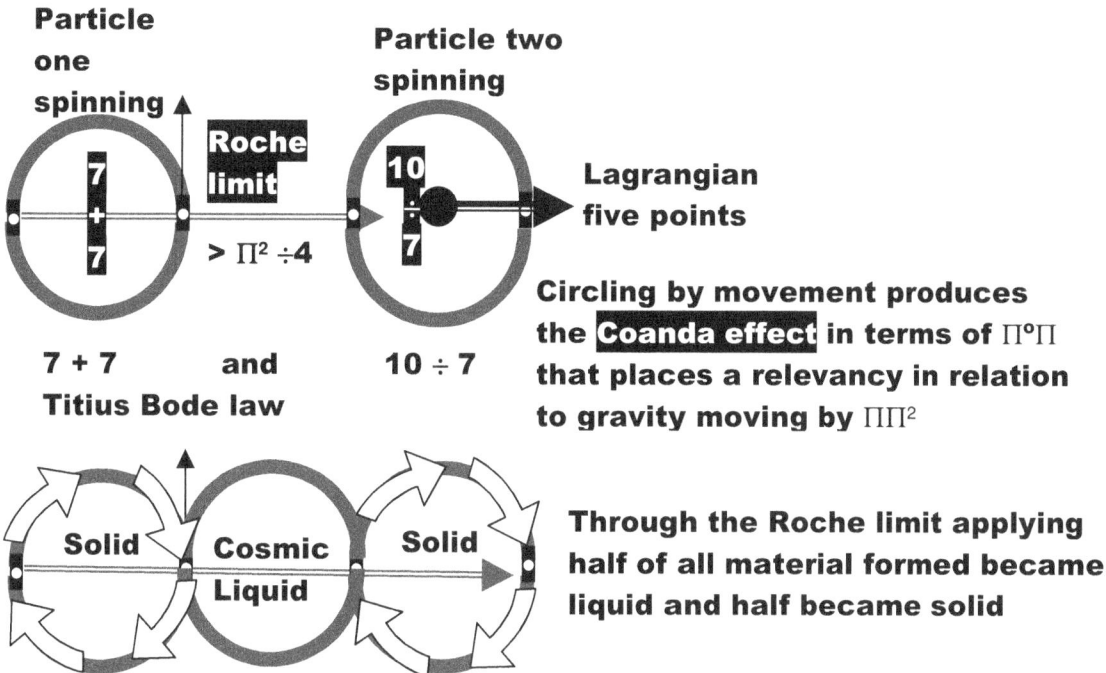

Circling by movement produces the Coanda effect in terms of $\Pi°\Pi$ that places a relevancy in relation to gravity moving by $\Pi\Pi^2$

Through the Roche limit applying half of all material formed became liquid and half became solid

The Roche limit > $\Pi^2 \div 4$ destroys relevancy coming about by directional movement by the division of four that produces the fifth point in accordance with the Lagrangian allocations and the direct positioning of material placing the material in close proximity is causing friction due to movement coming about by movement. Where the basic component that would eventually form material had $\Pi°\Pi$ with a governing $\Pi°$ singularity connecting to a controlling singularity Π, the liquid that fragmented as result of the Roche limit only had $\Pi°$ and therefore it formed a negative Π or Π^{-1}. This allowed that same structure being singularity form two categories where the one moves and the other does not move. In this there is another agenda, which I am not going into at this point. I shall say this much and that is this relevancy there is between that which has a controlled point serving singularity which we call solids and that which only holds singularity never controlled by movement thought of as cosmic liquid started the process that is gravity which is the Coanda effect. Between all materials moving there has to be a liquid keeping the movement apart or a layer of liquid formed by overheating will form. The process involved we call friction and the liquid we put between the solids to prevent friction we call lubricants, where lubricants always prevent the destruction of solids.

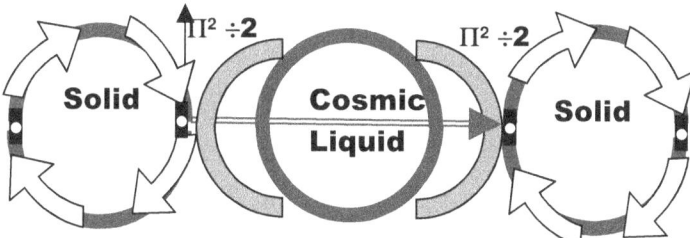

Where motion conducts electrical charging the process in which it functions puts electricity and gravity at equal terms. Electricity is equal to gravity because the charging requires the same process of conducting while it is the scale that differs. Both uses the charging of motion is to entice duplication of singularity, which places the flow of singularity in opposing poles as to re-unite singularity in terms of infinity and eternity. This is the basis, the heart and the sole ingredient of the Coanda principle that includes the Roche limit ($\Pi^2/4$). The charging of gravity $\Pi°\Pi$ to $\Pi\Pi^2 = \Pi^3 \div \Pi^2 = \Pi$ and the charging of space-time $\Pi^3 = \Pi^2\Pi$ is all due to the relevancy brought on by the Coanda principle. The value of motion came from singularity exciting singularity and that is the duplication while the duplication or motion presents the space.

But when this process was in place motion was not through space or motion did not introduce space but movement was purely to replicate. As what came about were incapable of motion it still required a tendency to apply motion that did separate $\Pi°$ from Π. This not only involved form but it involved all relevancies that did come or may in the future come about as a result of the attempt to commit motion. If mass was a factor contributing to gravity the cosmos would have frozen back to singularity without ever releasing singularity to relevancy because the mass at that point was limitless and the gravity was without measure. If mass did bring about gravity, the mass would have pushed all cosmic development into singularity and not allows singularity to duplicate.

Mass does not establish gravity. There is no magical graviton. In the beginning there was extreme mass and boy was there gravity! That which is infinitely cold parted from that which is eternally hot and in between a Universe developed. The only means that the cosmos could find a way to break from the grip of eternal eternity was to expand into relevancies. Such a feat can only go to task by forming opposing hot and cold. Becoming hot produces more of what is heating. That implies motion or a moving away from where it was by generating more of what is available. Only where hot released from cold could whatever was repeat once again and duplicate what was before into what then is more. Secured by motion T^2 in relation to a specific centre k^0 formed the first k which introduced Π and by Π coming about, this formed gravity by Π^2 and still applies from where singularity holds the Universe true to form. The k was an intention to place apart cold from hot and is immeasurable by today's standards as we have no means to measure and with the aid of our limited recourses we will not even qualify any worth as to be noticed.

The line that developed during this era was as flat as the singularity that it presented and as without space as singularity at the present is. This line is the line which light uses to flow and this line still connects relevancies through out the entire Universe. It allows us to see stars billions upon trillions of times more prominent that the Sun and it allows us to witness Black Holes that destroy light by the gravity flow it creates. Light holds a relevancy to time and light cannot be a constant of any sort because light holds space to time as relevant as all other moving objects.

The measure of movement is to have space flow in relation to time and where gravity forms movement in relation to light also forming movement there are two factors moving through space in relation to time. In the Black Hole the flow of gravity is so enormous it cancels the flow of light in relation to space backwards…then how can light be a constant. Moreover is the question that there is a period in place in space where time moves at a tempo it allows light to stand still and it then allows light to move backwards into the Black Hole. The light flowing away from a Black Hole is flowing so slow that the light is flowing reverse in terms of time lapsing within the Black Hole. The light is flowing backwards in terms of gravity or time flowing forward.

The spot formed a dot but since there was a total lack of space the spot formed again next to the dot that was and from they're a new dot formed on the place the new spot formed. This indicates a relevancy between space and time predating light to a point where light as a factor disappears from the Universe. The period I refer to predates light by longer than there are atoms throughout the cosmos. I said at first that if anything has anything to with gravity then it has to be Π and this is how it comes about.

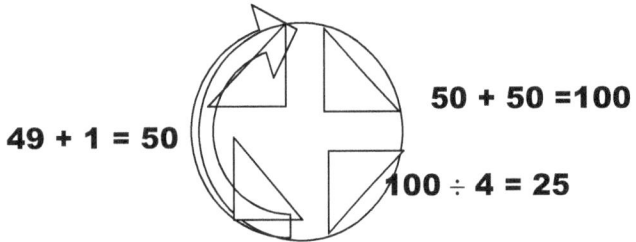

The Titius Bode law implicates 7 by the square holding relevance to 10 and that forms gravity as Π^2 positioning Π^3 in relation to time plus singularity expanding. The expanding forms as a half a circle worth 50 with spin which equal a double circle forming which then is 100 wile going through 4 sectors and that is a full circle forming 100 ÷ divided by the four is equal a value of 25 implicating a straight line forming where in singularity applying a straight line is equal to a triangle both being 180° but with Pythagoras again implicating the value against singularity positioning such allocated point then becomes the square root of 25 = 5 positioning singularity at 1^0

A straight line, a triangle and a half circle always have equal values in the singular dimensions and where singularity apply numerical values still had no place but was under development as the cosmos developed. I have very bad news for the mathematicians; mathematics did not produce the cosmos, it is the other way around, the cosmos manufactured mathematics when the cosmos came into development. The great mathematician's effort in trying to use the most complicated mathematics to calculate the start of the Universe with using enormous formulas is bizarrely mindless behaviour. The cosmos started before numbers were in use because the cosmos placed numbers in use. The cosmos started when a triangle and a half circle and a straight line had equal value because form in singularity is misplaced. As the straight line carries relevancy in elation to progress forming a half circle and this relates to the triangle and a common denominator is vested in the start from singularity. With the normal extending of singularity it will always form the triangle in a half circle whereby Π relates to the cube by 5 points to either side of the line singularity forms. Thus there are 10 standing related to seven and visa versa. By calculating the 4 squares in the circle with the dimensional changing of space the twenty five being square rooted becomes 5.

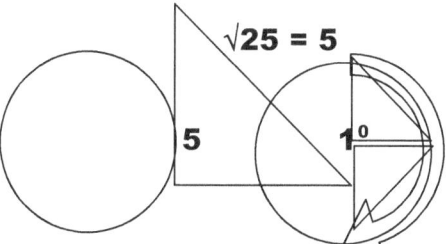

In that manner the Lagrangian points form another singularity marker at point 5 following the preceding 4 that divides the 100 that forms space into sectors and displacing the next sector to become a new allocation that presents singularity a position that sector would carry over an individual point serving singularity holding a dimensional value at 25 which by implicating singularity forms point 5. This then positions the point at 5 formed beyond $Π^2 ÷ 2 = 4.935$, which then would be point 5 that is -0.065 less than where singularity takes root once more.

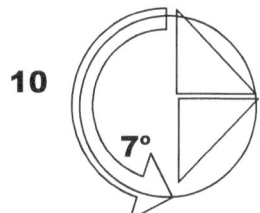

All rotation will be indicated as 10 over 7 or 7 over 10, which is the measured determining influence of rotation.

Where space comes into contact with the sphere the cube loses one of the six dimensions it has to the more dominating seven dimension of the sphere whereby the seven dimension in equilibrium will dominate the six dimension loosely connected by r bringing about that the cube then has 5 sides to the seven of the cube. This means that in the cube the "bottom falls out" and without a "bottom" to support objects they fall to Earth. Remember that a body "floats" in space, but at one specific point it starts to "fall" to the Earth. That is gravity and it is a dimension change much more than any force. I shall explain this last remark later on. This leads to the establishing the value of Π forming by movement expanding from the spot to the dot. By the movement thereof, applying the controlling singularity charges the governing singularity by the value of $Π^o$ while receiving the value of Π. But this value of pi or mathematically symbolised as Π does not drop out of thin air. It comes about as movement puts the factors in place from, which $Π^oΠ$ becomes viable in its forming of $Π^2$.

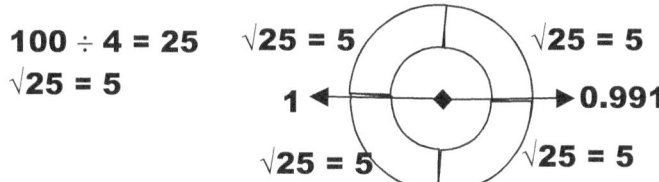

5 + 5 + 5 + 5 = 20
20 movements relating to singularity expanding is 20 + 1.991, which is the top part of Π.

The square of 25 according to the **law of Pythagoras** results in five points of movement carried through four opposing directional sides adds to twenty points in total that one revolution would go through. Adding the four fives would give 20 plus the singularity expanding from 0.991 to one adds a further 1.991 and in total that produces the value of Π = 21.991 rotating through 7o which then results in singularity forming $Π^oΠ$.

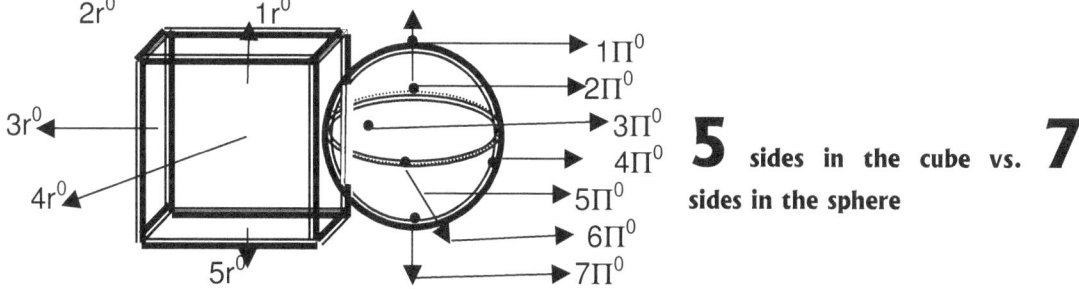

That too is the **Lagrangian** system with five cosmic structures holding relevancy to the centre structure where the centre structure stands in for seven positions diverting from singularity and the orbiting structures standing in for five positions in space.

In the **Roche limit** the space factor provides space to a solid structure and therefore the value of r is replaced by the value of Π bringing about a square in half of Π. The cube holding 5 to either side removes allowing the extending of Π to indicate position to space. Where Π extends to lock onto the next sphere's extending indicator, Π has to connect to Π forming the square of space and translating that to the half of Π being $(Π/2)^2$.

5/2
Five sides divided by two spheres.

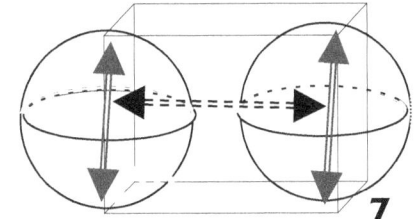

The Roche limit
$5/2 = (Π / 2 × Π / 2) = 2.4674$

7   Space-to-matter

The space between the spheres divide in half, but because of the extending of Π and not applying r as ordinary mathematics will suggest where Π replaces r the singularity extending from $Π^0$ will be half of Π in the square of $Π = (Π/2)^2 = $ **2.4674.** In this lies the dynamics why planets have a positional (be it rather a dimensional) relation of 7/10. There are many other borders that control space limitations such as Π forming the controlling point holding singularity.

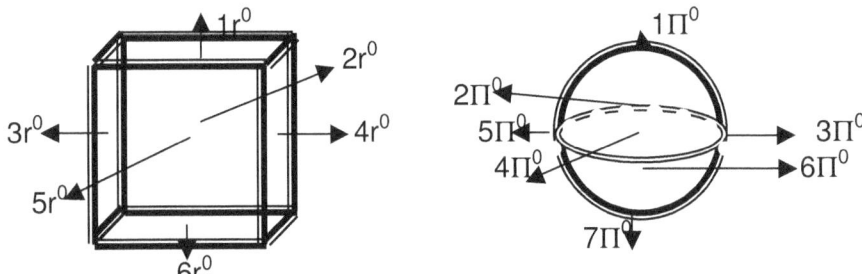

From this the Coanda effect takes president in the forming of gravity. The network of individual singularity not only provides spinning through governing singularity in the sphere but also provides spinning in the geodesic through out the cosmos linking all matter to matter in a network no one will ever come to understand in full. In the sphere the four squares forming the triangles linking the lines to the half circles holds space in time maintaining singularity of different assortments. In view of the matter-to-matter Roche factor where the factor consists forming a relation between particles occupying densified space-time of where $(Π / 2 × Π / 2)$ relating to the foursquare triangle the value of gravity $Π^2$ comes in position as $Π^2 / 4 × 4 = Π^2$. Every dot was $Π^0Π$ and every dot formed $Π^3$ in relation to $Π^2Π$ because of the expanding heat, which produced $Π^2$. However the expanding ran on a line concurrent with time flowing and no space came in place. With that a new relevancy came about forming a centre in between the four points of expansion that was resulting in time. But since the points were in themselves singularity, which is immovable and space-less, they still heated forming a cold motionless centre with the heat bringing about motion where this brought cooling and the movement produced a new spot on a new location that cooled because of the movement coming from heating and cooling afterwards. It became a repetition where infinity broke eternity by producing a centre because of space (or rather forms) forming the motion to enable the space to form in relation to the heat applying motion. This brought about a Cosmos being conceived.

As soon as the spot formed, the spot remained motionless since the spot cannot move. Then the spot overheated and by circling through four contradicting stages the dot formed a spot on a new location as the spot expanded. As the dot distributed in expansion by circling in four quadrants it went to a new location and because of the expanding the heat was more widely distributed, which brought about cooling. This allowed the newly allocated dot to cool and return to the position of a new spot. This process carried on as many times as there are spots and dots in the entire Universe.

The spot forms a full circle as it expands into a dot, but the line running through the circle is forever present because that is the future radius of the circle that will one day develop the circle, which is equal to the present diameter. The fact of the presence of such a possible line in such a possible circle dividing the possible circle into two parts makes the centre line equal to the half circle. The line forms the half circle but not only that the line presents the half circle as much as the line is the half circle. The line then is 180^0 and the half circle is 180^0 because in singularity the two factors are the same. The same value is of course $Π^0 = 1$. The issue of concern is to understand that singularity cannot move as much as the other part of singularity can't ever be motionless. One fact above all is clear and that is that singularity has no space. While singularity forms no part of the Universe we see and touch singularity also is not only part of the Universe but singularity is the Universe that forms. By establishing motion singularity has to be charged with the time delay we find space to be. The space is time taking a period or duration while moving from one point serving singularity to another point serving singularity while conducting the heat and the accumulation of heat that built up due to the retarding of the time to conduct the heat forms the space that is conductor to bring about the motion of the space.

It takes heat time to entice singularity and singularity can only entice b movement of singularity forming the control of the immovable governing point. Singularity cannot move and neither can singularity form space although singularity can add as space formed by movement. By enticing from one relevancy to another there is a bridging of heat by expanding and the expanding results in crossing over into a new point formed in a new time slot where this expanding forming duplication of singularity brings the contracting of cooling in order to send the gravity or the enticing or the relevancy to depart the space and reconnect the space to the next point serving singularity. Bridging all the accumulated various time delays that formed an accumulation of heat through time distorting brings us the space we see and have. However there is no true space in a stable Universe but for motion and it is eternal motionless space that puts a relevancy or a differentiation between motionlessness and moving. By this differentiation we find singularity charging space by the flow of time to provoke heat into forming space.

The development of the cosmos in the very first stage of development came into eras as the relevancies brought about new relevancies that spawned even newer relevancies that all remained in touch with the original singularity centres. It is these relevancies allowing light to flow that are allowing us to see by the distance space offers.

Every point that is serving singularity focused a new time position and by the intervention of space formed a delay in time in relevancy to a position singularity offered that eventually brought about space and every distortion of time brought more space. That movement through space by time concentrated as it activated singularity differentiation between singularity points that charged the points to form space by the movement of space. When the charging became overdue in some sectors it erupted in forming the Big Bang. By the time the Big Bang erupted there was such a huge backlog in heat and time corrupted and delayed the next result was the employing of space as a commodity in the Universe. The relevancy was C the gravity was C^2 and the space was C^3. That left what was inside atom still spinning faster than the speed of light applying the relevancy inside the atom of $k \geq C$ and on the outside of the atom space spun slower than the speed of light $k \leq\geq C$ where the electron applied the relevancy of $T^2 = C^2$ and that formed the atom which then became the cube of the speed of light $a^3 = C^3$. That left the atom at the relevant size of what the speed of light permitted at the time but since the Universe also expanded from that the relevancy the expanding brought on saw the atom grow in space to the extent it has now. The purpose of the star is to recapture the space the atom grew into and from there dismiss the space by spinning faster than whet the speed of light will be on the outside of the star.

All that are in place is confirmed in singularity. From singularity comes the motion of the space in relation to time and this we call space-time. Singularity is dimensionless, time less and space less and because of all this features it carries the value of Π^0. By expanding, singularity applies a relation coming about that reforms singularity from Π^0 to Π. Only when extending Π^0 to Π, the extending creates motion and the motion creates space that then doubles through motion applying which cuts the space in motion in half by matching the space as a duplicate. Motion creates another dimension or another level reforming singularity from Π^0 to Π or from Π to Π^2 or from Π^2 to Π^3

As said before we now know Π came about since Π is achieving form and not space. Only **r** can establish space as size will accumulate and as it had with everything else singularity had **r** covered by the measured value of one as in being **$r^0 = 1$**. By reducing r indefinitely to the tune of half each time, r would become infinitely small, in fact so space less it will be beyond human calculating means, however as mentioned in the case of the smallest dot holding one spot, r would become insignificant and beyond human comprehension even, but never will r reach zero and still Π would remain intact and dictating form. By reducing the circle radius **r** by half continuously this action will lead to an infinite small circle and an infinite number holding r would place **r** to the power of an exponential zero as a factor $r^0 = 1$. Then as a factor **r** would not contest any change when change is introduced into any future equation involving $\Pi^0\Pi$ but Π will remain because the circle as a form remains Π even being infinitely small. To amplify by dimension a value has to be set to find a value in r but if r remained covered by singularity all alterations that could possibly come about was in the form, which was Π. This made me realise that the cosmos has to contract by the same measure as it expands and always remain the same while it is so obviously growing in space. This means time is effected and time leaves space as the collected imprint of time and therefore it is time that deposits space and time is the "white hole" everyone is looking for all this time.

I came to realise I was facing a dilemma for which there is no clear answer. This expanding can be a problem one can wrestle with for one lifetime and never reach any conclusion. How can something grow without getting more of that what was before? Then it hit me like a ton of bricks. The answer is in heat but not heat, as we know heat. It is heat in getting relevancies between outer limits where it is heat that parts the smallest there can ever be from the largest there can ever be. Only heat could break the monotony of singularity. Thinking of heat in the way we now think of heat being in the form we now know heat as heat is now this statement does not make much sense. Since the Big Bang heat is material transforming from one state to another state. The change that took place involved singularity but singularity was 1^0 and being ^{0} it remained 1 and still could not grow. There was growth that came about and that is beyond argument. Heat rose from singularity, but if heat rose from singularity, it must be understood that singularity as a factor changed from 1^0 to 1^1, which means a relevancy came in place that no one could detect. It is true that 1^1 is still one, but one could then escape from singularity by producing factors other than 1 for instance forming 4^0. Heat came about but only as a relevancy to utter cold that formed singularity. If there is heat there is cold or if there is no heat there can be no cold. If there was a state of heat during the Big Bang of 10^{34} then what was cold?

Heat being 10^{34} without a bottom limit is meaningless. If one states a temperature it must hold relevance or it is without meaning. Space came into forming a relevancy that brought form. But since it is a relevancy and not a generation by accumulation, the form produced was Π. The spot formed a dot by heat and cold establishing relevancies and from that singularity was broken to allow all other points forming serving relevancy to come about. The cosmos did not start because of the lack of space. The cosmos started with heat and cold coming into a relevancy and in the cosmos there is no hot as much as there is no cold. The cosmos eventually broke into form we now think of as three dimensional as it freed from the confinement of singularity by establishing a singularity in a relation of heat and cold moving apart as it moves closer o once again form a mutual value in the end. The heat that at first came about was beyond measure because the cold that held the heat was also beyond measure. The immeasurable heat was on the outside of the dot that formed and the cold was on the inside of the spot that formed. The cold contracted because in nature cold contracts.

The heat expanded into a dimension of form and heat by expansion is in nature about motion. Motion is duplicating that which is and heat is what is duplicating by motion. But only heat by expansion was possible because in affect singularity cannot move. The motion became contraction, as the motion was the result of heat expanding which was forming four points in the rim of the dot. The expanding of the points created motion in relevance of a centre that formed because of the motion, which established an immovable centre as the Coanda effect, placed more dots in relation to more dots that formed. Then by filling the entire circle the spot grew outrageously large and spread the heat over a vast area. While filling the dot with heat the expanding and spreading of heat over a large area brought about cooling and once anything is part of the Universe, it has to remain in the Universe since it has no other place again to retreat to. Therefore by filling the dot and cooling it had to contract into a new spot and the new spot once more overheated because of lack of movement. The overheating cased expanding and expanding brought about movement and movement brought on space and space brought on cooling and the cycle repeated without ever stopping ever since it started. What proof do I have for that which I now state: the proof is in Π. The value of Π is equal to time being 3 with the adding of progress being 0.1416 times seven is 0.9911 and that is the point singularity grows from into 1. By spinning from the past (7° governing singularity) into the present (7° controlling singularity) and onto the future (7° newly positioned individual singularity) the value of 21 is established and (3×7°), but there is one additional value and that is what brings about the growth of time forming space which is 0.1416. The way that 10 + 19 + 1 forms also proves my point that in space 0.9911 grows from 0.9911 to become 1 and form ((10 + 10 + 0.9991 + 1)÷7 = Π.)

The growth of 7 + 7 = 14 in relation to 10 / 7 is the **Titius Bode law** input.

The newly allocated fifth point is the full circle completed with the new point (0.9991) being added and that is the **Lagrangian points**.

The **Roche limit** is the moving with time of material filling space in relation to space not filled with material forming space (Π^2) relating to the four opposing sides forming time.

The **Coanda effect** hold space controlled by movement forming material put in relation to space not controlled by movement filled with liquid. That is as simple as the explanation is concerning the four cosmic pillars. With all these pillars in place serving in the capacity of relevancies applying we are faced with the most important question of all and that is where the centre of the Universe would be located.

Arriving at the question about locating the space and time forming the centre that is forming the centre of the Universe, one has to realise the centre of the Universe is in every point establishing singularity that by the spin thereof is forming matter weather the matter surrounding such spin is big or small, in singularity size carries no significance. Big or small represents space and singularity is absolutely void of space. It is the impartiality of singularity that is claiming the value and not the differentiation of matter. One must realise that is the cosmos forming part of cosmology there are no big / small or hot /cold or near / far since every aspect holds singularity in relevance and without space there is only relevancies between matter and while matter could claim space and space is heat in a turnabout manner, from the vantage point that singularity holds, in singularity space is absent.

Every aspect in the cosmos is locked-in Universes that we named atoms, sealed off from other Universes by time differentiating that we call electrons and time from that point where electron spin time could be inclusive or exclusive depending on singularity holding relevancies relating to one another. The relevancies rely on inter dependence and inter linking that we find in how singularity applies, but as far as singularity goes, there is no differences according to human sizes or standards concerning big / small or hot /cold or near / far because singularity is void of space and only space defines such norms.

Accepting the principle of singularity bringing relevancies and not size unlocks the "so called mysteries" of the Universe and brings about clear understanding. It is all about accepting; acknowledging and interpreting the role singularity holding inter relevancies to pints serving singularity that maintains order in the Universe and forms matter. The value of this lies outside the Universe we have because from our view these points are so much equal that they are all the same and that makes the lot forming one dot. That is what happens in the Black Hole where the relevancy concentrates so much stability that all the instability of movement falls outside the realm of the governing singularity.

The concerning factor about time science fail to see is that in time the opposing factors of singularity is that with the movement of time everything that presently applies will oppose that which presently applies by the precise opposite. Following an ice age comes a heat age and we have left the ice age half a cycle or some $\Pi \div 2 \times 10 \times 10^3$ years ago minus the time still allowed coming to the turn about. Then it will take another $\Pi \div 2 \times 10 \times 10^3$ for the next half cycle to complete where the next turn about will take place. When we reach the top of the heat cycle then at that point the poles will alter and we will start going to the next ice age coming up. In considering the spinning motion in the fraction of time in the detailed instant every aspect of rotation will turn in every instant of change in time. The main characteristic about time is that although the points had the same characteristics only seconds before, they oppose the characteristics it had just before and just after the very second in which they are and to which they relate by similar points also in rotation. The fact of the graph proves my point in quarterly opposing dimensions and values. In considering the spinning motion in the fraction of time in the detailed instant every aspect of rotation will turn in every instant of change in time. The duration of the instant can change but the characteristic of time moving will always apply. Although the points had the same characteristics only seconds before, they oppose the characteristics it had just before and just after the very second in which they are and to which they relate by similar points also in rotation. The fact of the graph proves my point in quarterly opposing dimensions and values,

To introduce my theorem, in short I wish to bring a very short overview, before we start with the issues of the cosmic code and definitions I am forced to use to defend the point I make. I take the most common phenomenon on Earth and build a theory based on that principle. The basis of my theory is that everything is heat, be it solid, liquid, or gas. Culture tells us that hydrogen is a gas, or that gold is a solid. Our biggest drawback we can have is precisely such wrong cultural conceptions with no base of proof. Hydrogen is as much less a gas as gold is a solid because gold can boil and hydrogen can freeze. We connect a cultural conception to what we may presume as facts, but that does not make it a fact, it only makes our conclusions an elusion. What we have is heat contained in an atom that spins and the spin gives the atom control to the governing singularity. This heat is contained in a container called an atom and the spin controls the heat freezing the substance to become solid. The in between the solid particles we have heat that is uncontained since that heat does no spin and is therefore

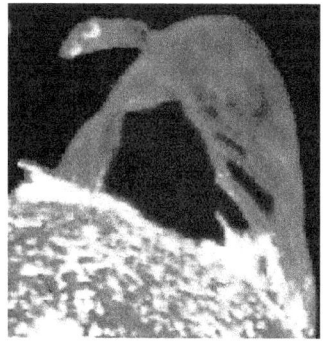

not controlled by motion. That is liquid in a dense form or it is gas in an expanded form but it still is heat. If we work with solids we work wit substance that favours the contained units called atoms and there is much shortage of the heat in between the solids making the structure rigid. If more of the uncontrolled substance comes into the combination the density drops and the mixture becomes less rigid in form and more liquid-like. If the uncontrolled heat becomes paramount and the solid particles are little in the compound we call this state of material a gas. A gas is when the expanding is high and a solid is when the uncontrolled heat is little in the mix with much frozen heat that makes the substance rigid.

All material is heat but some is cocooned and others are not part of the cocoon. Let us take this formula back to the accepting of the Big Bang and find sensibility amongst a lot of confusion that I can see. See the fluid push out of a bowl of liquid, spilling both sides as it falls into liquid. The inside of the Sun is not gas but it is fluid. The Sun condenses space, which is expanded by cooling outer space that carries the Ultimate in heat to form a condensed and liquid substance. In all of nature there is no NATURAL GAS as much as there is no NATURAL SOLID.

Look at the release of heat in the photograph showing an atomic explosion. The heat was contained in small particles called atoms but when the atoms overheat the space the atoms took up expanded by releasing all the heat that is visible in this release of heat. This proves that the atom is concentrated heat and by releasing the atomic structure this lot overheats, expands and the cosmic cocoon bursts open to release space which is heat in a condensed state. In order to release heat one has to overheat the container and get the lot to burst and release the solid that became a liquid, much like the liquid on the rim of the Sun. It is a nuclear explosion plain and simple.

If there is only liquid heat separating electrons, the electrons from different atoms will connect and overheat which will turn the inside solidness of the atom into an expanding, overheating and boiling liquid which is the process used to establish a fluid. This will turn the atoms into space-time. The process of linking atoms is better known to science as the process called fusion. The atoms melt together, they fuse, and they bond, but then they overheat, expand and explode into cosmic liquid. Hiroshima and Nagasaki as well as the Bikini Islands bear testimony of such an event. The cooling of the atoms is rendered passive and there is no more cosmic gas flow of heart to maintaining the atoms governing singularity and this upsets the relevancy that structurally forms the atom and keeps the atom dismissing space-time at a level below the cosmic relevancy of 112, and the whole atom turns liquid. The liquid forms energy as the liquid heat arrives at a higher spin rate than the frozen state will demand.

.Hydrogen 1	melts at -259^0 C,	boils at -252^0 C,
Helium 2	melts at -269^0 C	boils at $-268,9^0$ C
LITHIUM 3	melts 180^0 C	boils at 1300^0
BERYLLIUM 4	melts at 1287^0C	boils at 2770^0C
BORON 5	melts at 2030^0 C	boils 2550^0 C
Carbon 6	melts at 804^0 C	boils at 3470^0 C
Nitrogen 7	melts at -210^0C	boils at -195.8^0 C
Oxygen 8	melts at -218.8^0C	boils at -183^0 C
Fluorine 9	melts at -219.6^0 C	boils at -188.2^0 C
Neon10	melts at -248.59^0 C	boils at -246^0 C
Sodium 11	melts at 97.85^0 C	boils at892^0 C
Magnesium12	melts at 650^0 C	boils at 1107^0
Aluminum13	melts at 660^0 C	boils at 2450^0

No element is either a gas or is a fluid or is a solid. There are two groups always forming some mixture. We arrange the elements in such a manner, but that is only applying to the situation the Earth grants the elements.

When an element freezes it is solid notwithstanding…
When an element melts it becomes a liquid
When an element boils it is a gas again notwithstanding.

The Sun is the coldest place in the solar system and that is why it is freezing all the heat expanded to the limit in outer space to a liquid flowing around the Sun. You may think that at this point I reached a point of becoming ridiculous. Think about it …it is the spin of the Sun that freezes space to a liquid, which is visible to all, that wants to look without Newtonian brainwashing and by freezing the space even more, the space becomes a solid frozen inside an atom. By spinning the atom beyond the speed of light this movement is concentrating matter to a solid. By increasing the spin velocity to exceed the speed of light the matter holds less space in relation to the heat surrounding the matter and since matter is pure concentrated heat, the heat actually increases because the space around the matter is less dense heat than the density of heat the matter holds. Therefore, matter is the ultimate concentration of heat, because it is heat-frozen rock hard. By exceeding light it can become solid material, frozen to a cocoon we think of as atoms.

The colder anything is, the less space it requires and matter requires much less space than does outer space require space. If mater is heat frozen rock hard, then although having a higher concentration of heat, it also have a point within, that is much colder, and the flow of heat towards that point that holds the ultimate point that is cold, concentrates the heat as it condenses the space. To cool any overheating object, you increase the flow of air that is in contact with the overheating object. The air is space, and that increase in space brings about a reduction of heat by increasing the size the object has that is in contact with the heated space. That means the space -air ratio changes and the product resulting from increasing the airflow is heat reduction. However one has to look past the obvious to the factors not obviously seen. It also means by increasing the flow of heat, the heat factor reduces which means the object's size increases spreading the heat over a larger area and that will lead to a decrease in heat availability, with a lesser heat factor because the object increased in size.

The lesser heat factor will provide a faster relevancy to the flow of heat. That means the faster heat flow, the colder the object holding the heat becomes, because the more space offers the object a size increase and a size increase spreads the heat over a wider area there is to allow the flow of heat from a concentrated solid to an expanded air. Looking at a rotating object the relevancy of space reduces by the factor of four to the halving of the radius. If one draw a presumption that the same volume of heat flows through less space, the flow of heat must become faster by the square of the decrease of radius, and that reduction leads right down to the point of singularity. As that point ending space still connects to singularity, that point being without space becomes the point holding singularity as the reduction of space froze space out of existence at that point connecting to singularity. Touching singularity we have the flow of heat totally disappearing and going away from such a point serving singularity. The movement at the point serving singularity becomes "over-spaced", because the point of singularity is without space and in that it is without heat because it holds absolute movement. In terms of singularity, all things "over-spaced" are also overheating because it is spinning lesser and lesser as space increases and the ratio moves away from singularity. Therefore, as everything away from such a point is hotter, it will also hold more space to heat because the more space means the overheating factor is higher. That is the cosmos, the place hotter than singularity and therefore more space to hold heat. This applies in the precise way inside an atom. The atom holds more heat to less space and therefore it has to be colder on the inside than what it is on the outside. The electron must be the hottest and the proton covering singularity must be the coldest.

This is correct, but it is also relevant because what may be frozen to our view is still liquid to nature. Even freezing an object to a temperature of −273° only means that the heat surrounding the atoms is at its limit of expanding as it is forming a gas. It is at its limit, but still it is a gas. If it was more than a gas frozen by the electron to form a liquid that surrounds the atom, the electrons will touch, because the electron is the state of heat where heat becomes so cold it turns to liquid.

Energy is the release of heat to gas, forming space, and the space creates the nuclear winds, accompanied by radiation and light. Radiation is only intense spinning gas and light is a heat droplet. Therefore the heat becomes condensed heat in the form of heat "Vapour" and heat "drops". This is the closest we can come to see what is on the inside of the atom. The rest our minds must tell us, and not our eyes. When heat cools, it becomes more condensed, holding less space, and is therefore it may seem hotter but according to science principles it must be colder. Behind the electron, where the neutron takes liquid form and provides the atom with a fluid substance the movement of heat is faster than at the point that the electron holds, because it spins with an ever reducing radius towards the final solid proton, therefore as the radius reduces to the atom core, the spinning of the heat holds four times less space for every time the radius halves in length. Having an ever-increasing spin rate, the heat moves faster, condensing even further as it progresses to the proton core. That makes one realize that inside the atom the heat contracts, holding less space and by holding less space it must be in a more frozen, or colder state. This is about the same as feeling the core of a radiator of a cooler unit, and thinking if there is that much heat on the outside, how much heat will be on the inside. Well, the similarities may not be that correct, but the principle is very much corresponding. Matter is frozen heat because it is spinning through less space.

I shall make a statement, that will surprise every reader, I think. Space has no value, because space is the product of time in singularity, a part of singularity, it is a by-product of time in singularity that has no cause to be except to hold heat to the relevancy of $\Pi°\Pi$ that positions light and light alone to a specific point where time formed space in terms of cosmic development at such a precise moment. When looking art water boiling inside the boiling water one finds bubbles coming about in the water. No one blew bubbles in the water, it obviously could not enter from the top, because it will eventually rise to the top and it did not enter from the sides or the bottom of the pot, the heat that forms the pot is watertight as much as air tight.

Yet there are newly formed vacant space, that came from nowhere and will disappear into "thin air" as soon as it leaves the pot through the top of the pot where the water ends in space. No one can detect from where the bubbles forming space came but the space is there all the same formed as vapour in the water and between the water vapour there is no space, because there is only heat. If it was not heat, the water will be less hot, and be a liquid. The bubbles formed as heat coming from the plate, expanded into space (not air because air would immediately escape) and eventually the bubbles will rise to the surface. This proves that space has no value, except for the heat that forms the space. This is another point made, which proves that outer space is actually less dense heat and matter is more dense heat.

Stars expand and contract through singularity development claiming heat from space and developing matter in relation to the progress singularity has. This is a securing value in relation to the governing singularity. The expanding of space is represented on all levels but along with the expanding is the diminishing of space as well and such diminishing shows relevance to growth by indicating how the reducing of space forms a density and the increasing of the mass in those truly massive stars.

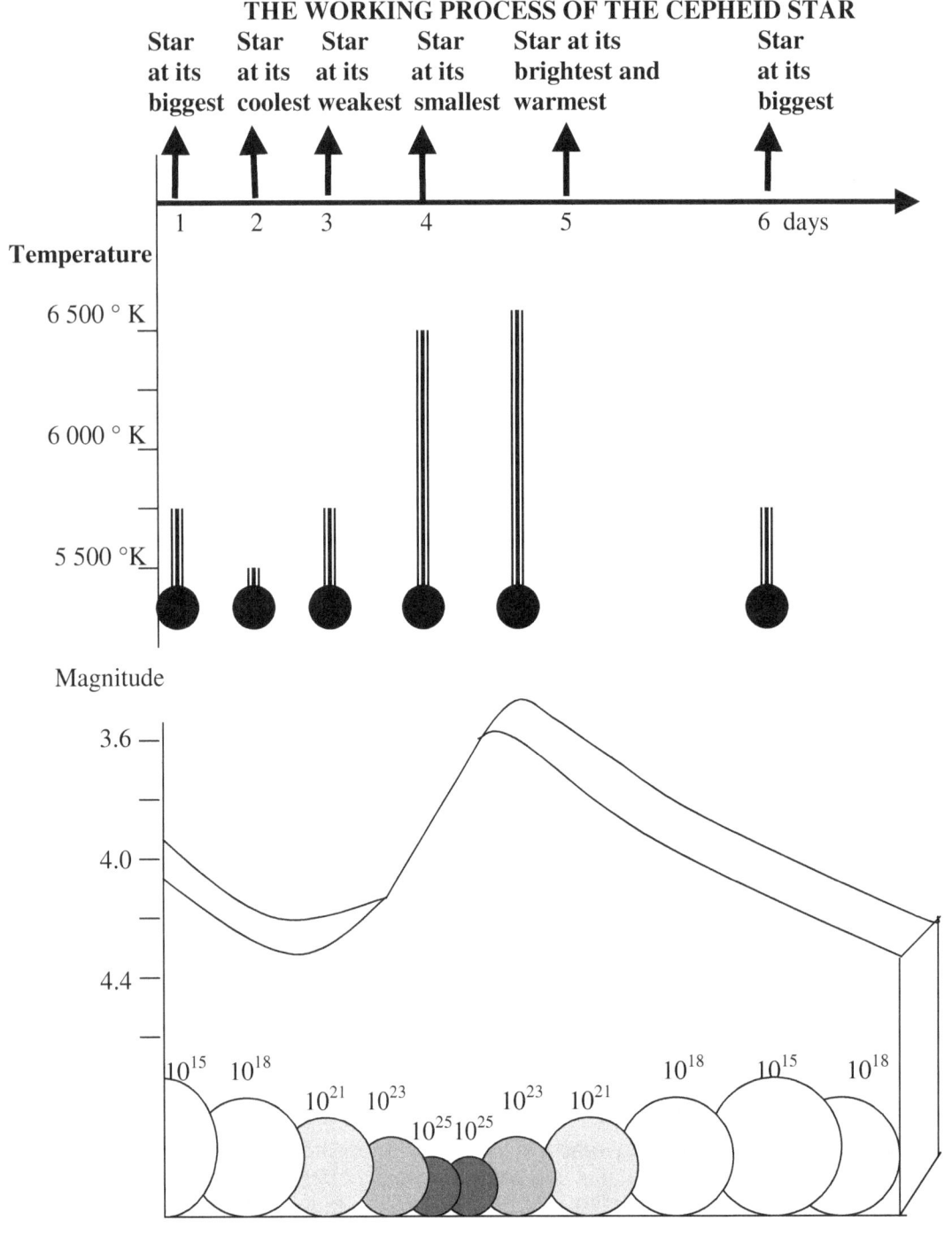

In this example I follow Newtonian tradition by allowing culture to prescribe heat specifications used in the above sketch through which I show how time slows down as the pulsating star expands and contracts. The values I use is so low because the heat levels I Implicate is totally inadequate as to what we will find within such stars and the levels of heat go far beyond than what we have the ability to comprehend and therefore I use some values we might associate with given our small capabilities as humans trying to understand Creation. We will never have the ability to fathom values applying within these stars except in cosmic relevancies according to the comic code, which is the system I try to introduce in this book.

At this point I arrive at the biggest bone of contention I have with Newtonian science. Newtonian science has it that everything depends on mass and in truth it is just not that simple. It is not mass that produces gravity; it is the density of matter and the way that solids interact with liquids that produces gravity. When one cubic meter of water form a vapour it will be a cloud with just as much mass as that which is equal to the cubic meter of liquid water but the vapour will float in mid air. It will hold all the mass in mid air because it holds more space, therefore there is less dense heat than there would be when the water was liquid because the water has less space in a liquid form. The mass of the vapour is the same as the mass of the water, yet gravity applies less to the vapour because of the abundance of space allowing less density. Gravity has much less forceful pulling power on the vapour. It is not the space holding the solid in place that one should calculate, but the density between the heat and the solid within the heat forming a unit within the space in question. All space holds time therefore it is space-time. Time commits space to an allocated place. Space does not apply and can disappear, if it was not for singularity presenting heat within that space and it is the movement placing singularity at a worth that allows space to be. Space does not exist without heat, and gravity concentrate and allows for more substantial or denser heat, therefore space can solidify, and it all depends on the movement of the solid. Gravity concentrates heat by reducing space. That brings us to gravity and time. What is time? If some of the Newtonians wish to bend time, they should at least find out what time is before they can start bending it in all forms and shapes.

I have explained that gravity is the duplicating of material by the expanding of overheating material and by doing that the process then is followed by the dismissing of heat and in that process of expanding the heat the heat is distributed over a larger area, which then leads on to the introduction of cooling. We find this process very well expressed in the cycle of Cepheid stars and especially in the way that Pulsars pulsate. These Pulsars and giant Cepheid pushes back the envelope of time to a point where all the atoms within the star form a governing singularity that holds the entire star in singularity custody a point wear time starts or ends space where at that point time forming space only grants pulsating motion as it comes to the edge of the Universe. The above-mentioned stars retain time to space at a level where singularity implicates the measure of space altogether directly. The removing of space is due to the cooling process applying and that cooling coming about leads to the concentration of heat making heat denser which is the liquid we see around the Sun. The Sun is cold because of the incredible movement the Sun performs and by moving that dynamically the Sun contracts the expanded heat forming outer space which liquefies the cosmic gas to become cosmic liquid which we call plasma or light or electrons or probably a thousand another names to bring about total Newtonian confusion about what Newtonians are unable to understand. Gravity is a process of overheating space that expands that is followed by the cooling down of space to form a liquid. The liquid admits high volumes of heat from the dense liquid form to the highly expanded form we think of as outer space. The Sun through spinning concentrates the heat that it then stores the dense but cold heat in concentrated space around the edge of the Sun and further freezes space to turn the liquids surrounding the Sun into solids formed further down in the very heart of the Sun core. By duplicating the protons bring about cooling that freezes space-time into the single dimension of singularity by removing space-time altogether. Science holds culture much more prominent than what science holds facts prominent.

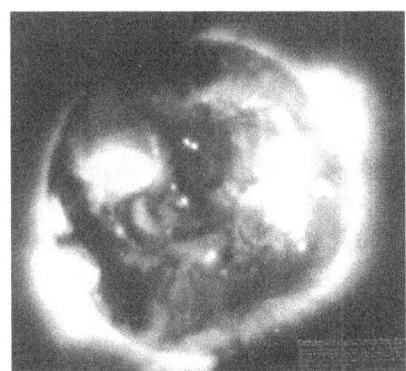

It is said that the Sun is hydrogen and in the same trend it is said the Sun is a gas giant. I have shown how liquid squirts from the surface of the Sun, which then is called plasma and prominences in order not to recognise the true meaning of the substance squirting from the "gas" Sun. This is said that the Sun is gas because culture teaches the Newtonian mind that hydrogen is gas. If the Sun is hydrogen and the Sun is 6500° then the Sun must be hydrogen and gas because it is impossible for the Newtonian mind to collect all the implications of 6500° and hydrogen being a gas and the Sun being hot into one collective giant concept. This means that notwithstanding what Newtonians physically see in relation to what the Sun tells them about liquid squirting from the inside of the Sun, Newtonians still insist on telling the Sun what it has to be and if their Newtonian culture press them to find the Sun to be a gas ball, then that is what the Sun shall be. The truth is that hydrogen is as much a liquid as iron is a gas and neon is a solid. It depends on the element relating to the space/heat in the circumstances surrounding the substance at that very precise instant in time. We have to stop telling the cosmos to show us what we wish to find and start accepting what the cosmos is telling us to find. The culture that I am referring to is all about nothing. Newtonians have this idea that they can tell the cosmos because Newton told the cosmos that the cosmos is contracting by the force that mass provides and the cosmos has to obey Newton and is idea of mass pulling and on that trend Newtonians tell the cosmos what should apply and how the cosmos should react and what the cosmos be made of. The governing singularity places a freezing value on the controlling singularly and in that way the Sun applies all mutual singularity within the framework of what the governing singularity controls by. Proof of this again is the Roche limit, the indisputable presence of the Titius Bode law applying, the Lagrangian points positioning being present and the Sun acting in relation to the liquid versus solid of the Coanda effect bringing about gravity.

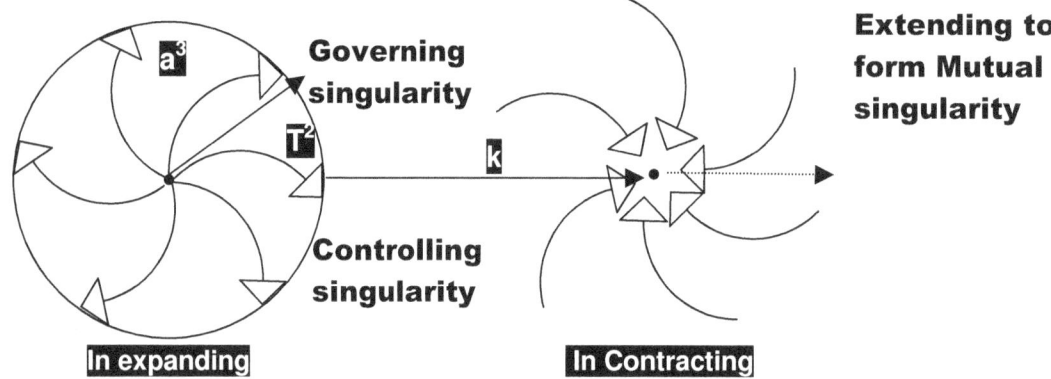

In the first picture on the next page one must be blind not to see that it is raw cosmic liquid that spills into outer space as it squirts from the surface of the Sun. This shows clearly that gravity is the result of contracting heat and by attracting heat it is condensing heat onto the surface of the Sun and into the Sun. When there is contact between the liquid already condensed and the hot liquid coming in from outer space this uneven heat will cause an overflow the liquid cosmic heat as it will expand under certain conditions bringing aspects about and some will flow back into outer space. The entire scenario of gravity is that the star cools outer space, which is in a gaseous sate down to a much colder substance being cosmic liquid, and this is the result of movement. This process of cooling is regulated from the centre of any star that is forming the governing singularity in terms of the controlling singularity.

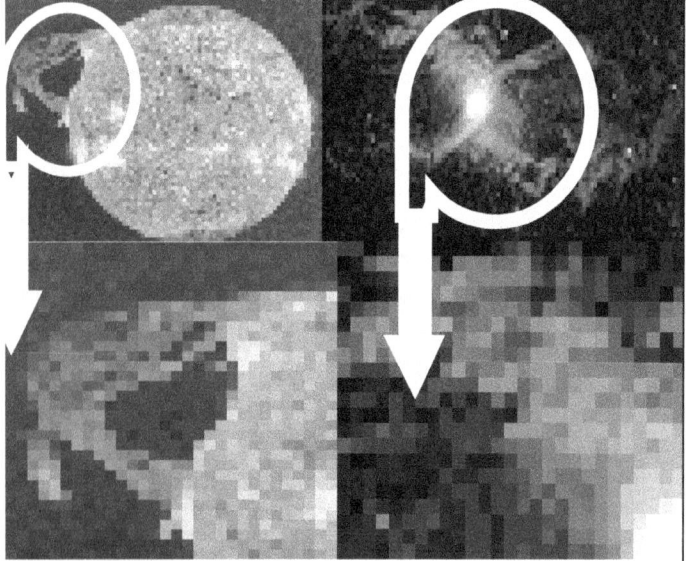

In the next scenario presented in the pictures the overheating core is prominent in relation to the cosmic liquid as the core holding the governing singularity is much more compact than the liquid flowing out to outer space and that brings about that the heat will flow to a less compact but hotter area from a more compact but denser region. In this case the star is overheating and with that it can no longer protect its individual singularity in cooling. The pictures show clearly the difference of a star NOT overheating being "normal" with liquid pouring from it and then become a gas as it evaporates into what we call light. The star was unable to charge sufficient movement to cool the structure in terms of the centre and the star overheated by spilling its content into outer space.

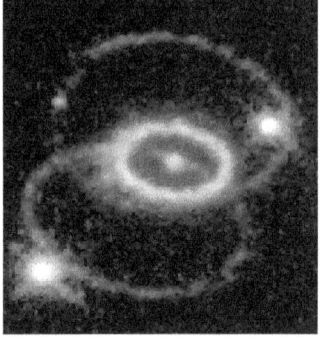

Looking at a star overheating it is obvious how that part which covers singularity breaks into heat when control of heat is insufficient and overheating of the liquid forming the star erupts into uncontained cosmos liquid. By demolishing singularity it means Π demolishes the very point holding singularity and which is the point serving reference to gravity. The question then to be answered whenever asked is why would stars overheat? We can blame pressure, but pressure would not bring about a star disintegrating from the centre, as the star depicted here clearly does. Pressure would be when something is pushed into a container artificially and whatever is pushed into the container is trying its best to get out again and that too is not happening.

Nothing is pushed into the star, the star has no containing wall and nothing tries to escape. A burst from pressure should blow the sides out and there is not even sides constructed that can burst. Stars we call Super Nova has blowouts. That much man has known since before writing began, but since of late this phenomenon becomes more and more seemingly misunderstood. If stars blow as we can clearly see from the picture just above, then the explosion is happening to the star because of eternal heat becoming too much in the cooker. Then we must go on a hunt to find what is behind the heating of the cooker and why is the container under all other normal conditions blessed with the ability to control the heat.

Our conception of what is hot and what is cold in the realm of what science laws put forward as principles bringing total confusion to the concept about the principles of cosmology. What we think of as being hot

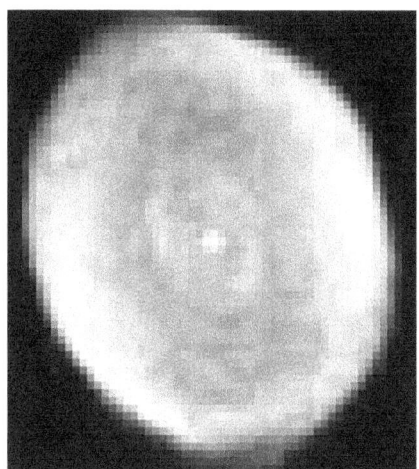

can't be hot because that heat is getting rid of the heat, which means it, is distancing itself from the condensed heat within. Only when it is fully expanded does that area accept the heat and then it could be regarded as being with heat. Every one knows that a gas is one dimension HOTTER than liquid as liquid again is one dimension HOTTER than being solid. If the star is liquid on the inside, and the liquid evaporates when coming into contact with outer space, then outer space is the hottest, notwithstanding whatever boundaries and values we humans attach to the dimension. Our human standards have to change to accommodate the rules laid down by the cosmos and not apply the cosmos to suit our rules of hot and cold, big and small, near and far. In the picture I present there is clear distinction between what is reminisce of the inside of the star and the rest of outer space. In the case of the Super Nova, something prevented the liquid to turn into gas, therefore overheating. One can clearly see the liquid remaining a glued substance in contras with the gas forming

outer space. The liquid froze as a liquid becoming a cosmic lollypop. That which prevents the overheating turned the layers into frozen identities and still not overheating, therefore it became a liquid outside the star. This star was turned into a miniature galactica, sustaining billions of individual singularity, because the governing singularity did not destroy, but the singularity of every point serving as controlling singularity is still in support of one another. Again I have to press the thought that it is singularity determining space-time to form through conditions that bring about the state between what forms matter and what does not form matter. We think that matter can be solid liquid or gas, but it is the condition of the space-time derived from singularity that places the form and conditions valuing the form of the elements. Hydrogen can be as much a solid as gold can be a gas.

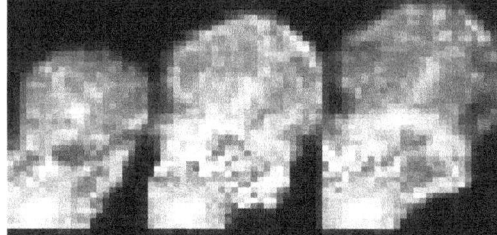

When heat surges and becomes too high, it turns into space. This can only be when what is cooling is not cooling sufficiently enough any longer. What then happens we call an explosion! It is frequently seen, yet never acknowledged by science. When heat reduces, it relinquishes space in the producing of more concentrated heat, this process we see as cooling. When heat becomes hotter it expands into exploding

What ever the terms used there must be a recognising of the inter relation between heat and space where the reducing of the one will lead to the increase of the other. The star does not apply pressure to bring about fusion, it freezes the elements into fusion by applying millions and even billions of degrees Celsius. What is applying to stars inside the galactica centre is applying to particles inside the sun.

Science sees the nuclear reaction but do not recognise and therefore do not admit that the nuclear reaction is three different phases. At the beginning of the process all the heat is solid, placed in a container by nature and the container has a human name called the atom nucleus. In the atomic explosion there are three ingredients that are distinctly apart. The solid has to overheat. When the solid melts down, it becomes a fluid. The fluid we gave the name of light. There is not enough space to explain the detail of the argument, but light is not a gas, it is a fluid. The first step of the nuclear explosion is converting the solid to liquid. In the liquid state the star does not overheat. The overheating becomes part of the second phase. That phase involves the turning of the cosmic-fluid to a cosmic-gas we call space. Space is heat overheated creating space by expanding, as heat is space concentrated creating a fluid or liquid not yet correctly named by concentrating heat into a denser but smaller space.

I do not think that I or any other person is at liberty to try to calculate any on goings with in the star but from what is clear from the outside one may come to some measured idea of the stars position in space –time.

The fact that it can freeze heat to liquid surrounding hydrogen while holding a temperature of 6500^0 C should be an indication it is not what we seem to acknowledge as normal. The governing singularity controlling the Sun places under gravity totally different conditions applying within the mutual singularity of the Sun than we humans could ever understand. Every star or if you wish to accept the validity of planets, then even planets form a Universe independent of the rest by the measure of the governing singularity controlling the mutual singularity through the guidance of the individual singularity within the structure. The sun is freezing hydrogen to a dense liquid at 6500^0 while space is boiling (expanding through overheating according to the Hubble Constant) at -273^0. Science have to review there thoughts on relevancies because what seems to be hot is cold under certain circumstances and what seems to be cold to a point of freezing is boiling hot. There are no standard issue and fit all through out the Universe. Every singularity attaches different criteria to borders controlling the space-time within. What fits humans on Earth does not even suit conditions on the moon, yet science cannot appreciate that the moon applies very different standards to that of every structure and every structure is a cosmos on its own turf, supplying its own turf.

It is not the specifics that are of importance because the specifics changes from considerably to completely when taking into account that hydrogen remains in a frozen state at 6500^0 C within the Sun spinning therefore it is obvious we have to look at other clues to give some indication of what is in process. On Earth in the time we have as a duration we find hydrogen freezing at 269 ^{0}C as where it freezes on the sun at 6500^0 C, which implicate the reduction of space to an enormous increase in time duration.

It is just as important to realise the existence of the two material formed by singularity as it is just as important to realise that heat is another form of material and a separate form of material. The one type spins in confined space of the atom moving faster than the speed of light and the other being slower than the speed of light does not spin and is outside the atomic confined space.

This puts the one at a point serving governing singularity Π^o at a point forming controlling singularity Π and the other at a pint serving singularity as time moving $\Pi\Pi^2$. The two forms of singularity are the same that developed on an equal basis and came as a result of the other. The one was produced to save the other and what the one produced saved the other. The one principle brought the incentive for motion while the other took the incentive by providing the motion. In this Kepler's formula prediction of $a^3 = T^2k$ is validated. The one produced what the other captured and the one retained what the other delivered. The motion initially did not bring the required relief and another form had to be devised. By overheating and increasing space it counteracted overheating and by removing the expanded material and retaining it onto the contracting of the other, did the two forms form a synopsis where by all received benefits in the form of cooling. Only when further requirements developed, did the need arise for more within the cosmos to be made available. Then form had to become part of the cosmos. The first demand on motion asked no further changes because one change brought on satisfaction to all that suited all.

The second was more general and on an ad hoc basis that was established to fit the need of individual places and not groupings in the broader perspective to fit individuals at large because there were two forms to cater for and not the extensive range we now have involving space by three dimensions. At first the establishing of motion set a trend that brought on required results by moving in parallel time lines that formed but afterwards demand for more drove development to find new avenues into which development will gave to move. That made space a requirement and the Big Bang became the space that formed the space vacancy required, in which to move. Space became a demanding issue as the heat levels raging out of control. The heat had to be stored in space by becoming space to retain heat for later consumption. The number in ratio that produced the heat providing particles that offered to release their form in contribution to have those that retained form, do so to save those others retaining form. But those at the time that Creation had on offer became those ones that became the danger of destroying Creation instead of saving Creation. From this a code was born that would put in place formulas that would govern Creation in the manner we now appreciate. At the time of development only singularity was in place and that is still the true part of Creation present. Then from this the development of space became an answer and space spawned by the dimension of Π. Then the cosmos as we now know Creation to be formed as $7/10\ (\Pi^6) \div 6$ came into place and I will explain this shortly because at this point where space was born a lot of other things also prove that space is forming the history of time.

This form came about when only form was present in the cosmos. It was in a time era where form featured in relevancies that would lead to one day becoming the atom. The atom forms a dual purpose of duplicating as well as dismissing and some proton layouts prefer the one better and the other find the opposing qualities better suited. This relevancy came in place when time was not time forming space we can see and space was only holding relevancies that ignored form. Time is forever eternity being interrupted by movement in infinity to bring about eternity ticking as infinity ticks. Before the space we now see came in place, singularity took on stages in forming relevancies between duplicating and dismissing space-time, which incidentally was not yet truly space-time in the sense we think of as space-time. At first a dot moved from the spot leaving the spot but taking with the spot as part of the dot to remain in the dot. The two never separated but the one allowed the other to be.

As the dot confirmed a discrepancy between infinity and eternity by defining infinity as an interruption of eternity cold and hot parted a union.

The dot that formed was not space but a relaying of time to form a new point of singularity where eternity was interrupted by infinity. Time took form from 1^0 to 1^1 or from Π^0 to Π. It brought form into differentiating between interrupted eternities with infinity doing the interrupting

Only after this time of single dimensions came to value which has the true distinct relevance, form in the concept of dimensions came about that positioned a time differentiation outside the realm of time by four. In this realisation we can assume that space had some meaning at this point and the formula used to investigate suggests just that.

The lagging of exciting one point in relation to another point takes time. It takes time to send the message across to get singularity at that point excited. It takes effort to bridge from the dominating singularity to the independent singularity and that effort slows time down.

In all it is not mass contributing to gravity but gravity establishing mass. Mass has no influence on gravity but mass is the creation of gravity.

Nitrogen 7	melts at -210°C	boils at -195.8° C
Oxygen 8	melts at -218.8 °C	boils at -183° C
Fluorine 9	melts at -219.6° C	boils at -188.2° C
Neon 10	melts at -248.59° C	boils at -246° C
Sodium 11	melts at 97.85° C	boils at 892° C
Magnesium 12	melts at 650° C	boils at 1107°
Aluminum 13	melts at 660° C	boils at 2450°
Silicon 14	melts at 1412° C	boils at 2680° C
Phosphorus 15	melts at 44.25° C	boils at 280° C
Sulphur 16	melts 119° C	boils at 444.6C
Chlorine 17	melts at -101	boils at -34.7 C
Argon 18	melts at -189.4° C	boils at -185.8° C
Potassium 19	melts at 63.2° C	boils at 760° C
Calcium 20	melts at 838° C	boils at 1440° C

One will find that whatever group one chooses there are gasses and there are solids and being whatever does not involve mass. If mass was attracting mass by the force thereof then the strongest mass must be attracted to the largest mass in the atom and the least mass must float in the air. All the "light" atoms must be floating as gas and the "heavy" atoms must have no buoyancy at all. F = G (M.m) r^2 hardly can even begin to explain the fact that there is a gas that is more massive than iron but floats in the breeze just as hydrogen which is the least massive element.

It is elements that is volatile or the lack thereof that holds the key to buoyancy that holds the key to floating or falling to the Earth. Excluding Argon, which is 3 x heavier than carbon and is a less dense carbon? I mention carbon because carbon is one of the densest materials there is:

Scandium 21	melts at - 157° C	boils at -152° C
Titanium 22	melts at 1670° C	boils at 3260° C
Vanadium 23	melts at 1902° C	boils at 3400° C
Chromium 24	melts at 1857° C	boils at 2665° C
Manganese 25	melts at 1244° C	boils at 2150° C

Ignoring these facts, Mainstream science will hardly answer the problem we do not understand and such ignoring brings strong doubts about the quality and sincerity of science.

Every element has a specific function in the Universe and plays a role according to the displacement figure and it is that that has the key in the role it plays by contracting or by expanding heat.

Krypton 36	melts at 1539° C	boils at 2730° C
Iron 26	melts at 1536.5° C	boils at 3000° C
Cobalt 27	melts at 1495° C	boils at 2900° C
Nickel 28	melts at 1453° C	boils at 2730° C
Palladium 46	melts at 1552° C	boils at 3980° C
Silver 47	melts at 1412° C	boils at 2680° C
Cadmium 48	melts at 321.03° C	boils at 765° C
Xenon 54	melts at -111.79° C	boils a -108° C

In stars some elements have to be volatile to collect heat and others have to be constraining to dissolve heat. Stars have to have an Iron core and the iron core has to spin within copper holding cobalt as an intermediate. The focus pf elements are to unify what is infinity with what is eternity and that it does by movement.

I am going to show and prove in this book that electricity and gravity is the very same thing and atoms are small pumps pumping liquid to the inside of the stars just as a centrifugal pump would do. Some elements have to reconcile with infinity favouring contraction while others have to lean towards expanding which releases in directionally in relation consoling with eternity. The overall function is to unite heat that is in eternity within singularity within infinity.

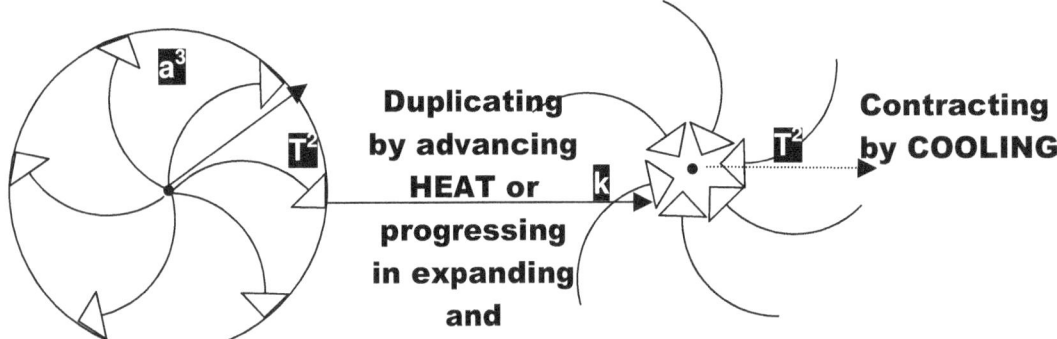

All material has a relevancy in forming duplicating or advancing the spread of heat comes about where the linear travel or the space established plays a domineering role and in others accumulating heat or the contracting or the dissolving of heat by cooling plays a absolute overburdening factor. Thus means in the one $\Pi^2 = \Pi^3 \div \Pi$ is domineering and in the other $\Pi = \Pi^3 \div \Pi^2$. The one is much more volatile than the other and the other is much more stable in its approach to space-time.

The crossing of the divide is material forming space by pushing time forming space into duplicating material progressively or by dismissing heat into the ranks of singularity confined to the control thereof. When time brought in a five points to the four points it took time to be, that fifth point became more than only form, it became space because it was one point outside the Universe of four or of form.

1 Hydrogen 1   melts at -259^0 C   boils at -252^0 C,
Hydrogen depends on interacting as duplicating

2 Helium 2   melts at -269^0 C   boils at $-268,9^0$ C
Helium 2 or Deuterium depends on some duplicating and some dismissing

3   Lithium 3 melts 180^0 C   boils at 1300^0
Lithium 3 depends on more duplicating than dismissing

4 Beryllium 4   melts at 1287^0C boils at 2770^0C
Beryllium 4 depends much more on dismissing than duplicating than interacting

5 Boron 5   melts at 2030^0 C boils 2550^0 C
Boron 5 depends much more on dismissing and very little on duplicating

6 Carbon 6   melts at 804^0C   boils at 3470^0 C
Carbon 6 depends as much on dismissing as it depends on duplicating making carbon most unique.

7 Nitrogen7.   melts at -210^0C   boils at -195.8^0 C
Nitrogen 7 is the epitome of duplicating by violently expanding and in fact Nitrogen returns all stored heat back to movement by applying movement as little other can repeat.

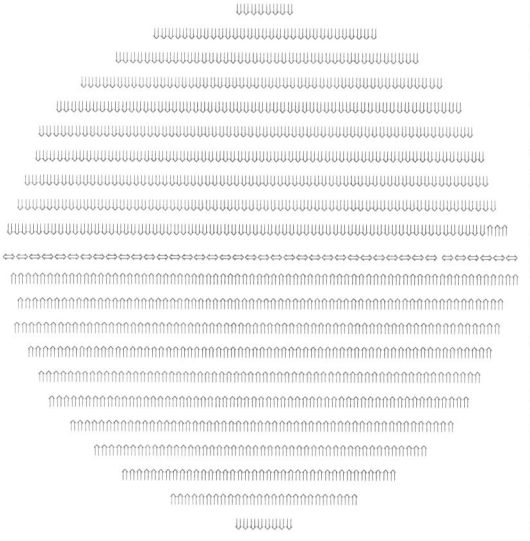

Since the star is the total configuration of the atom's characteristics, the atoms will tell us what we should know about every layer from what is applying in such a layer to what characteristics such a layer would show when it provides the function of what it has to for fill within the star. It is a case of looking at the function of the atom when dealing with the balance applying as the atom is either liquid prone being volatile or not moving in respect to liquid and therefore contracting liquid.

Every atom holds (I am guessing) as many dots or points serving singularity as the Sun has subatomic particles per atoms and that would still be a very conservative guess. These points serving singularity is the true Universe activating gravity. Every dot is a controlling centre selecting a governing centre where every governing centre controls movement of space from a centre. This goes on as long as there are spots forming groups as individual singularity being part of the mutual singularity and therefore being unable to survive independently. Outside this arrangement are liquids forming singularity with "broken" dots. The dots form groups to survive and as a group the survival depends on how fast the group has to spin to remain cool. In another book I reserve one chapter to explain the phenomenon what I called the Lagrangian atom. These dots arrange in a manner that they could either favour the space duplicating aspect or the space dismissing aspect.

The Earth is mainly about duplication of space much more than dismissing of space and so is every structure in the solar system. This means the earth is solid

This can only be the result of the fact that even in the case of the Sun the centre is almost entirely liquid heat and the liquid heat produces sufficient space to dismiss by the centre that holds the heavy metal particles, which is doing all the dismissing. The liquidity provides motion while the solidity removes motion in the centre of the star. The dismissing going on is in the space factor where the space leads to a denser heat within that space because there are insufficient material to accommodate all the heat by the dismissing factor T^2. In that case motion far outweighs dismissing $k>T^2$ but a time comes in every star that the dismissing takes absolute charge. $k<T^2$. That is when the star goes dark. When the gravity exceeds the speed of light the star can admit no more light but in a violent flash. There are many names connected to this process.

The atom restricts dismissing of space by the containing structure to the atoms relevancy being Π^0 in singularity bringing on Π relating to $(\Pi^2+\Pi^2)(\Pi^2\Pi)3 = 1836.1181$. This is the relevancy there is between the electron and the proton. Science cal this mass but it is a dismissing capability. This we will classify as normal applying structure values the atom has in outer space or in structures with very little atmosphere. Please note there is no pressure involved because the motion involved creates conditions naturally instead of unnatural pumping that causes pressure. There obviously cannot be pressure within a star because the star is not surrounded by a containing wall and there are liquids flowing towards the star.

What the Coanda effect proves is that the rotating motion acclaims a centre that exemplifies all phenomena in nature as we use nature to our advantage. All of nature including gravity and even our thought we generate uses the same method of motion forming a circle going around Π in rotation Π^2 and in the centre of the circle a point of no motion holding no space comes about. This is what Kepler taught us when he taught us $a^3 = k\ T^2$. With the Coanda effect forms the basic principle of all natural phenomena we can see from that, that the motion of liquid in the presence of a solid this movement forms a centre that excites as it establishes singularity. From that rotation, space flows to a controlling centre but because of the lack of motion in that centre there is a lack of space in that centre. Even the electron is performing on the grounds laid down by the Coanda effect. In some cases the liquid takes front stage and the volatility is prominent making the element forming as a "gas" and in other cases the solid part is predominant leaving us thinking of the element in terms of a solid. Therefore there is proof of a flow towards such an established centre and there is control from that point of singularity. In every case the singularity controlling space-time setting standards for space dismissing in relation to space duplicating.

The duplicating factor stands in regard to the movement by repositioning the solid while moving the solid in terms of the liquid in expanding while the dismissing factor is the contracting of the liquid towards the centre of the atomic singularity. Again I put it that mass but mass has little influence on the scenario. There is a balance between the duplication in relation to the dismissing of space and the relation extends to the number of atomic elements present which then creates the balance applying within the atom forming the bonding there is relative to liquids and to solids. As the liquid heat subsides towards the centre of the star and the heat density is dissolved by the dismissing-prone elements the motion or moving ability of the star flowing by route of the neutron the liquid as a factor fades away because then the flow of heat within the atom exceeds the speed of light making it solid.

Everything except life as an energy is heat. Life is energy but is distinctly not part of the cosmic form of energy. It is all a balance where the flow of heat started at a point and pushed in a direction. This brings me to my first definition. What is energy? I shall prove that **energy is the interaction between heat creating space and by demolishing space, heat concentrates. All energy in the Universe holds relativity to this, no matter what.**

Everything in the cosmos is moving, either by own individual accord, or under the influence of some other singularity in dominance. In explaining we return to the top. When the top is in a state of motionlessness on own accord it is everything but motionless. The motion it adapts are synchronised with the Earth in harmony with the solar system and according to the greater picture of the cosmos. When an energy source not related to the cosmos called life intervenes and energises the tops motion, the singularity in that top suddenly jumps to life. By adopting a rotation energised to an unnatural state of energising because of life's intervention, the singularity of the top is not in charge but as it applies more and more energy, it will begin to find a means whereby it can escape and apply individual singularity as the top starts to separate from the singularity the Earth holds. The singularity holding the Earth would then allow the singularity of the top to rotate within a specific band where that a specific band of being active before the earth's singularity will start to destroy the singularity in rebellion.

The top on the other hand will try its outmost, when the singularity it holds gets by individual spin is too strong to remain be in domination of the Earth's singularity. This proves there are singularity producing conflict and singularity performing in harmony with one another. The terms bring about conflict and harmony. The motion of the top is an attempt to begin applying an individual singularity space-time defying and standing apart from the earth's gravity. That action we see as the top starts rotating in a manner where the top does not align with the Earth's singularity.

With the adding of spin, the time the top holds becomes unrelated to the time the earth holds and the top will start a campaign too escape from the singularity domination the Earth has on the top. When the time or spin of the top exceeds the limits the Earth places on the top, the top would emerge by trying to escape from constrains placed by the Earth.

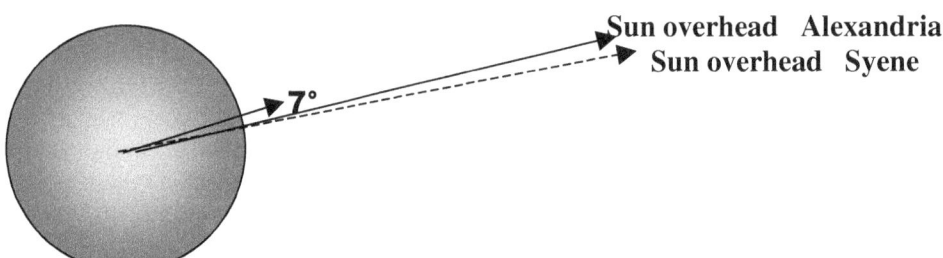

It took Eratosthenes of Syene (276 – 194 BC) a Greek astronomer who in the year 240 BC made a discovery that the earth has a profile of $7°$. Since then no one ever did anything about it. When any singularity wishes to disconnect from the earths singularity, specific pre-calculated laws would have to comply to allow the lesser object to divorce from the larger object. I indicated how the dimensions of 10/7 and 7/10 interact to form (Π^2).

The view I represent at this point is known to science for almost as long as science knows mathematics. Not long after the law of Pythagoras was understood where Pythagoras introduced mathematics Eratosthenes of Syene made as big a discovery as Pythagoras did. But in the one instance the world took notice because the world could see and understand and the other instance the world disregarded the findings because the world did not see what the implications were. The value of Π is known to science before science knew what planets were. Yet, in all the time since then all along during Newtonian discovery not once did one person try to reconcile the value of Π with gravity or for that matter with the circle and see how the concept of gravity harmonises with Π as all the values forming Π interacts

The Titius Bode influence in a manner that on the one side holds the matter-to-matter relation of 7+7/10 whilst on the other side during the same time holds the space-to-matter relation of 10/7 forming equal and opposing values. From this the orbits of cosmic structures are always oval favouring the singularity dynamics of the one structure at one point and switching the favouring to the other structure on the opposing side. Because the structures can never be equal in size (singularity will not permit that where the Roche principle will intervene) the shape is always "off centre" as well.

The Curvature of space-time is as common as space-time

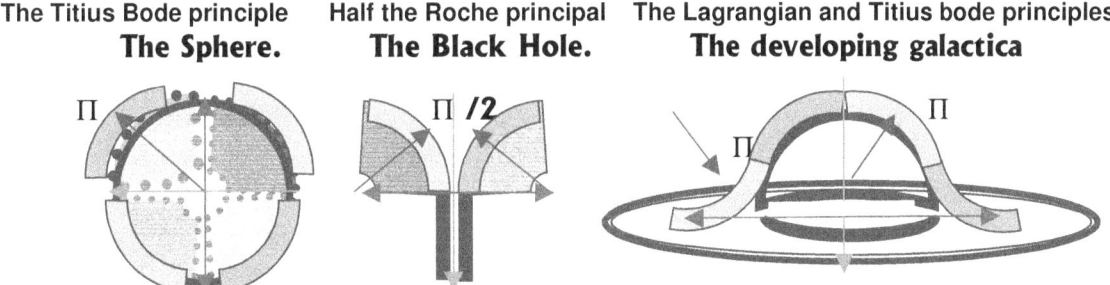

Bubble or normal Π   **Inverted Π**   **Double inverted Π**

The reasons for this applying I explain explicitly in detail.

With singularity being where it is and in a common place as it is the curvature of space-time also becomes a common factor but not yet commonly distinguished because singularity changes the profile too suit space-time.

In the movement we have the contraction of dismissing space where the movement consist of duplicating what is there from the past $Π^2$ to singularity + and again duplicating $Π^2$ by movement to the future. That reduces space-time in heat to singularity by distributing material into the realm of space.

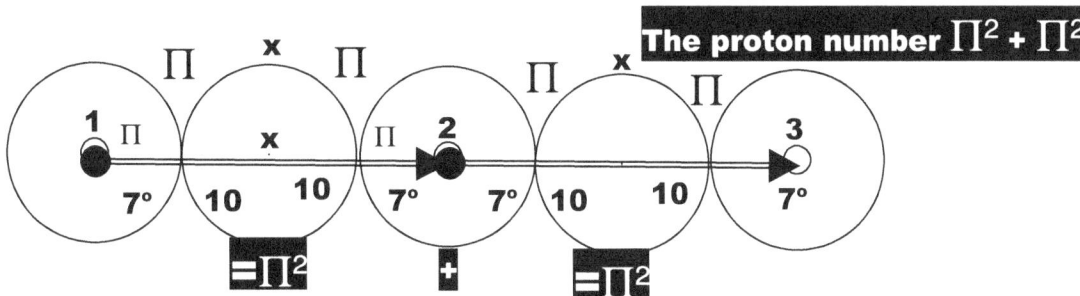

Within the limited boundary of the atom the neutron plays the liquid part in applying relevancy in relation to the Coanda effect. The neutron puts singularity attached t movement by forming a sub-atomic liquid.

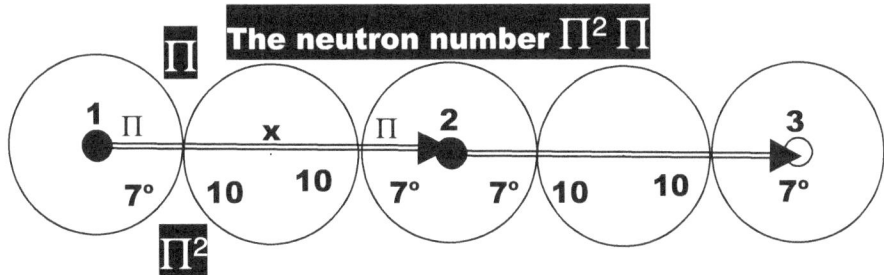

The electron puts time in relation to the Universe where every atom is a Universe locked in time. By the motion of singularity's inability to move we find the past relating to the present flowing to the future.

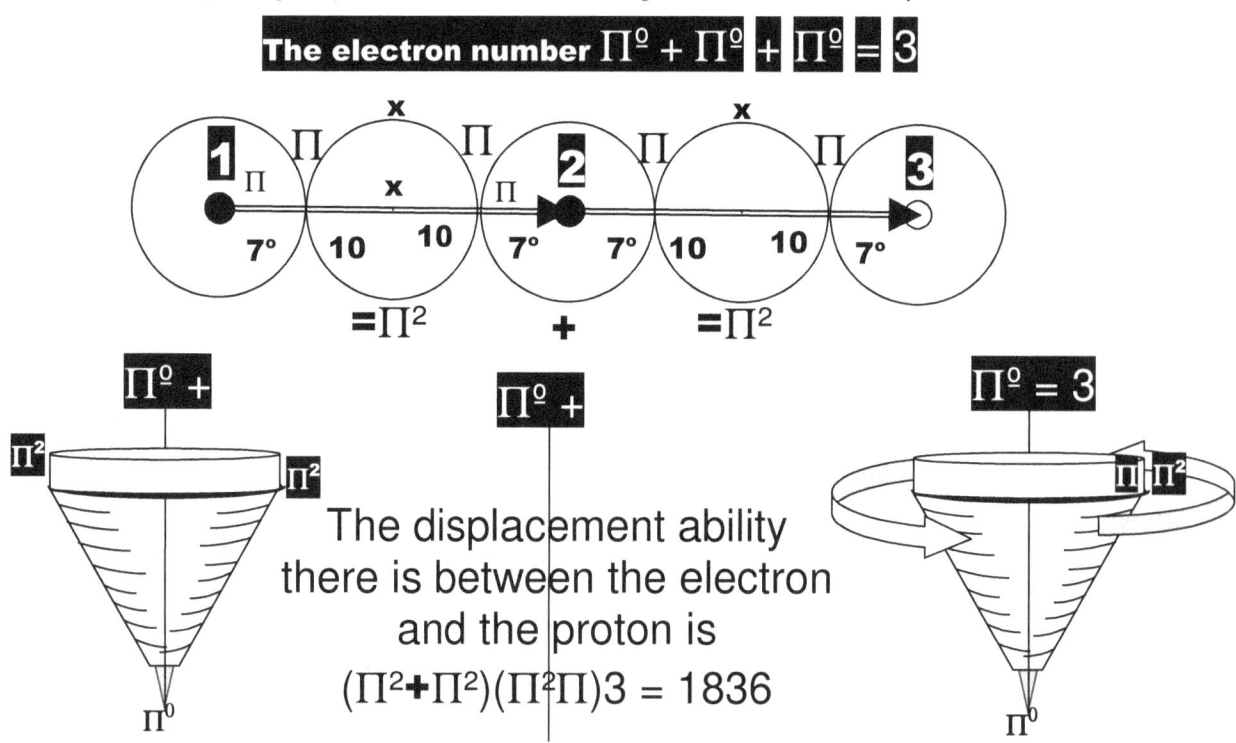

This explains true gravity as applied by singularity. Let us investigate and try and find a way by using logic how a star applies gravity. I hope it dawned on all that mass is the biggest con anyone ever pulled on so many persons on Earth. Therefore it is not the space the dots are surrounded with that holds a worth, but the number of dots forming a chain serving singularity that is important. It is not the size of the number of dots occupying the position or the size of the space the dots occupy that is prominent. All the dots are equal in such a manner that all the dots are one dot in essence. It is the relation in the dismissing of space and the duplicating of space that becomes important. It is the relevant position according to singularity applying that carries the worth. The less space there is the more the favour will be to reduce the space because of the advantage the dots have in securing space-time that will prevent overheating. On the other hand the more space secured will also prevent overheating and therefore those will opt to duplicate space in order to find space to secure and prevent overheating.

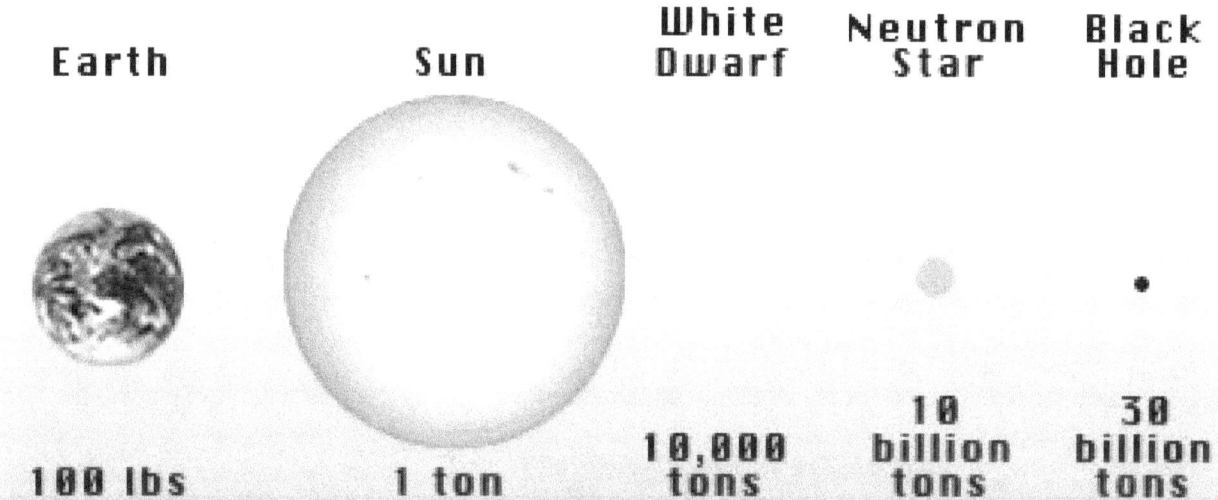

Since the Earth has no singularity demand that is much better developed than the Universe sustains in outer space, we find on Earth an atomic relevancy applying of Π^o to $(\Pi^2+\Pi^2)(\Pi^2\Pi)3$ which to us humans is adequate. But in units holding more demand on the controlling singularity the space-time displacing relating to space duplication the relevancy out there presents much more demands on atomic structures occupying space within the star containing through set boundaries. In the stars serving more developed governing singularity there is much space filled with atoms occupying much less relative space. In the stars more prominent but holding lesser space such as Pulsars and Black Holes the atoms must also hold lesser space but they also hold more protons by number serving countless more atoms in the much reduced space.

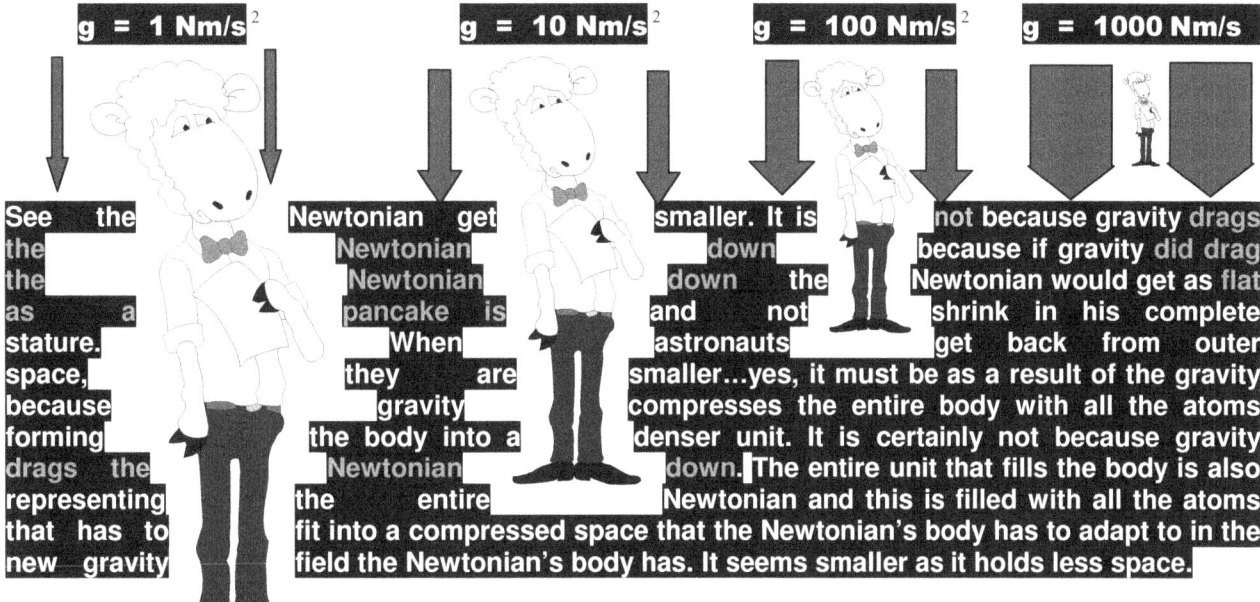

See the Newtonian get smaller. It is not because gravity drags the Newtonian down because if gravity did drag the Newtonian down the Newtonian would get as flat as a pancake is and not shrink in his complete stature. When astronauts get back from outer space, they are smaller...yes, it must be as a result of the gravity because gravity compresses the entire body with all the atoms forming the body into a denser unit. It is certainly not because gravity drags the Newtonian down. The entire unit that fills the body is also representing the entire Newtonian and this is filled with all the atoms that has to fit into a compressed space that the Newtonian's body has to adapt to in the new gravity field the Newtonian's body has. It seems smaller as it holds less space.

It is obvious that when "gravity" increases the space the atom captures decreases in relation to the gravity increase because the gravity reduces the space the body holds and in that the density of the atom increases many times over. If the density increases the heat level has to surge but since gravity is the cooling of material by applying movement we find the heat levels remain on par. This however does not translate in the atoms remaining at equilibrium and therefore with the relevancy of the governing singularity applying to the controlling singularity can't remain even.

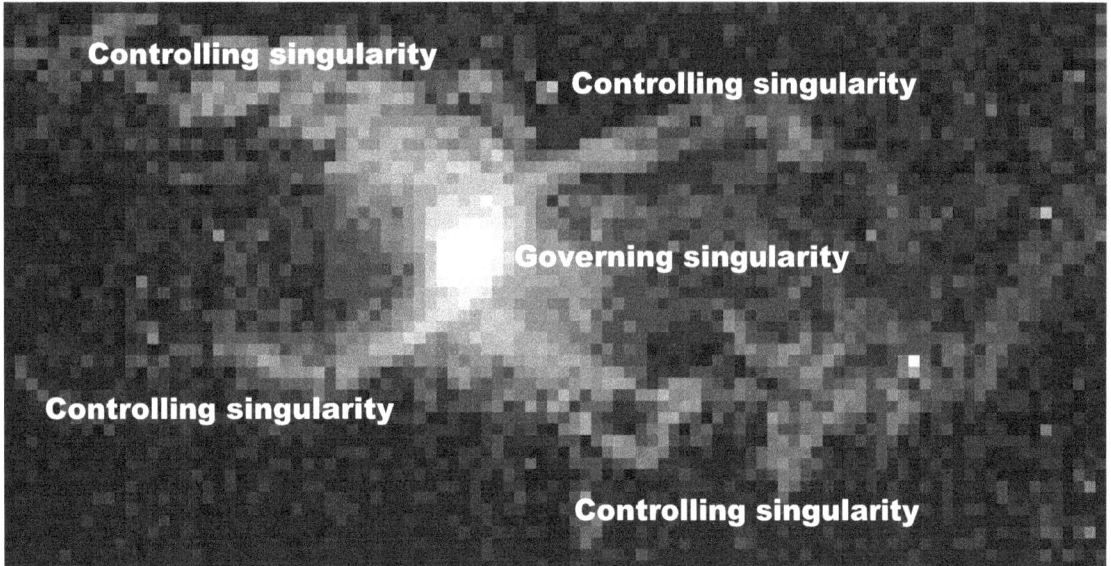

In the photograph one can clearly see how the relevancy between the governing singularity and the controlling singularity changed and by extending the controlling singularity in terms of the governing singularity the resulting expanding brought about a slower movement in spin. This then resulted in less rapid duplication of material and less rapid duplication increased the levels of heat within the star by reducing the degree of density in the heat compactness. The lead to overheating that then resulted to an "explosion" or a better term would be a release of heat. That is all that a Super Nova is! It is a star overheating and expanding by renewing the relevancy applying between the governing singularity and the controlling singularity as far as the occupying space-time that the mutual singularity will determine.

The governing singularity allows a more prominent controlling singularity to influence all the atoms forming the entire body the Earth holds because of a reaction coming from the accumulated value unifying the atomic individual singularity of all the atoms forming a unified governing singularity within the Earth and this works down to every atom forming the bonding value of the mutual singularity within the controlling singularity field. Therefore in something as impressive as the Sun notwithstanding how small the Sun is as a star, the relevancy has to change from Π° to $(\Pi^2+\Pi^2)(\Pi^2\Pi)3$ to be something much different because on Earth as it is in outer space, the electron has a mass of $9,109 \times 10^{-31}$ and the proton has that incredible bigger mass of $1,673 \times 10^{-27}$ and it makes the proton an 1836 times more massive or as I would put it

The Absolute Relevancy of Singularity   Page 172   used to explain THE COSMIC CODE

more compact and denser than the electron. This difference may apply to one of the feeblest fields existing and producing mass but when this ratio is in the Sun it would be a complete different kettle of fish to boil. The entire ratio would change to something totally different and it is in that sense that the Cosmic Code comes into play.

We have to remember that this is where Newtonian "mass" applies. All the atoms form a point serving a governing singularity Π^o and the volume of this movement positions the governing singularity in terms of producing space by producing movement. The more the movement is, the less space will apply. The atoms place a value of 1^1 in relation to the star's singularity of 1^0 and to us this is the same. But on the flip side of the Universe where reality is the key we have a difference between 1^0 and 1^1. To us however, it is the same and seems equal to us living on the side of the Universe that constitutes of light forming a history of time and that is what holds no reality but only space. The effect of time or gravity is to dissolve the history of time, which is written in light by light and that is filling eternity. By gravity applying contraction the movement between infinity standing still and eternity moving around infinity wherever infinity is and by that action dissolves time in the form of light which is space into the singular spot holding infinity and this process was ongoing ever since when the last perfect moment was unperfected moment came about. I call that moment –Alfa.

This is the instant when that which can reduce no more disconnected from that which can increase no more and infinity parted with eternity…but this information as to how this came about I present in a much more complicated book yet to be published. All space is forming the history of time written in heat and gravity will dissolve all of it. Even the heat that fills the atom is subject to movement in the history of time and will dissolve into infinity as time progresses. The flow of such heat from eternity towards infinity causes a centrifugal flow of space contraction that places a relevance of $\Pi\Pi^2$ onto any body that is secured say to the Earth. It is when the "body" touching the Earth surface forms Π and connects with the Earth centre holding the governing singularity Π^o while the Earth centre provides the gravity Π^2 that the chain of singularity runs from the Earth's governing singularity to charge the body that has "mass" with the controlling singularity Π and only by such a link in singularity does "mass" become valid as the movement of gravity Π^2 generates $\Pi^o\Pi$.

The space the particles hold is directly in relation to the particles the containing structure duplicates. The atoms form the star and the governing singularity is responsible for binding the star by the movement of all the atoms in the star. The more space that is relevant to the structure that the star duplicate by motion is then in turn once again relevant to the space the structure dismissed or displaced by proton action in space less units. The reason why the proton enlarges and reduces is because the proton feeds the point serving singularity with heat in maintaining temperature. The more space the particle claims in relation to the space the container holds that relates to the space the container duplicates is relative to the space the containing structure destroys. From that mass derives a factor value. As individual occupying space the atom is an individual container by own merits and as such duplicates space in this regard within the specific confinements of atoms. Atoms does not function on mass but in regard to density of that specific proton layout as it refers to liquid heat in accordance to the Coanda effect.

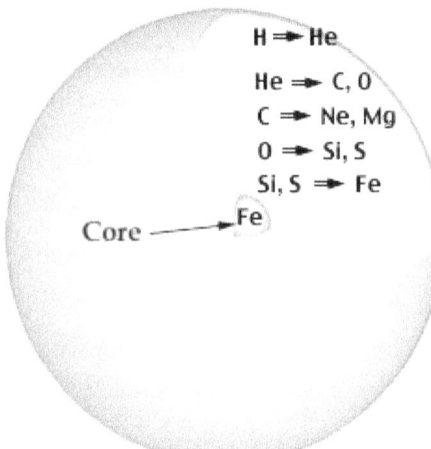

Every layer holding different atoms within the star has a different function to fore fill in regard to the dismissing of space-time. The fact that the stars have iron cores is not coincidental but is to generate gravity by spinning through a specific space in relation to other space. Every layer holding a different atomic number positions the qualities of the atoms of that type in arrangement with the purpose and the function they have to perform. The function of the star is not to "die" and float in never-never land forever and ever because that thinking is as pre-historic and outdated as all the other Newtonian backwards concepts. Stars have a cosmic function and due to that stars can't "die" because stars have never been alive. Stars have the function to re-unite infinity with eternity by eliminating space which is the history of time written in light.

By contracting space into a denser form the space becomes so dense that by motion the space becomes solid and doing that solidifying of liquid to solid is the purpose of the atom while the function of the star is to collect the expanded heat we think of as space.

The Universe started from something I named time in singularity formed by infinity parting their mutual allocation position with eternity. This is where hot and cold became relative in opposing values. Time in singularity is the only constant because there is always a direct line of contact to time factor in singularity Π… And time in governing singularity is the end product of everything that is cold, frozen beyond space while outer space is the limit in expanding space and therefore is the hottest space in all of space. We view the ultimate freezing point to be outer space at a temperature of minus 273°C but thinking in such terms prove to be fatally incorrect and quite the opposite to what science shows to be true. We say that heat will always move, from the hottest area to the coldest area but in terms of hot and cold that is incorrect. Heat will move from the most condensed area into the most expanded area and the most condensed area is the coldest while the most expanded area must be the hottest…just because cold contracts and heat expands. If the outer space were the coldest point there is, the Earth would be frozen solid, because heat will never arrive at the Earth from outer space.

We would have a continuous flow of any heat the Earth may hold to outer space. Such is not the case. We may view that heat flows from the Earth at night to outer space and that is correct. However there is no flow of heat that will bring about temperatures to fall to limits even close to outer space. Nothing on Earth can reach −200° without the interaction of human life. That means the Earth will always be a place colder than the Universe in outer space, just because we have more heat condensed in space on Earth than in the Universe in outer space. If it was not the case, the Earth should be at least as cold as the Universe because all heat will flow from the Earth to the outer space. To condensate heat one has to concentrate heat by removing space and the only way to remove space is to cool the heat to a liquid form.

See how the liquid they call prominence and also call plasma expands from the confinement the Sun offers to the expanded space outer space offers. To do that the heat has to increase in order to expand that dramatically. However again in typical Newtonian fashion it is science that wishes t tell the Universe what the Universe should be. If the Sun freezes heat to a liquid at 6500° we will tell the Sun that the Sun is hot because we feel the heat. We feel the heat because we are the factor that is the cold because we hold material at a specific temperature in order to sustain life. If we wish to see the true picture we must view what we see in terms of what applies to the cosmos and not in terms of what applies to life, where life is totally alien to the cosmos. The duplicating of the enormous Sun (compared to our human standards) moves around as this giant structure pulsates. The pulsating of the Sun is beyond our visibility but it is there nonetheless for otherwise it would not be able to move. This plasma I call cosmic liquid is contracted towards the Sun in the same way as what a centrifugal pump would pump water towards its centre holding the governing singularity. The controlling singularity cools outer space down to 6500° from the totally expanded -273° and in human terms this makes no sense, but please remember Human sense makes no sense in terms of what applies on the Sun and in the Sun.

The governing singularity applies conditions we do not understand and never will understand and the atomic relevancies playing out in the singularity the Sun applies goes far beyond what we would never understand.

Why is it that the Earth would appear hotter, but is colder than the Universe? It is the fact of gravity, but it is the terms connected to gravity that I reject because of the connection science apply to gravity being a force. If it was a force then something not visible or beyond explanation that is in the realms of magical is pulling or pushing and since we do not believe in spirits creating magic forces, and no one can detect anyone applying any force, we have to dismiss the connection to a force or forces. For the convenience of culture and in order not to confuse any reader in the introduction, I remain using the term gravity but in doing so, I dissociate myself totally from the implication that science connects to gravity because gravity is movement and time is movement and therefore gravity is time moving space along. Gravity is what contracts space and by contracting space gravity produces denser space and denser space is more concentrated heat, which in terms of science is contracted space where contracted space is cold.

When space expands it is the result of heat because everything that heats up also expands while everything that cools down also contracts. I realise this clashes with every aspect of cultural science and with anything ever believed in science but one has to stick to science to value the true picture. It is because the Sun is so cold that the Sun can absorb a very limited value of heat and must therefore reject what it does not need or use. This happens to all material. Light bounces off material because the material only uses that much it requires and rejects the rest. That does no make the atoms a provides of heat but the atoms remain a user of light but as the atoms can only use a very limited amount it has to reject the rest. As the heat concentrates the heat loses space to gain condensation thereof and as a result the condensations is the result of space cooling by allowing heat to be more per cubic whatever of space. By the Earth moving the Earth reduces space and the result is the Earth is cooling the space the Earth is within because the Earth duplicates the space more often than outer space does. Thereby it spreads the heat it is in over more space that results in less heat per space per any volumetric unit.

We feel heat flowing from the Sun and therefore we associate the Sun with heat. Let's put it in terms of light. We see an object being a specific colour and then we associate that object with that colour. We say a rose is red but the rose is all the colours there is except the red we see it rejects because the rose rejects red and therefore we see red flowing from the rose. Any colour of any object is not the colour the object portrays because the specific colour in which the object shines is the specific colour the object rejects. It rejects the colour we se and we believe it is because it disassociates its colours with that specific variation of red. We then tend to say it still is that colour just because the object is shining away that colour that it rejects but that is a wild guess on our part because what is colour. Colour comes from a specific vibration that associates with the colour and therefore the object is not that colour but merely vibrates at that

frequency whereby the object rejects the colour it disassociates itself with. Now we take this very same scenario to the Sun where the Sun is shining on us every colour we might imagine as a form of heat.
One may say that our ability to feel heat is the effect of the Sun shining on the Earth, and one may be partly correct, except when considering the wider picture. Measure the space forming outer space that the Sun pours light into equally as it shines on the Earth. If one would divide that area the Earth holds into the area covering the vastness of what we call outer space then the Earth stands apart from the rest of outer space. I truly think no ordinary pocket size calculator will provide a realistic reading when stating the difference in concentration there is between the area the Earth occupies and therefore heats and the rest covering outer space. Yet the Earth is many times hotter than the rest of space, if it was merely the Sun shining on the earth, the Earth was no factor to consider in relation to the space out there. That alone cannot account for the difference there is in temperature.

This brings us to gravity, (not being a force instigated by mass). My definition about gravity is that: **Gravity is the reduction of space to concentrate heat, by the rotating movement of a solid in the boundaries of a fluid** therefore the more gravity there is, the more heat there will be. The Earth or any cosmic structure claims heat from outer space by concentrating heat. You may shout that fusion is the blaming contributor to the heat in the sun, but if there is something like fusion then fusion has to come as a product of freezing or no fusion could ever be possible. Even if there is something such as pressure which I dispute totally in more comprehensive works, then pressure leads to heat and heat leads to expanding because pressure is the restricting of space and restricting space is condensing space which is a trigger for things o expand or to burst or to explode.

Somewhere close to the beginning I started off by saying everything started at a point that was so cold, space froze to singularity while the other part burst into space while still remaining on the time line. Let us test this statement: Only by heating two parts will the two parts part from a joint point. When the object becomes colder, the space it holds reduces until such a point that the two parts will unite onto one point. The less space per atom there is the colder the object must be. Anything can freeze rock solid, as everything can boil and burst into gas. This means all elements in nature, is neither gas, nor liquid nor solid. It is the space that is between the elements that allow the elements the form they hold at that moment. By reducing space, space has to concentrate because to concentrate is to reduce the solution. In that concentrated space would be more heat per unit of space but to get more heat concentrated into a unit per space the heat has to cool to be contained in such a volumetric space. By reducing space, we find more heat. However, one should take this following as a cosmic law: heat is space reduced and space is heat expanded but the two are flip sides of the very same coin.

If you find more space it can only be by increasing heat, and you find more heat it will be the result of space reducing, and in a state of concentration will heat seem more because the heat is colder but in a state of expanding the heat will seem less because the concentration of heat per volume is less. If this is true then heat and space is the same thing. When the Universe was the size of a Neutron, the Universe had unlimited heat distributed in space the size of a neutron because the movement that had to apply back then exceeded the speed of light because according to the GUT theory everything outside solids was as dense as thee electron. Today everything is expanded limitless and the expanding produced a heat we think f as cold. The earth does the same with the spin the Earth produces.

The further one rise above sea level the colder it becomes and the deeper one goes into the Earth (not the sea) the hotter the atmosphere becomes. The earth spinning forms gravity and the gravity is condensing air or cosmic liquid or whatever you wish to call it to a liquid state.

By establishing gravity we find the conduct of singularity interacting and therefore establishing the value of Π. However at the point Π is validated the Universe is still flat. The explaining of the following is a little bit extraordinary but so the understanding of any of the factors forming cosmology. By using singularity the use of singularity not only involves the equal measure of singularity but in principle being singularity means there can't be two being equal. Being singularity means there is one point holding singularity where it is the same singularity because space-parting singularity is multiplying the very same singularity since singularity is 1^0. Therefore space can be multiplied but not as it involves singularity. Multiplying one with one result in achieving one $1 \times 1 = 1$ and therefore one is not just equal to one but is the very same one $1 = 1$ notwithstanding whatever space forms between the points representing the same singularity because the space forming is also singularity. By multiplying one with one result in achieving one $1 \times 1 = 1$ becomes the repeat of one ($1 = 1$) and not the duplication of one (1×1).

Looking at the overall picture we may find that 7 spins around 1 but since 7 spins around the same one it is spinning the same 7 that then creates a dual in 10 where ten serves as a value that forms space.

The star is in place to contract what expands through time development and the atom is in place to unite what separated as time parted infinity from eternity to establish space. The star gathers space-time from space by pumping space-time to the atoms within the controlling singularity of the star and through that movement gravity contracts the liquid into the atom and the atom is at task to re-unite infinity with eternity as it once was.

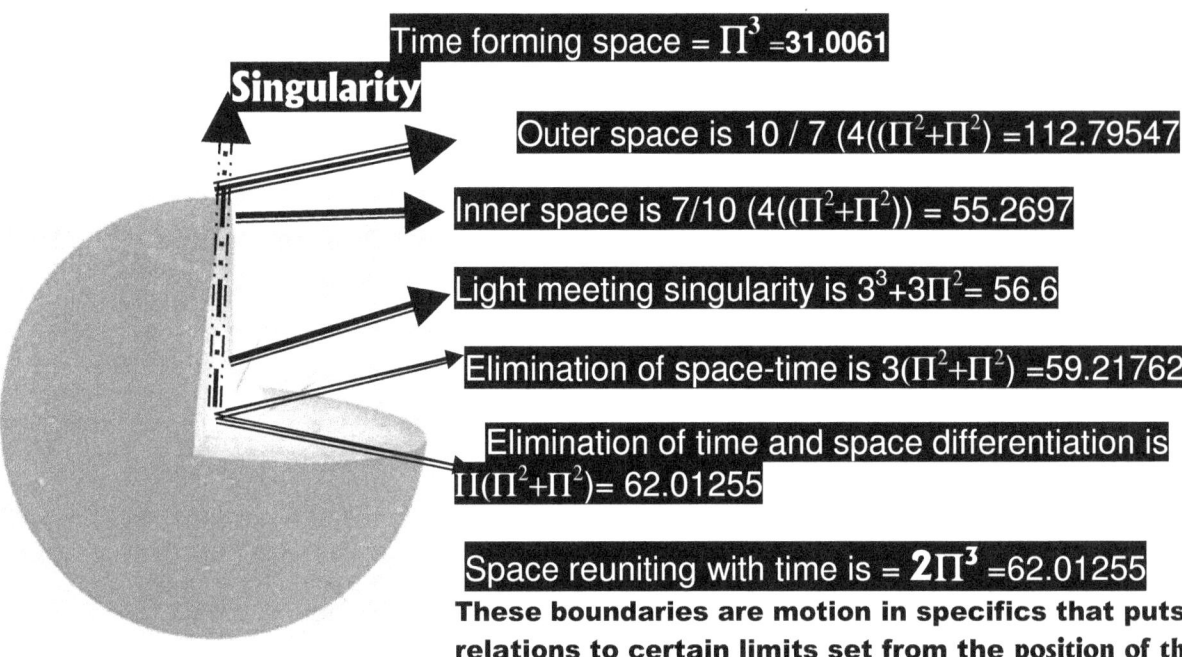

These boundaries are motion in specifics that puts relations to certain limits set from the position of the point serving the governing singularity outwards. It proves that there are dimensional implications all around and that the dimensions are valid. The same implications are validating other principles in the cosmos such as in the case of the Titius Bode phenomenon by implicating the Coanda effect and others.

Every star is as much a small red dwarf as the star is a Cepheid giant as much as the star is a neutron star and a Pulsar that would eventually become a Black Giant. Even the Earth and the moon are stars under construction and in development. The star has every layer that has to develop as the star progresses through many phases in the range of gravity applying in that period in time. Jupiter has a lot of Moons that will become a lot of planets when Jupiter is a full grown star such as the Sun is at present and by that time the Sun is no longer a part of the Milky Way but became a part of the Blackness we call outer space, grown to something only visible through long telescopic lenses from on of the Planets surrounding the Star we use t call Jupiter. In the same manner as Jupiter develops by time forming space, so would every satellite orbiting Jupiter develop and grow as s[ace grows until every satellite becomes a planet that would in the future to come also become a star. Every time the instant is renewed, a new relevancy comes into position placing Π in terms of an allocated position according to Π^0 by the measure of gravity Π^2

Every layer in the star represents one more era of development as the relevancy changes that apply to the star. As the Universe expands, so does the star shrink and the star never shrinks as much as the Universe never expands. It is a changing of relevancies. In the same way as the Universe becomes less dense, so does the density if the star increases. In the same way as s the gravity of the Universe subsides, so does the gravity of the star become overbearing domineering…it is conditions contributing to changes applying that refocuses relevancies. It is all in development as time produces space by relinquishing space. Every layer in the star represents another phase that the star has to go through in development of the star and every star is a time cell under construction. The star pumps by gravity's contraction gas to a liquid state from outer space to within the star where the atoms take the liquid and by contraction develops the liquid into a solid that then transfers as heat to the governing singularity in order to maintain the temperature of the star's governing singularity that will charge the movement in accordance with the mutual singularity depending on the independent singularity holding the controlling singularity in position of gravity.

The cosmos is in a developing stage, which is in a cycle where galactica has then role as to affirm material in crystals called atoms and groups called stars. The galactica is there to preserve singularity in the role that singularity plays in relevancy. While time in outer space expands as it becomes less dense we have material also expanding while becoming more dense. This puts movement to singularity where that which holds infinity and can never grow, then grows extensively while absorbing from that which can never end holding eternity to someday end in closure. I wish to put what I now am about to explain in more commonly used terms although according to my way of thinking it is less accurate and that means I go not fully agree with the way I explain it in terms of terminology used. One example is that science think of outer space as space and I think of outer space as time and I have not enough space in this introduction booklet to explain why it is that I refer to outer space as time but I do so in one of my yet unpublished letters.

BACK THEN when the Universe was new

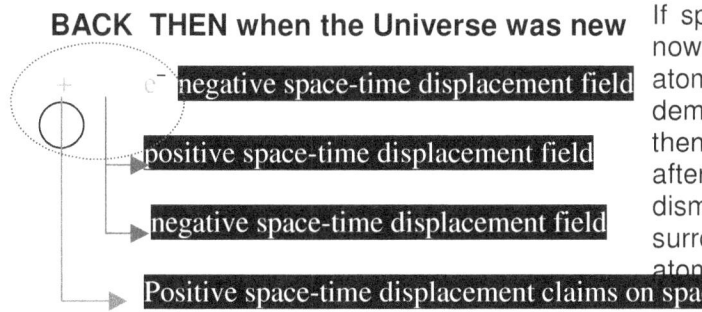

If space outside the atom grew to where we now see the Universe the space within the atom also grew substantially and if a star demolishes the space it has, such a star must then also reduce the atomic space because after all that is what the star is all about, dismissing of all space including the space surrounding as well as the space inside the atom.

PRESENTLY we refer to the sizes we find space has in the Sun as quantum meaning they are inexplicably big

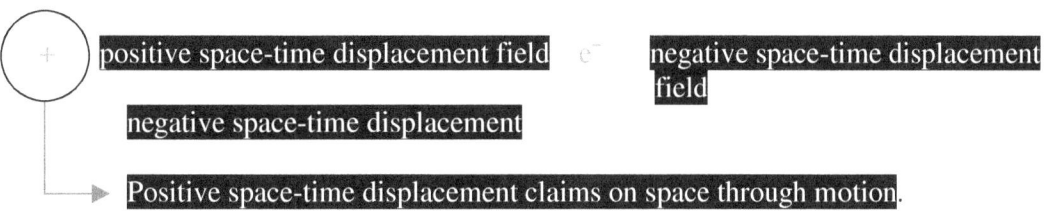

IN FUTURE TO COME things are going to get a lot bigger than now in the present.

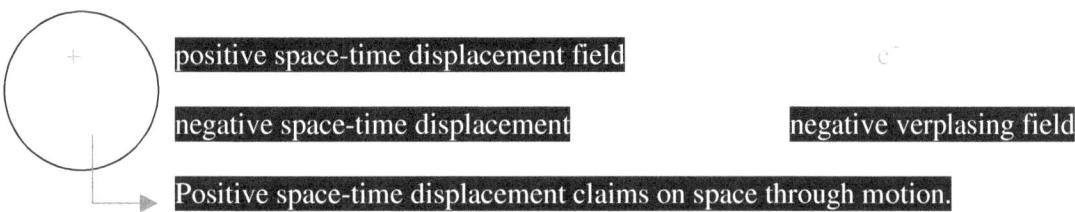

The manner, in which the schematic layout presents itself as follows.
ATOM NUCLEUS   ELECTRON

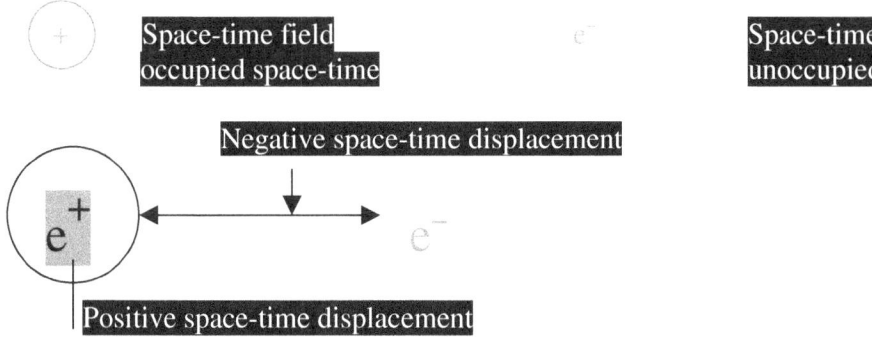

As the atom expands, it pushes all space into expanding because it takes the heat from space where the heat in the space retains the growth of the space by allowing the distribution of space to progress in favour of space and against the density of heat losing its applying relevance. The density of heat reduces as the space increases. That is on the outside of the atom. On the inside of the atom the complete opposite applies. By relieving the space of heat on the outside (outer space) $k^{-1} = T^2/a^3$ the density of space grows (atomic inner space) inside the atom (or star) $T^2 = a^3/k$ by comparable measure to the density heat loses in outer space, therefore the Universe is growing in space by the measure the Universe is losing space. This is the effect that the Coanda effect will leave as imprint on the cosmos. The Coanda effect is the transfer of singularity $\Pi^0\Pi$ in relation to gravity $\Pi\Pi^2$ that re-affirms space occupied Π^3.

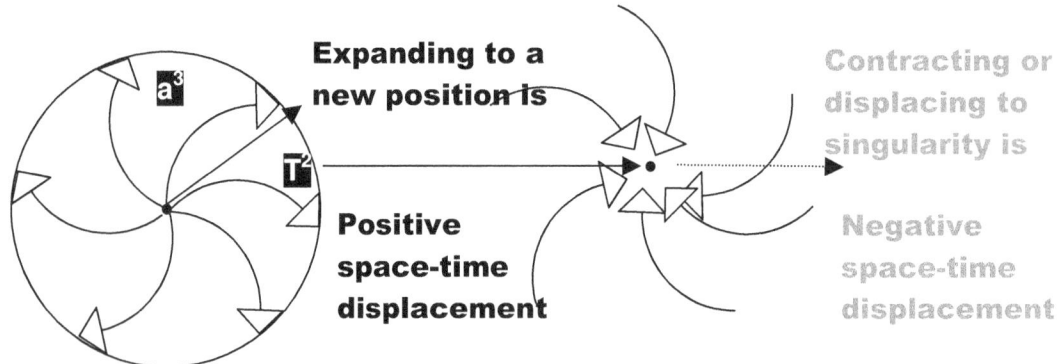

The true expanding is within the atoms where singularity confirms the controlling singularity in terms of the governing singularity as every point serving a individual singularity in terms of the mutual singularity applying at such a moment. The terms of relevancy changes as the Universe are constructed through development of singularity. As the atom grows in space, which is time, the controlling singularity will re-affirm a new state of gravity in terms of the governing singularity. We see the space expands and we see the relevancies part by forming space.

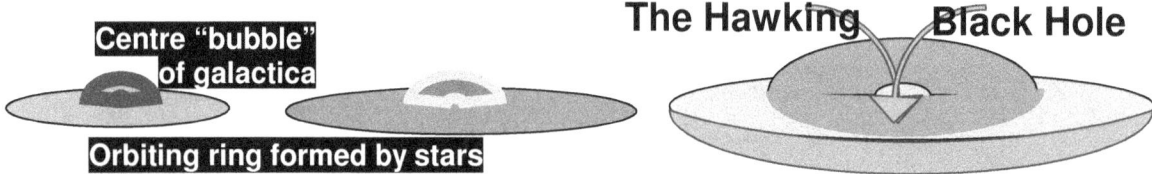

This brings about a reaction in the unoccupied space-time of the relation between the unoccupied to occupy to densified space-time in an ever-altering ratio.

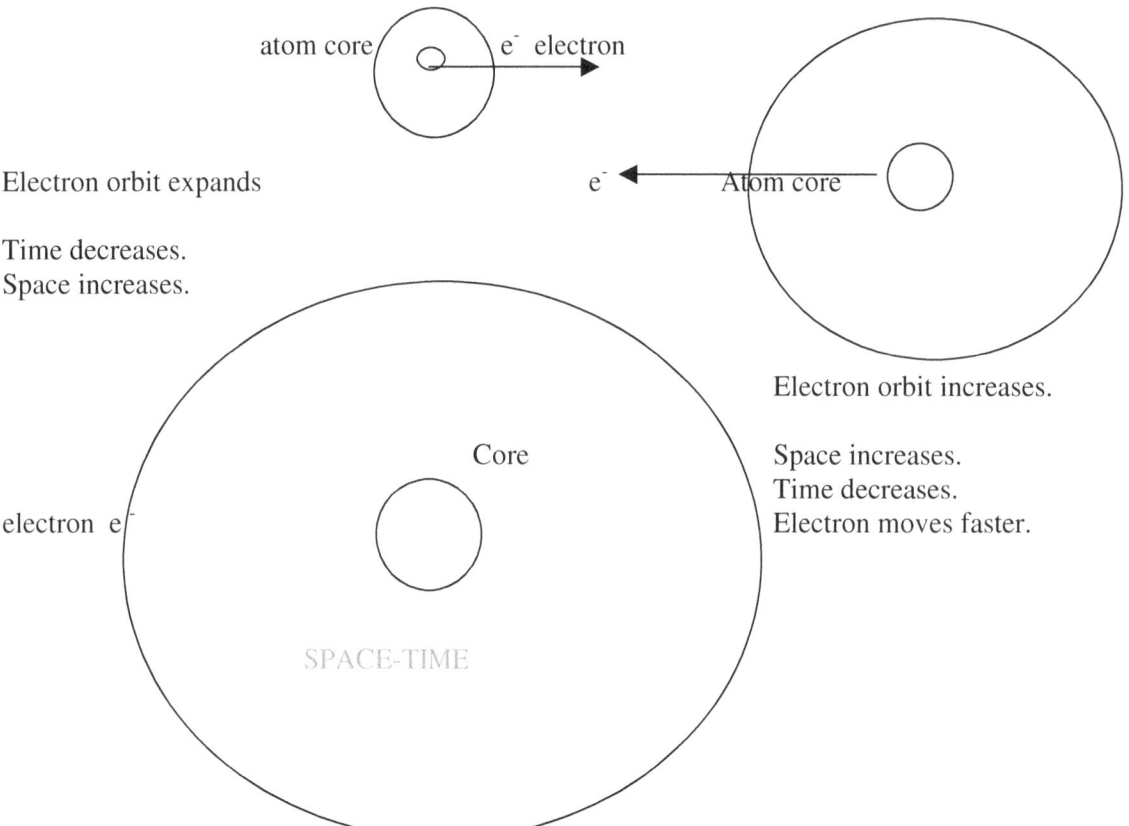

From singularity come the motion of the space we call space-time. Singularity is dimensionless, time less and space less and because of all this features it carries the value of Π^0. By expanding does singularity apply a relation coming about that reforms singularity from Π^0 to Π. Only when extending Π^0 to Π the extending creates motion and the motion creates space that then increases through motion applying which puts greater emphasis on the space in motion by matching the space as a duplicate that results in relocating positions. Motion creates another dimension or another level reforming singularity from Π^0 to Π confirming Π to Π^2 or from Π^2 to Π^3. The motion comes as a result of different motion claiming space within space in relation to individual positions they hold relating to singularity. If everything started off small it must include everything and not part of everything.

We have to regard the changing in its full capacity. How big was the atom as a unit when a star was the size of an atom? The relevancies apply all the way through and not just when science needs them to explain certain aspects gone array. Gravity to us in the way we experience gravity being on Earth and part of Earth yet also apart from Earth we find the connection we have called gravity to be $\Pi^3 \Rightarrow \Pi^2 \Pi$. Explained it would read that it would be the space we are in Π^3 that we claim within the Earth is confined to the space the Earth claims Π^2 extending from outer space Π to the centre Π^0. But the rules applying to the roles changed a little since Π^2 applies to the motion singularity uses to retrieve space and Π indicates lateral shift.

It is not only outer space that grows because **k = a³ / T²** is as much the cosmic value as the value within the atom. That means that **k = a³ / T²** is also in place within the atom and that shows the space within the atom grows as the Universe grows because the atom represents the Universe that is in growth. However at this point I have to state that there is a point where the governing singularity takes control and then the atomic space reduces. In the beginning of a star's life cycle the mutual singularity is dominant and the individual singularity forms the supreme partnership. Then as the governing singularity grows, the control moves from the mutual singularity to the governing singularity and at that point the governing singularity starts to form gravity that exceeds the speed f light. This gravity that exceeds the speed of light would at first be in the very centre of the star, but this gravity that exceeds the speed of light will progressively extend as it starts to demolish the layers one by one from the outside of the star going inwards.

First the hydrogen layer would disappear and then the helium layer and the star will divulge layer after layer until only a centre core consisting of iron, cobalt, nickel and copper remains that all have the ability to form gravity that exceeds the speed of light by dismissing space-time through gravity at a rate being higher than $3\Pi^2\ 3^3$ which is the relevancy of the speed of light and at such a point in the star's cosmic development the growth of atomic occupied space declines. As soon as the governing singularity's gravity extends a gravity ability to advance beyond the speed of light, the dismissing of space-time to form atomic-space increasing in space stops. The borderline is the speed of light, which is C or $3\Pi^2\ 3^3$. As the governing singularity forms gravity beyond C the governing singularity takes control from the mutual singularity that holds a respect on the individual singularity and the gravity halts space forming or space increasing within the atom.

Then the density increases as the space diminishes. By taking control of relevancy charging gravity, the governing singularity charges a gravity in movement that enables a decline in space growth and it retracts the controlling singularity within the star as it then moves the controlling singularity within the atom not towards outer space but towards inner space or in the direction of the governing singularity. The space the star claims reduces as much as within the star moving the controlling relevancy towards the governing singularity until a point is reached where the star's governing singularity is the same as all the atom's governing singularity and the atom as a factor disappear within the star. This we then call a Black Hole. Gravity is the contraction of space while the speed of light is the expanding of space and this process started with the Big Bang occurring.

The contraction of space flows towards the star while light flows away from the star and when the contraction of space-time flowing towards the star exceeds the relevancy in movement the speed of light can and does go negative. It is fantasy to think f the speed of light as a constant because in strong gravity fields the field bends the light meaning it is strong enough to divert the direction and in very strong gravity fields it puts the speed of light in a contracting phases and in that sense the speed of light is negative to what it was before or say which it is in the gravity of the Sun. The truth underlying the matter is that all the atoms form an accumulating movement that renders the governing singularity the possible movement to bring about gravity and the gravity is totally dependent of the individual atomic movement to sustain the balance in heat control by the movement of every layer. The Universe expands by way of increasing space as a factor of growth and as the star's contraction capability we think of as gravity is going beyond the speed of light the atomic structure diminishes in space occupied by material while the "mass" increase exponentially.

By forming gravity beyond the speed of light this brings about space-time reduction from the centre of the proton and this halts all cosmic growth coming from the centre to the atomic proton cluster. As the atom expands in space-time by the accumulation of growth of space the proton can also grow dimensionally bigger through the neutron growing in stature. It expands in captured space–time by pushing the electron walls to allow the atom more space to occupy. It is pushing the electron to achieve a distance every time in the same manner that the body lets nails and hair grows. There are three factors of space-time where space-time is released. Cosmic unity and space and heat parted as singularity released the space heat holds by forming motion which produces time to set boundaries and relevancies applying.

In that we have to place every particle holding singularity as a universe separated from other universes by singularity allowing spin and having space-time in three major dimensions. In the atom every line connecting the electron with the neutron with the proton forms a relevancy of $(\Pi^2 + \Pi^2)(\Pi^2\Pi)3 = 1836$

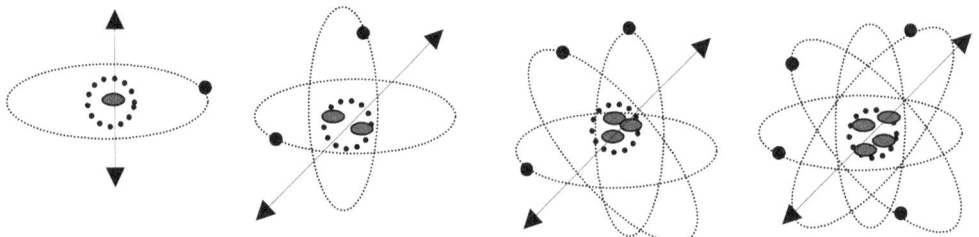

The neutron position ($\Pi^2\Pi$)
The neutron has no mass. The neutron holds no detectible mass. This means the neutron is a fluid that surrounds the proton and being the fluid of equal proportions it forms no density and only serves as a transmitter in heat flow towards the proton. It doubles as a heat transformer to the protons. The Neutron is the cosmos because the neutron forms the movement of space in a solid state that the proton represents. The proton is $k = a^3 / T^2$ the neutron is circular motion which is $T^2 = a^3 / k$ while the atom is $a^3 = T^2k$.

The electron position (3)
There is no electron in a physical capacity. One cannot catch an electron to go and show it to your Ma and ask whether you are permitted to keep it and to use it for decoration of your room. You can do it with the rest of the atom, but not with the electron. You can't use it as a free electron to drive your radio. It serves as a doorway to reduce the flow of heat from atmospheric value to the speed of light and we call this process electricity. The electron is the whirl that one find that is similar to the whole water creates as the water accelerate in flowing speed when going through the hole in the bathtub.

There is a hole...yes that is the neutron but the whirl by which the heat restriction is accelerated, the hole as such is not a physical thing and therefore there are no free electrons flying around waiting to be captured. There can be doorways created serving as electrons any place and then those gateways each will become an electron. The electron has the outer edge of the atom to fill, that barrier between that which predates the Big Bang and that which post dates the Big Bang which also is that which is on the outside of the atom that is far greater in volumetric size than the inside of the atom where space is at a premium compared to the outside. Even just looking at a picture indicating the atom should tell everyone all there is to tell about gravity. There is an electron serving a neutron that connects to a proton and this cluster could be one or many but this line causes a flow where each has an identifiable place in the line of flow. Some with many proton/ neutron/ electron holds volatility and others with small number forms relevance as a solid. It cannot be size or numbers of these within the string and the numbers forming the string in quantity that implicates mass and therefore it must be the relevance of the string fitting the Coanda effect on gravity that produces the solidity or the volatility. Referring to the massive potential size discrepancy that became synonymous with the quantum factor, the name used as a quantum state is indicating unbelievable space that is beyond any explaining the size the atom holds and space going the way of the atom from the electron down to the proton. The name alone given to this state of space became synonymous with incredibly big, unexplainably huge and beyond measure of understanding

The proton position ($\Pi^2 + \Pi^2$)
The position the proton has within the circular atom cannot promote any idea that the proton can be that more massive than what the electron is yet it is a ratio of $(\Pi^2 + \Pi^2)(\Pi^2\Pi)3 = 1836$ not more massive but

more dense. If we think of more massive in the Newtonian sense however the thought that comes to mind is bigger in volume and there is just not enough room down there in the centre of the atom. Therefore one has to search and see what there is down there that would allow the proton to be that more massive. The only thing in total abundance where the proton is, is the absolute lack of space and the only way that the proton can be that much more massive is if it destroyed 1836 times the space than that which the electron manages or on the other hand if the electron can duplicate 1836 times more space than what the proton can achieve. This means that the electron duplicates space 1836 times more than the proton or the proton reduces space 1836 times more efficient than does the electron. This atomic relevancy between the electron and the proton that we at present hold did not always apply and as space grew in progressive development the atom had to adapt in relation to the requirements of singularity. This relevancy brings on atomic displacement by movement and there is not always displacement.

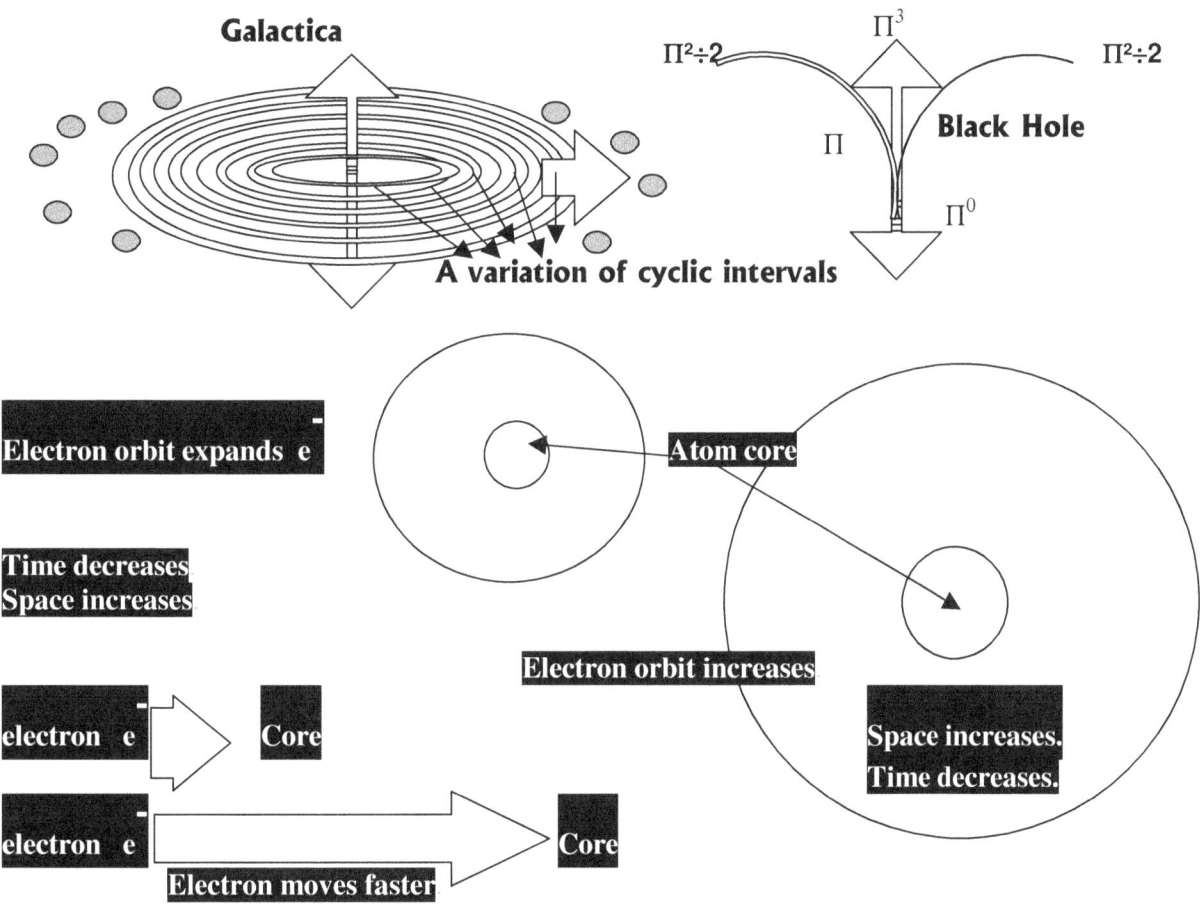

The only way a star can grow is to mimic the growth pattern set by the atoms in the star. The only way a galactica can grow is to mimic the growth patterns set by the various stars in the galactica. The atoms form growth that influences the star, which influences the galactica, and in this manner the atom is the driving force that finally shapes the cosmos. The star is just a cosmic atom forming a united front for all the atoms within the star. In that sense the atom is the Universe that forms $\Pi^0\Pi$ in the relation it forms with the governing singularity positioning the controlling singularity and in that the controlling singularity provides the spin Π^2 which puts a boundary and a limit on the edge of the atom as a form Π^3. This pattern is a continuous extending of singularity growing from a point that still holds space although the space referred to is still not being located within the region of the cosmos we view as having space as a legal value and hose point not being space is still being deposited by time as space forming no space but by the light deposited at that point during that particular instant that remains as a reminder of what was present when time deposited the space at that point. It serves as a reminder and has no other value or function and in that we can travel through it because it has no credibility but by forming a mirage of what once was.

The one part of creation the official verdict and mine is in agreement on is that it all started small, but in saying that I go one step further by saying the Universe was at the same time eternally large. It is all about relevancies forming as singularity applying matter in relation to the overheating it started to combat. It started with singularity producing matter and the matter changed in relevancy to one another by becoming solid or liquid in relation to each other. Space still was at a premium because the space we know and we see as gas, was not yet part of creation. Since the Big Bang the fluid heat is in a process of converting to space enlarging the role of time as the universe still systematically overheats.

The reality about this diagram is that every aspect of the cosmos relates to this diagram.
Since the Big Bang the atom increased because all the relevant space in the Universe increased as the balance shifted. With that being the case the reverse has to apply as the star is about contracting and reducing space by motion producing gravity.

The cosmos cannot be if the cosmos do not share with everything else in the cosmos but the sharing is always producing relevancy to the position of another factor forming the Universe.

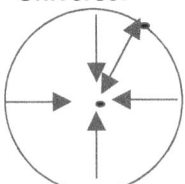

The inner space is applying positive space-time displacement in relation to the object in rotation while relocating the entire atom within the mutual singularity confinement as positive displacement.

The orbiting object forming the outer ring is in a negative displacement in relation to the inner centre. It is the centre forming the proton that puts a controlling singularity at a point where the electron supplies the divide that the neutron confirms. The entire idea behind the atomic chain is to reduce or concentrate heat. This it does in contraction by applying motion it is securing space

The heat the inner structure secures also prevents the motion from applying to the object because it became dominant enough to reduce the space towards the heated centre and in doing that it is producing space to secure space through applying motion to the space. This means when the speed of light applies the electron is valid but beyond that the neutron applies

What makes the cosmos is the variety of structures forming what we think is the Universe formed by all participating objects of all sizes while only using the same singularity the differentiation singularity maintains in relevancy. There is no big or small because that which has the biggest control in the cosmos is also incidentally the smallest there ever can be in the cosmos. In fact it is so small it cannot even directly claim a any part within the cosmos. To make sense of this lot to me being somewhat of a dimwit, I placed two opposing motions in relations to each other and the one will always show one or two relevancies in relation to the other. I named duplication positive displacement and contraction negative displacement.

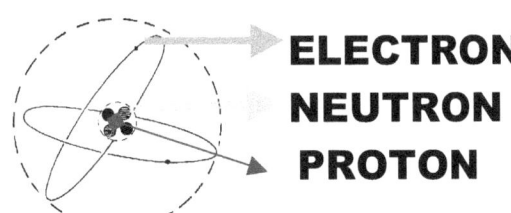

ELECTRON is about confirming space
NEUTRON conforming space
PROTON converting space

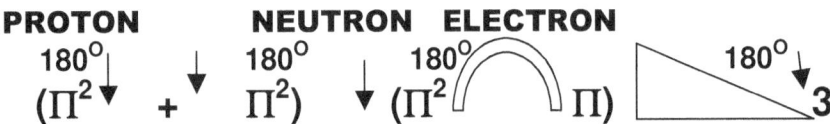

If one looks at the transmission of sound, it too depends on the relocation of matter, but to a very small degree, and in this process lies the transmitting of sound. The sound goes both ways $(\Pi^2 + \Pi^2)$ forward and backwards as well as up (Π^2) and the sound also goes down connecting to the earth (Π). To make the error of judgment in confusing the process with the breaking of the Doppler rings are quite understandable. In the Black Hole the atom completely forfeits all relevancies. The biggest activity there can be is in a Black Hole forming space-time confined directly to singularity where that pushes time eternal and by doing that places al motion activity into outer space as it renders space in infinity alone. The Black Hole placed an entire galactica worth of material that used to be a star within singularity having no space inside but only the moving time on the outside to become singularity.

It is about confirming space $(\Pi^2 + \Pi^2)$ conforming space $(\Pi^2\Pi)$ and converting space **3.**

The Universe comprises of the atom and a star is just another atom holding lots of atoms within its own space where the galactica holds lots and lots of atoms in lots of stars within its own space. Therefore the Universe is the atom that is stretching out indefinitely. This comes about from the fact that all particles are the same formed by singularity within the Universe since all particles serve singularity. From singularity the size space-time takes up is unimportant because from singularity space-time is merely principals connecting singularity forming energy as gravity or antigravity and presenting space through the relevancies forming the motion of time. That is particles and atoms that is surrounding singularity, protecting singularity, maintaining singularity and securing the surviving of singularity. This service of space is done by motion in duplicating space or extending space.

By accepting the Big Bang theory and acknowledging Hubble's evidence science has to have a look at the way they cling onto Newtonian views. There is no chance that by purely using the diameter, one can calculate and determine the gravity of the star. If that was the case then how can science ever try to explain the pulsar or the Neutron star?

Even by bullshitting in the face of contradicting evidence there is no manner in which to explain any star by birth or by death because a star cannot be born in the manner science tries to prove as much as a star cannot die in the manner science promotes. I truly cannot see how science can stick to the notion of the speed of light being a constant which they measure in the gravity field of the Earth while they also know that gravity slows the speed of light down to become a minus and still claim the speed of light as a constant when all evidence proves the very opposite. If their theory of a constant applies then how can the Black Hole break the constant of the speed of light? Before the light tries to escape the motion formed which the gravity of the Black Hole introduces the light goes into reverse by falling into the Black Hole and to come to that reverse the light will first arrive at a point of being very motionless and therefore the light will lose the space it holds in that specific position in that very moment. When surging towards the centre of the Black Hole the light will go slower and slower until it comes to a standstill from where it then turns in direction and go into a reverse as it then accelerates faster than the speed of light down the gravity funnel towards the centre of the Black Hole. Remember that where gravity leads to contracting into a smaller but condense area, light does the opposite by going towards expanding into a larger area. In that way we can say the light is the very infamous and very sought after "anti-gravity" every Newtonian has been looking for since God knows when.

The limit that caps the atom is the speed of light, which is in place and governed by the electron. The electron sees to it that the gravity the atom provides can't exceed the speed of light. However, everyone knows the more comprehensive stars charges gravity that exceeds the speed of light.

In the centre of the star where the governing singularity is vested and still being without space, a point is situated that is visible nonetheless.

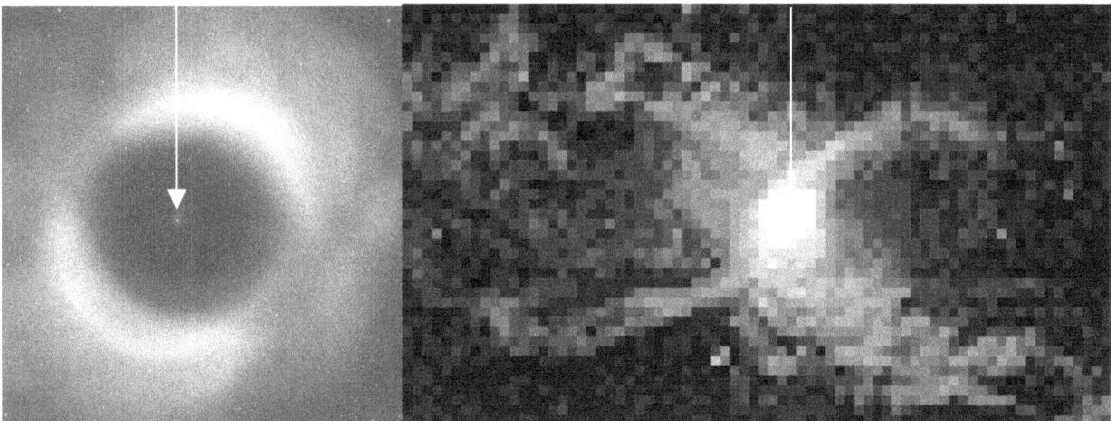

All the time the light holds an even relevancy but since the relevancy overpowers the light speed of $3\Pi^2\Pi^0$ which is the ultimate antigravity possible, that relevancy can change into 29.6 km / sec or it can accept the factor of one as it rushes down the gravity funnel of the Black Hole. When the factor of movement is one it will show no individual motion going into expansion and away from the star and therefore the light photons then relinquishes the entire claim the photons had on space by advancing in the direction of gravity and accelerating at a pace towards the inner atom at a rate where the gravity is exceeding the speed of light. If the atom did not have gravity that is quicker than the speed of light, the atom would not be able to attract light by casting a shadow of darkness. That makes every atom a future Black Hole in waiting. In that event where the star would have gravity that draws light towards the star then all the light the star generates will go dark. At that pint the individual atomic governing singularity relinquishes control to the governing singularity that the star generates.

The electron seizes to be valid within the star, as all heat then is frozen solid as it enters the star. This is not that unusual because in the atom there is also a certain absorbing of light and the rest releases as colours casting light to expand away from the atom. But during this process of forming "anti gravity" and expanding into space the relevancy factor remains $3\Pi^2\Pi^0$ until the stronger Π^3 relinquishes the space during the time by and in that the atom is rendering the light or the expanding motion no relevancy. This will be when $\Pi^3 = \Pi^2\Pi$ whereas light is $3\Pi^2\Pi^0$. This means that space held by material will always apply stronger contraction of gravity than would light apply by expanding but also this is limited to the atom or stars gravity capability $\Pi^2 = \Pi^3 \div \Pi$.

Wherever light flows light will hold the relative motion of $3\Pi^2\Pi^0$ in relation to singularity being a proportional factor to all changes in space-time that may affect space-time and wherever singularity applies Π^3. A constant derives from a law and the cosmos does not break laws. Breaking down the relevancy of light would be 3 that forms time by moving Π^2 in relation with gravity performing as a result of singularity Π^0.

In the cosmos the only filled space is where the atom occupies space. It is the same ingredient that fills outer space but rapids motion spinning beyond the speed of light solidifies the substance by controlling the substance. The atom fills the space inside stars and the atom must be the space that the star ultimate will deplete. The atom depletes the space outside the atom and conform the space to controlled solids where singularity then depletes the substance by uniting eternity to infinity. The speed of light is just another part of atomic space because science is aware that once at the time at the beginning of the Big Bang the outer space comprised of a density stronger than what the electron was.

The complexity of the Universe rides on the balance that singularity introduced and the entire Universe invested in movement placing relevancy between that which can never stop expanding and that, which can never reduce further. In the centre is singularity standing still. That will remove all space because there is not motion that produces space. But the standing still of singularity produces space through expansion by overheating and the overheating constitutes of motion where only motion will provide the required cooling. But it is both the occupied space holding the seven relevancies to singularity that expands and as that expands the ten relevancies are holding occupied space within unoccupied space that produces the other expanding factor.

By starting to explain the cosmic code we have to start with the most obvious of explanations. The cosmos starts where the periodic table ends or starts depending on the angle one looks at it. Space-time dismissing converts space into a sphere and this process starts to show where the cosmic space-time dismissing brings about a form that ends expanding and show the start of the atomic forming decline of space- in time at **7 / 10 (Π^6) / 6 = 112.162**. This is not the point where gravity begins because gravity starts at a point somewhat just above that at **10 / 7(4($\Pi^2+\Pi^2$) = 112.79**.

In this we find the following relevancies forming singularity:
7/10 Mutual singularity becoming individual singularity

$\Pi^2\Pi = \Pi^3$ governing singularity referring to Kepler's formula that states $a^3 = T^2k$

$\Pi^2\Pi = \Pi^3$ controlling singularity referring to Kepler's formula that states $a^3 = T^2k$

$\div$ 6 points in the cube forming the Principle singularity referring to Kepler's formula that indicates to a pint $k^0 = a^3 \div T^2k$

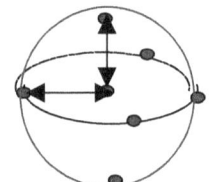

In the sphere there are never only one direction implicated in movement. Movement are always in relation to the centre position because as a line goes up it also goes in or out. When a line goes north or south, it also comes towards the centre or going away from the centre.

There is always relevancy present in movement. As this moving indicates direction it also apply Π^2 for indicating value forming the time factor. There are the four forming the circle that rotates around the three that forms the line better known as the axis in the sphere and this makes for 6 points that represents a value of Π, thus giving a compliment of Π to the value of 6.

In the sphere there is no radius but only the extending of Π from the centre Π in six opposing directions relating to one another by the square but remaining Π because of the unity the matter holds in relating to space. It is not possible to draw a precise line that would form a precise ring and not cut some atoms in parts. Because there will always be an atom disallowing the precise positioning of the circle the circle continues on a solid basis holding Π as a positional reference and not r. In every sphere there then are the seven Π relating in precise dimensional and positional equality forming equilibrium to the centre Π as well as to one another by 90^0 and 180^0 implicating the dimensional positioning. Therefore the sphere holds $_7\Pi$ and the cube holds $6 \times r^2$

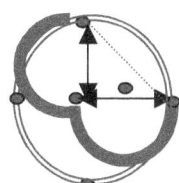

Because every moving line represents one quarter of the sphere in relation to the rest of the sphere and the line also indicate the relevant position between the point indicated and the point in the centre it is a relevancy of singularity in progress. By connecting the line, as Pythagoras will suggest the singularity within the sphere become a specific value indicated representing one half circle.

There is one substance forming the cosmos and there are two forms formed by one substance forming the cosmos where the one is controlled by movement sealed in an atom and the other does not have movement except for the expanding of space.

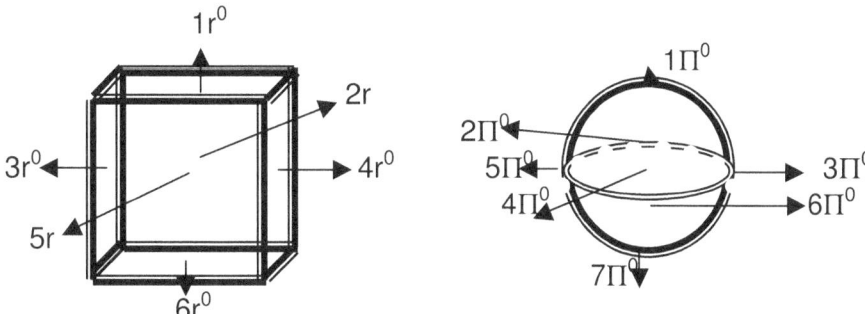

We have a circle forming by a circle forming leaving six dimensions worth Π that all serve one centre holding singularity. There is a circle running from top to bottom and there is a circle running from side to side and back to front. All these points forming Π are directly in relation to one specific pointing the centre forming singularity. The entire worth of the controlling singularity is Π^6 because of the six points forming one combined value that places the controlling singularity at a measure with the centre governing singularity. In this we find the value of Π^6 and this is very simple to understand but also so is the rest very simple to grasp.

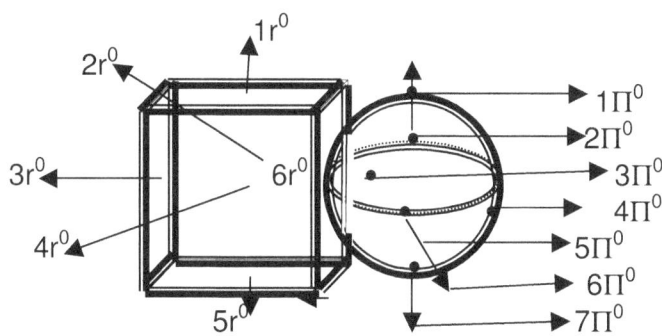

As previously indicated the space surrounding the sphere holds a value of 10 in relation to the spinning 7 that forms the diversion of the line into a circle. Every 5 points serve two spots as the movement is from 5 points to five following points making it 10 points.

Front = $4r^0$

Back = $2r^0$

Left = $3r^0$

Right = $6r^0$

Top = $1r^0$

Bottom = $5r^0$ or close enough it makes no difference

The value of $7^2 + 1^2 = 50 \times 2 = 100$ by the root thereof is 10. That is the one scenario the Universe uses to build a cosmic space and that is mathematically proven. But also we have the five points forming when the fifth point follows the four opposing spinning value of $4\Pi^2$ which then forms a completed circle. Then as a result of this five points forming a square in movement we have 25 forming as a result of 5^2 and since this doubles by movement we have 25 duplicating to fifty which is the one half of the 100 that forms space. That too is the Lagrangian system with five cosmic structures holding relevancy to the centre structure where the centre structure stands in for seven positions diverting from singularity and the orbiting structures standing in for five positions in space.

With singularity placed in infinity within the centre of every rotating object every atom and its relation to its surroundings including other atoms form space-time diverting from the point holding singularity as far as rotation goes because every object holds three relative positions in as far as where it was, where it is and where it will be in relation to singularity providing time. I elaborate on this else where.

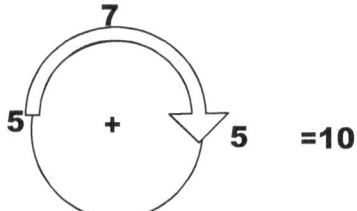

There are **5** moving by the diverting of **7** to another five forming **10** that stands related to **7**. All that this implies is that rotational movement is taking place and it is concentrating either space or form.

The important part in this relation is to see which goes into a relevancy with singularity. In other words it is important to see by division which value will end as 1^0. In that we can see whether space or form is compromising with the movement that occurs. It is a question of what moves where because if there is movement there is compromise occurring either in form or in space and one of the two factors stand to gain as the other factors stands to compromise what it represents.

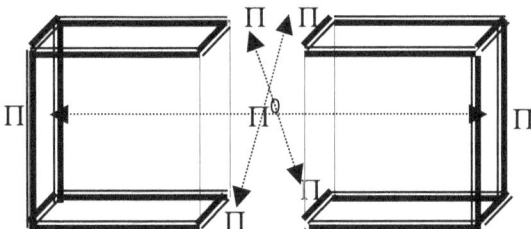

Then this lot forming the sphere spins within a cube. The sphere holds controlled heat within moving space where the space is strictly related to the centre. This we call material and it is singularity in movement. Then every point holding the controlling singularity Π is connected to the governing singularity Π^0 by the movement parting the sphere holding the boundary of Π^3 controlled spaces in the relation of gravity Π^2. This defines a Universe we think of as an atom or a star or a galactica or anything going bigger or smaller that spins. This then puts whatever spins in a division of 6 sides to the sphere holding Π^6 to the exponent of 6. $7 / 10\ (\Pi^6) / 6 = 112.162$

$a^3 = (T^2\ k) = a^{3 + 2 + 1 = 6}$ is representing a sphere within a cube with the sphere holding a centre that is presuming the position of singularity Π^0 within the centre that connects to a border formed by Π. This connection of the border to the centre is part of $k^0 = 1 =$ **singularity**. Einstein proved that at the point where he claimed that space reduces to form a flat Universe and such reducing reaches a point where space as a factor in the third dimension disappears into the single dimension (space going flat) and at that point gravity is overwhelming. Einstein interpreted this, as that the complete Universe is going flat with all the space instantaneously disappearing but while it may be true that the Universe is going flat, that can only be within the region of space deprived singularity, since singularity represents the Universe that is as flat as anything could get. However there is an illusion side we see and that illusion never goes flat. It is the relevancies inter fazing between space contracting with heat concentrating and singularity connecting infinity to eternity. This action is much the same as electricity in alternating current, which is just the same as gravity. We have two forms holding the Universe which are running parallel to each other where the one forms singularity and the other resulting in space that turns and moves while the other is flat with only singularity representing value.

Humans' (including Einstein) interpretation of the Universe is faulty but the faulty aspect does not include the fact that the Universe is going flat. The misunderstanding is only about which is part of the Universe that is going flat and to which part of the Universe the flatness refers too. According to Einstein he proved that the Universe is alternating between going flat and holding space but his lack of studying Kepler prevented him to see what truly happens. By only using mathematics it brought about his spontaneous misinterpretation collected from our culture where we wish to tell the Universe what it should be according to Newton. Einstein saw half of the Universe and that lead to his incorrect interpreting of what the Universe actually is. We all have a faulty perception of the Universe because not only he (Einstein) as an individual scientist but all humans throughout history have also never asked the Universe what the Universe is. Kepler did and the Universe answered using the mathematical equation $k^0 = a^3 / T^2 k$, which when interpreted means singularity in position is placing space-time is the Universe. No one ever thought about this statement Kepler introduced in sincerity because from a Newtonian aspect it seems silly. But rethink the silliness presented by the Newtonian Universal centre and compare that thinking about what the Universe told Kepler then decide what is silly. Newton's never acquiring the effort to do a study of Kepler's work withheld him (Einstein) from reading his very own mathematical translation accurately because apart from Newton, Einstein must be considered the second most important Newtonian ever. What Einstein saw was that space disappears and he then jumped to the conclusion that the space he saw in his mathematical equations was outer space referring to the space falling outside the parameters of the material occupied space secluded by dimensional borders. In the sphere placing the borders that the sphere holds there are deliberate and very distinctly placed edges or points forming a specific distance from the centre. The centre is also proven beyond any debate.

The centre of any sphere has to be at the very point where space completely falls away. It is at the point where all the points of line centres meet by the crossing the centre of their individual connection coming in to contact as a group. In that way one may assume that the lines connecting the controlling points on the other end are crossing on a centre point that all that is participating in the constructing of the sphere is democratically electing such a centre. Please note this conclusion very well because this forms the heart of the Coanda principle. That will put that position where the lines cross which in itself is centralising all space in the sphere at that point, such crossing point will become very distinct and controlling where that point forms in the single dimension and singularity is the single dimension. But Kepler also solves another riddle that truly got Newtonians unstuck. This, to which I now refer, is what is referred to when they refer to the Hubble constant.

Einstein read his calculation correctly but his interpretation thereof wrong because the Universe we observe does not disappear into obscurity within singularity by going flat. We always have space and the Universe is not disappearing into singularity. It is time that takes space into relevance of singularity as time goes into singularity. The growth we see in the Universe is an adding of space in every cycle completed by every cycle, which all the protons complete.

The adding is the smallest addition that can come about in the shortest period of repeating by cycle rotation there can ever be. This growth of space-time next to singularity confirms the growth of singularity as singularity recalls the space it uses to grow in the time it grows. The margin of growth will be by the extension of Π in the formula $\Pi = \Pi^3 / \Pi^2$. Every cycle completed in the relation to space by the initial value of Π. $\Pi = \Pi^3 / \Pi^2$ leave ultimately Π^1 extending as space or as Kepler chose to indicate it as k^1. But that too has to be compensated by the duration of time reducing the time aspect by the margin that the space expands. This confirms what is evident in the Hubble Constant. The further one looks at time the more time seems to race because time has the invert properties we give to space.

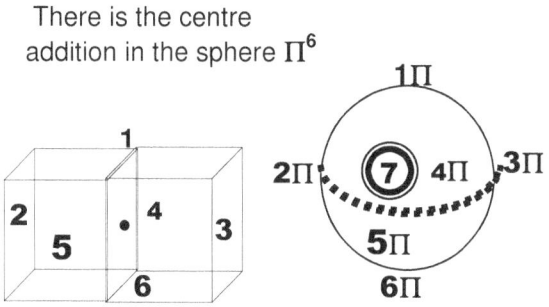

There is the centre addition in the sphere Π^6

Taking the outlook from the point the sphere is holding from that centre out into space there are ten points connecting to the centre. In that are the dimensions of singularity connecting to space where five connects to space in the second dimension of singularity, and five connects in the third dimension of singularity. On the other hand the cube does show a very different characteristic, which involves only six sides (at least) connected.

Kepler's formula also indicates that a sphere is within a cube that is holding a sphere

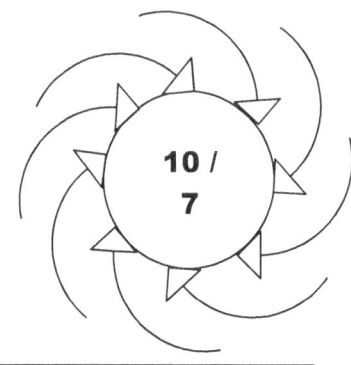

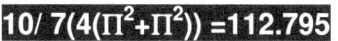

$10/7(4(\Pi^2+\Pi^2)) = 112.795$

$7/10(4(\Pi^2+\Pi^2)) = 55.3$

Kepler said that the space there is, $a^3 = T^2k$ is in motion and therefore there are no space any where except space in motion forming space-time. There is only space duplicating during any specific time duration and only space-time can relate to a specific point where singularity establishes control. There is no space only because it is all space-time, which means space in motion by duplication thereof. Therefore when one removes motion from the space factor, space will collapse. That is what a Black hole is all about! In relation to the mindset of modern man what I am referring to is unthinkable not to be absolutely part of common sense

The motion of space-time is about space duplicating by filling the following space in the following time in a relevancy where the front space will be followed by the space just behind that and that will be followed by the space just behind that.

Material within stars (and outside stars play the game of follow my leader) because the material follow in a flow where the one behind captures the space it will occupy as well as the space it will represent in the flow directed by the direction of time. The material flowing into an object is connected to material flowing out of the object by motion we talk about as momentum.

When creating a solid barrier to stop material flowing we destroy the space by preventing the motion to continue the flow will also stop space from forming, and then we stop the movement cutting off the flow with some intervention blocking the space-motion but by stopping the momentum we destroy the space duplication. We refer to this as a collision between bodies but what it is, is that we fill the space which the space in motion was supposed the fill, therefore we destroy the form of the space in motion and every time we destroy all space by creating a point that can't hold two different values of space filling heat bringing about a colliding destruction that produces more space in the end. There is occupying space and there is vacant space and the occupying space removes the vacant space and it does not fill the vacant space but it replaces the vacant space with filled space. The more space we supply in relation to the space in motion the more space in motion will be stopped where the space in motion will become a flat object as all the material in motion share the same space. Space a^3 then becomes more represented by T^2 at any point serving singularity as k.

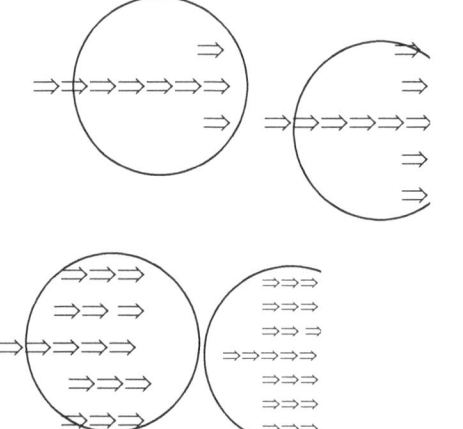

Stars in our Universe are controlled by singularity that defines the outer borders as $10/7(4((\Pi^2 + \Pi^2)) = 112.795$ on the out side and channelling the flow through the value of $7/10 (4((\Pi^2 + \Pi^2)) = 55$ to an inner border of $(\Pi(\Pi^2 + \Pi^2)) = 62$ in the centre of the star. That means the space envelope is $10/7$ and that is confined to $7/10$. That produces the gravity in the stars

As the space in motion is occupying less space due to the motion duplicating and reducing the space, the space in motion will need less space to duplicate, thereby then create motion. Smaller objects can apply faster motion since smaller objects require less space to duplicate and therefore less time to do the duplicating of the space.

Science should become serious about the task they chose to perform and not going flat out covering up mistakes by denying mistakes their Master made. I am aware that science is aware of the contradictions Newtonian science advocate mainly because no one this far acted surprised or bewildered when I mention it when they then have to realise it for the first time in their entire existence. Every Professor Doctor Very Educated flashes their accomplishments an the number of degrees they hold when telling me I do not understand "classical mathematics and physics" but not once did any one of them showed me which is the part I do not understand! It is I that can and that do show Every Professor Doctor Very Educated what it is that they with the many degrees don't understand and not one did once take my challenge to show what it is about "classical mathematics and physics" they understand which I do not understand. They know exactly what I say when I say it and with that they are performing acts so shocking about self-protection and self-preservation. When Roche presented his findings they should have realised there is something missing with the way they see things. When Hubble presented his findings they should have put Einstein to task about finding the mistake they made and not a manner to cover up the mistakes they made about the Universe contracting. Newton's arguments mathematically are about contraction and Hubble proved that the entire Universe proved without doubt that all indications contradict Newton by expanding. By them then going more overboard and going further away from the truth by covering the misconception even further and to order Einstein to measure the Universe was frankly an act of madness! Why do they not test?

Newton said a sphere is $a^3 = 4/3\,\Pi\,r^3$

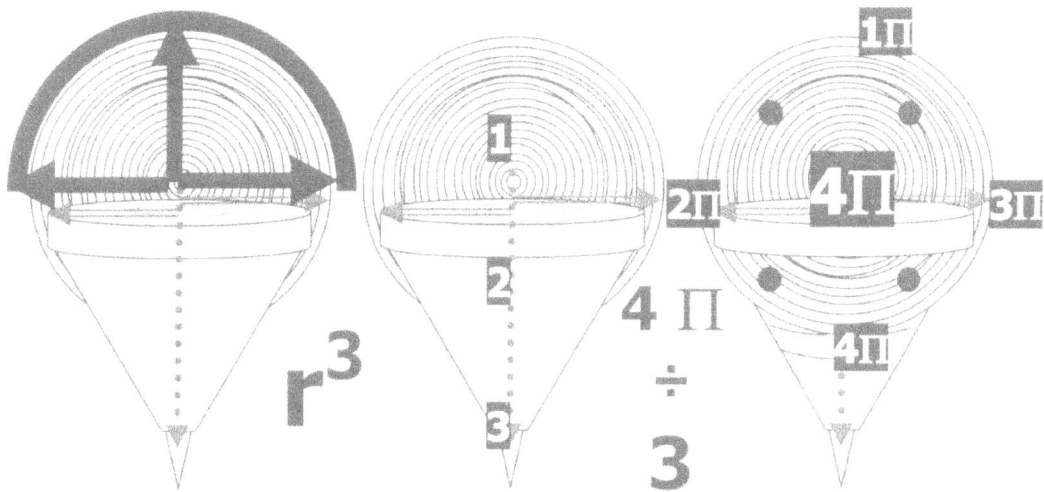

Kepler said a sphere is $a^3 = kT^2$

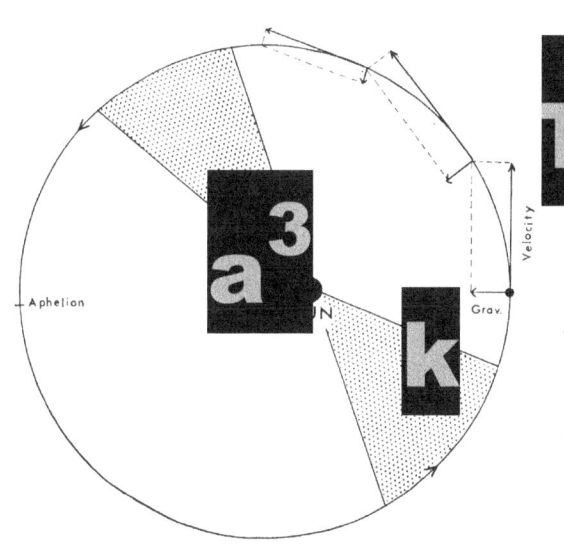

Our instincts, our logic and our calculating process all indicate that the sphere holds a centre point from where six evenly positioned point's position matter to be. Using The formula $F=G\,(M_1.m_2)/\,r^2$ it indicates to a force pulling objects closer, where each force is coming from each centre point the body in question has. Where every atom dismisses space as it is spinning, the spinning find refilling from outside coming as it flows inwards directly in a circle to the centre (because there is always a double connection to gravity in motion) of the cyclical sphere. At one point in the precise centre, the dismissing of the total ability of all the components within the structure, finds a peak and at such a peak the dismissing forms the biggest influence on all points at the outside. That places the border forming of the group selecting the unit by motion and such reducing becomes inherent part of the form of the sphere spinning to create cycles. In nature the only form provided is the sphere forming a relation within the cube.

The contraction that causes the reducing of space must commit the two bodies towards a point in each case being spot on in the middle, notwithstanding what direction the force is applying, the body will draw to the centre when being part of the unit. Only when the heat ratio promotes more duplicating than dismissing due to the spiral cyclic relation the elements hold with heat will that relation with heat counteract the dismissing and neutralise the overall effect of the collective dismissing.

Newton's $F = G\,(M.m)/r^2$ from figures Kepler left us and see how far did planets shift closer, taking into consideration the sophisticated measuring apparatus they have. Prove what is lectured to millions of students about Newton's contracting or pulling gravity. Where such observation did take place it rather proves definitely quite the opposite because the distance between the Earth and the moon is growing and it is not shrinking. With me openly criticising Newton and Newtonians lecturing none applying information across all of the Universities across the world (I guess) will, once again make this book as successful as the others in the past and that is they ignore it flatly. The Academics get very hostile attitudes towards me when I blame the Academics to their faces that they are not about gaining knowledge but about conserving the past misguiding through protectionism. Universities protect their own without any willingness to test that which it protects. All evidence should be clear in confirming that the basis on which the entire world of science forms a union is founding their policies and beliefs on incorrect principles.

Space-time holding a six dimensional Universe is :

$7 \div 10 \, (\Pi^6)/6 = 112$ Outer space expanding to the ultimate

$10/ 7(4(\Pi^2+\Pi^2) = 112$. Gravity fully expanded starts contracting at

$2(10\Pi^2+10\Pi^2) / 7 = 56.6$ (light ends movement)

$(\Pi^2 / 2)(4(\Pi^2+\Pi^2) / 7 = 55.6$ Roche in cube: Condensing heat

$7/10(4(\Pi^2+\Pi^2)) = 55.2$ Contracting heat to an Inner space limit

$\Pi \, (\Pi^2+\Pi^2) = 62.0$: Ending of space in the dimension perspective

Everything in the cosmos is moving; on two accounts by own individual accord, as well as under the influence of some other singularity in dominance. In explaining we return to the top. The two opposing motions are inseparable and always in relevancy but never to size because it is always to motion. In explaining we return to the child toy in the spinning top.

We use the most infinite to view and formulate what we think is going on in the Universe. In the accompanying picture one can see lines that show $\Pi°\Pi$ manifesting and there are innumerable more spots or lines representing Π as a relevance to singularity. Being Π sets us in the centre of the Universe. We take so much light for granted, never thinking for one second how impossible our relation with light truly is. This totally extraordinary relation we have with light must be one of the reasons why we humans put our position we have in the Universe in such a pivotal place. The fact that we as life carrying individuals especially we in person that are all blessed with the ability to only use our hind legs to walk on, on the surface of the Earth has the idea that the Universe was created especially for us, us being those holding life. Think again and such an idea supports everything everyone thinks of his own importance. Such an idea is absolutely bizarre.

All the questions, but mostly the unanswered questions about what is more nothing and what is less nothing that forms outer space being between the Sun and the planets drew me to the realisation there can be no such a quantity in space as nothing because even space has to be something. Clearly as it is for any one to see a nuclear explosions creates space. In explosions Academics portray the winds as shock waves, but what is the shock wave other than new space coming into prominence and rearrange the structure in relation to the new space just created by liquid heat unleashing the created space as well as the space volume that came in place. In that way it is clear that releasing heat brings about the expanding of the radius r as part of the sphere forming space. Hubble proved the Universe is expanding. Then by backtracking we have to set about reducing the sphere constituting the expanding Universe. If r in the circle is growing we have to reduce r to backtrack. When the circle reduces, the value allocated to r will become implicated because r determines specific size. Not so in the case of Π, because Π in the true sense only indicate that the circle is a square without corners and therefore Π dictates form and not size. By reducing size only r comes into contest and will point to such reduction. By reducing the circle radius r by half, continuously will lead to an infinite small circle but Π will remain because the circle as a form remains even being infinitely small. To find the value of the centre we have to go to the smallest point there could ever be.

One should not try to focus on an image of such a spot or dot because there is no image. The line dividing the cosmos that runs through every particle, no matter how large or small, is beyond our vision. Such a small line, so small it is not even noticeable and is so small it is not part of the cosmos we can see but is large enough to part the cosmos into sectors. It splits the biggest there is into particles and we are not even able to notice the precise location of such a split. In truth there is no top or bottom that we living in 3D can see. We shall have to use a general conception brought about by intelligence. Your intellect tells you about such a spot, but that is all indication anyone will ever find because that spot is on the other side of the Universe (quite literally). From the centre of the dot there is a top and a bottom spot. From those points there is connection with four quarters. That produces six connecting points that are all aligning to the centre. Because it serves big and small, hot and cold equal and alike and it is the smallest cutting the biggest into equality, size is of no issue. Size is what man makes of it. In the Universe there is no size in hot and cold, large and small. And that dismisses all prominence of what we ever wish to give mass. For the smallest there is, singularity is serving the largest there is equally.

If the Universe spins around a centre point holding singularity, and singularity confirms the centre of the Universe, then every particle holds the centre of the Universe making the number of universal centres immeasurable many, and every atom and sub atom particle presented outside the atom in smaller bits, are all not pieces of the Universe but they are a Universe surrounded by many Universes. If every atomic particle no matter how small is holding the centre of the Universe, then the gravity is coming about from that point because that is where the gravity applying in the Universe are applying contraction. There was a beginning that saw a radius between objects so small the size will never again repeat. The diameter of the particles were also next to nothing but that should not be a contributing factor surely…the main focus point is that particles were as cramped as it shall never again be repeated during this Universe cycle.

The value formulated in $(\Pi^2+\Pi^2)(\Pi^2 \times \Pi \times 3) = 1836$ is an atomic relevancy that projected from the atom to the star directly without loss of any translation. The formula of $F = G (M_1 \times m_2) / r^2$ only apply in a very specific range, and at a very determinable point the formula does not effect objects in the air. After such a point one will find satellites able to orbit, be it art a definite pace that matches the rotation of the Earth. Still…below such a point (B) orbiting objects will come crushing down to the Earth. On Earth a person could apply mass as a specific relevance placing an object in context with the rest of the Earth in order to give some understanding about physics and the concepts physics form but then giving mass on Earth is just another relevancy applying and is not the key that unlocks the cosmos. This giving mass and awarding mass can't be taken out to the rest of the cosmos. The key to the cosmos is in relevancies applying in relation to singularity and singularity is one aspect that goes beyond mathematical calculations.

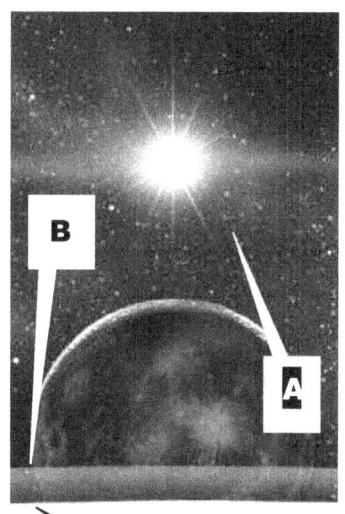

From point (B) to the Earth Newton's formula applies and from point (B) upward Kepler's formula applies, but my pointing this out brings about all sorts of annoyance concerning academics. It must be clear to all persons that there is a big difference between the applying of Newton's $F = G (M_1 \times m_2) / r^2$ and Kepler's $a^3 = T^2 k$. When the objects reach some point they will drop to the Earth and when that happens, mass do not play a part in the speeds they come to reach.

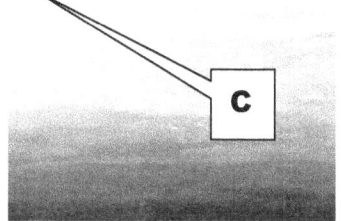

Space goes flat within the centre of the star and in all stars this is present. All stars are Black Holes in the making because as much as space expands, by the same margin does stars contract and these poles move apart as time find a way to unite eternity with infinity once again. Black Holes are not about material escaping or not escaping but it is about motion of space not being able to support the dimensional space in which we live. But Black Holes are just the peak of an evolutionary development that can be explained so easily when using the cosmic dimensional code. I say the star is one huge atom and this star has no mass at all because mass is part of Newtonian imagination. The star is just a large atom!

When looking at photographic images coming from the Sun we can clearly see that the fluid pushes out of a bowl of liquid and the telescopic images coming from the Sun via the camera lens shows that there can be no doubt that the Sun is a bowl of liquid sloshing like a boiling kettle of soup (from within stars, spilling both sides as it falls back into the liquid pool forming the sun). The inside of the Sun is not gas but it is fluid and it is cosmic gas cooled to the point of being cosmic fluid. In all of nature there is no **natural gas** as much as there is no **natural solid**.

Hydrogen is as much a liquid as iron is a gas and neon is a solid. It depends on the element relating to the space/heat in the circumstances surrounding the substance at that very precise instant in time. We have to stop telling the cosmos to show us what we wish to find and start accepting what the cosmos is telling us what is out there that is what we should look for and that is what we should find. The fluid state and the gas state is expendable material that stars remove through development but then so is all space-time and material. It is just more o the same thing where one is moving towards the other that is spinning towards a very specific centre. Using the location to explore these books that I announce will indicate as to how singularity came about to form space-time and commanded motion by creating space. In that exploring we may find out that the Universe is already contracting as much as it is expanding and it is contracting by expanding because it is through the contracting that it is expanding; the answer comes about from $a^3=T^2k$. My effort with the criticizing of the Academics was never to attack the world of physics and I never had it in mind to destroy any work made by them. I can ill afford enemies and even less enemies as power full as the Academics are. But on the other hand I cannot go on praising work I disagree with when I aim in full knowledge in the mistakes I see about their work.

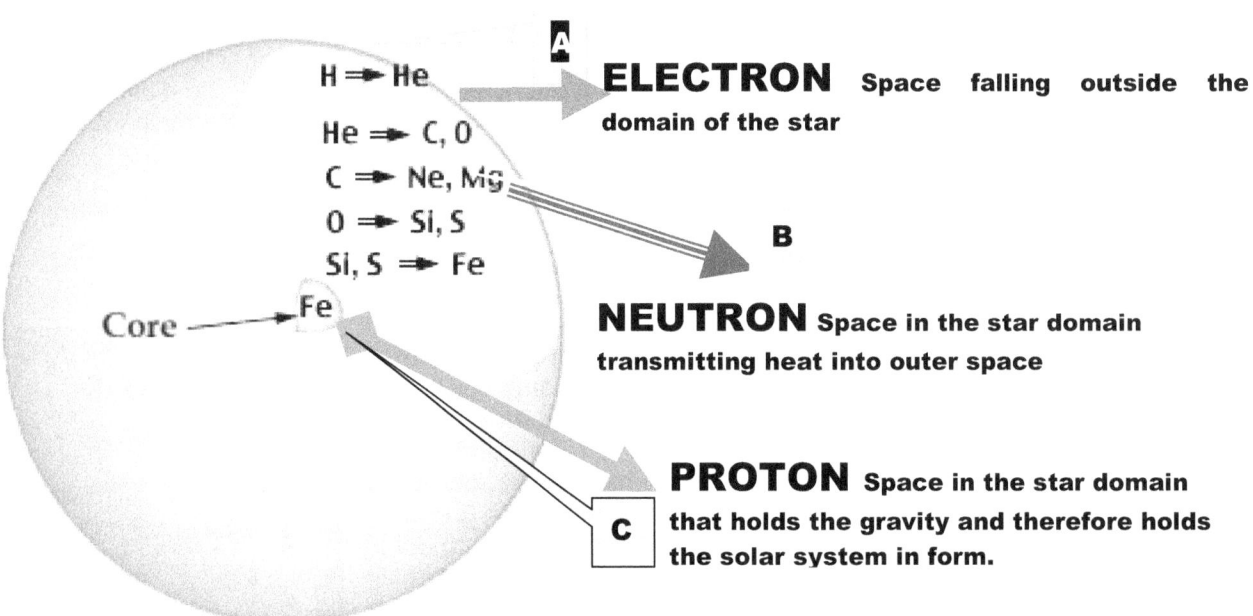

$A =(\Pi^0 + \Pi^0 + \Pi^0 = 3)$

Translated to the atomic part that focuses on the collecting and accumulating heat in space by transforming space into condensate heat by concentrating the time and establishing general differentiating between the outer space and the inner space of the star

$B= (\Pi^2 \times \Pi)$

Represents the neutron part of the star where the motion maintaining of the star is confirmed in the carbon nitrogen oxygen space-time accelerating within the star to meet in the silicon layer.

$C = (\Pi^2 + \Pi^2)$

This part represents the proton phase and the proton part of the star is where all space-time is confirmed to singularity by dismissing of space-time and establishing growth in securing singularity maintaining.

It then is the atom in the most centre part where space and time meets singularity that is representing the idea Einstein found where he said there is a Universe collapsing to a single dimension. The demise of space and the flatness obtained in this is an atomic contributing factor in every atom. Where the proton numbers exceeds $(\Pi^2/ 2)(4(\Pi^2+\Pi^2)/7=55.6$ reaching $3(\Pi^2 + \Pi^2) = 59.2$ at a point post of the proton where gravity initiates in according with the proton dimensional configuration **of** $(\Pi^2+\Pi^2)(\Pi^2 \times \Pi \times 3) \div \Pi^3 =$ **59.21.** The relevancy of the atom proves the three forms of material there are in the cosmos. The solid structure is represented by the proton relating to singularity in the form of $(\Pi^2+\Pi^2)$. The neutron is the liquid part representing the motion or movable part in the cosmos as $(\Pi^2 \Pi)$. Then the final stage is the space era represented by the 3-part in the atomic relevancy where the liquid formed space. The last part representing 3 is the Big Bang era where heat changed to space. The evidence of this statement is so clearly visible any one must see its presence.

This is a much better solution to the Newtonian misconceptions of mass inexplicably pulling other mass along nothing forming space. Using Newton's formula the radius in the square dividing the shared and combined mass of the particles the relevant mass of the particles rises by the square as the radius reduces. If the radius becomes infinite, the relevant mass that the particles will produce goes up eternal. No force in the world would keep particles apart drawing on each other with such an applying force where such a force is divided by an infinitely small separating radius is producing force as big as the Universe is capable to produce. That was the Big Bang. The Universe always flicker as a solid being in a state of duplicating which is held by a flow of gravity that can never repeat once segregated. The question in need of the answer that will bring in the light is then: "What brought on the break?" This is a recipe for joining and not dividing. Still according to the Universe I am able to witness the dividing that became enormous and the joining distance being practically irrelevant. The gravity was more than words can describe, the heat was able to melt it all in one structure, but that did not happen. It split into an innumerable number of billions of individual atoms.

When examining the case where two balls drop vertically, gravity, as a force does not apply and therefore gravity does not come into effect because there is no difference in speed or duration.

With out any apparent reason the formula is substituted with the following formula:
$g = G(M . m) /r^2$ where:
G = the gravitational constant,
M = the mass of the body,
M = the mass of the lesser body
r^2 = the radius between the two bodies.

Whatever this formula needs it is desperately lacking a foundation of basic necessities to provide substance about basic fundamental logical questions no one ever dares to ask. The slightest provocative questions leave this application high and dry. There is a dot needed in a spot where r needs some defining of proof.

Should one ever wish to paint showing that atoms formed as a result of the one mass groping the other mass as the two were clutching one another till dooms day arrive bringing coupling, then such a venture has to start by providing the incentive to initiate a centre from where the clutching may originate and why such clutching will originate. What brought about a centre r that brought about mass one groping mass two and why did mass one as well as mass two form because the first and foremost forming of mass one and mass two before the groping started had to involve a centre beforehand? Before being a mass it had to have a centre to secure that mass. What brought about such centre into the cosmos? Stopping at 10^{-43} sec there had to be a centre to start whatever mass was groping another mass. By not providing a centre one drops into a basin fluid with nonsense since the initial explaining of what and why and where the very centre arrived at that brought about the groping in the first place the whole notion is a loose unstable not very well thought through blubbering of ridiculousness in the realms of Little Red Riding Hood. Why was there one centre because the moment you say mass, some centre had to establish the mass notwithstanding the size or nature of mass?

Mass is contained, so what contained the first mass before there was any mass to contain around a centre? At least I can bring a centre to the table as I prove with my reasons I give for such a centre forming before even any mass came groping and snaring other mass. I can indicate with a fair amount of logical accuracy why such a centre will come into place. I can show how and why and where motion brought on this very first groping and clutching if there is groping and clutching to begin with. I can show what phenomenon is responsible for this centre to become practical and I can show the phenomenon positioning and locating of such a centre. I can show in the shortest manner by using accepted mathematics in the simplest of forms why space is linked to time and all space revolve around a centre using time to do so. I show what natural occurrences produce motion because producing motion provides the backlog of heat with the space it establishes by delaying time. Space forming to accommodate heat expanding is providing motion that brings about cooling through duplication that is the result of motion. Cooling establishes duplication which provides reducing as the product is reduced by duplicating and in that duplicating the doubling of pace is halving the heat in the space. It is a natural process that can be tested anywhere in the cosmos. Heat something and the space grow bigger whereas cooling the something will naturally reduce the space it holds. It is no diverting of nature by creating anti whatever that disappeared and comes back just like the Phantom does in the comics.

Newtonian science should not become annoyed with me criticising their thinking but not only that, they should show evidence immediately to back their claims as to show how far did the solar system-structures move closer, if Newton's gravitational pulling proves to be correct. From that we then can calculate what collisions we are waiting for and how long before the big solar clashing will begin. The absence when they're just mentioning such a possibility confirms to me they know as well as I do there is no evidence of the moon reaching the Earth, with no evidence of pulling or tugging and the Universe is in synchrony more than any person may ever be able to prove.

There is a position that is in motion that is forming the very edge of the outside. To be in motion the position must be in relation to a point from a centre. From the centre there must be a specific allocated space ending at the object in motion and starting from a centre that has no dimensions. The object in motion determines the one limit and the centre with no sides and no space, which is standing still in singularity, determines the other limit. By that we can see there is only one way of looking at what we can observe and that is from the outside in.

Where we have space forming the point of motion and the sphere standing still we find that space returns to singularity and this contraction is in relation to material and material is supporting singularity. In one case we have outer space or comic gas forming singularity as it is expanded to the maximum limit where gravity begins conforming space $10/\ 7(4(\Pi^2+\Pi^2)) =112.795$. Then the flow of heat is concentrated by the movement of iron $7/\ 10\ (4(\Pi^2+\Pi^2)) =55.3$ in relation to copper $\Pi(\Pi^2+\Pi^2)) =62.01$ and at that point the sphere that forms disappears into singularity as the proton $(\Pi^2+\Pi^2)$ that covers the governing singularity it represents unites with the controlling singularity Π

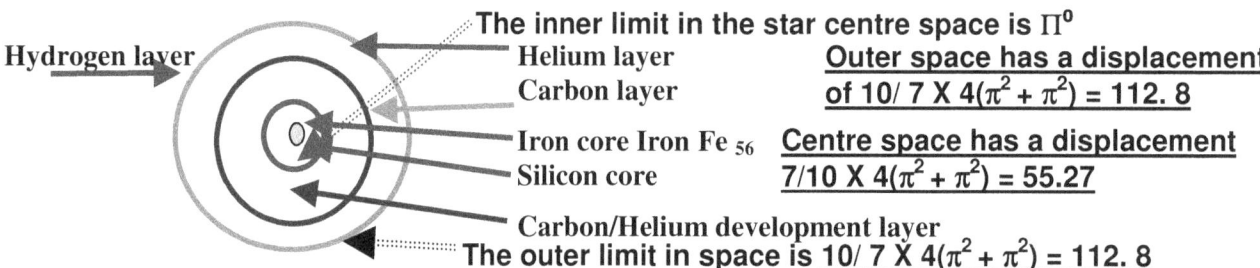

$10/\ 7(4(\Pi^2+\Pi^2)) =112.79$. The movement of gravity from the most expanded position @ 112

$7/\ 10\ (4(\Pi^2+\Pi^2)) =55.3$. The movement of gravity charged by an Iron $_{56}$ inner core, which is the same process as charging electricity

$\Pi(\Pi^2+\Pi^2)) = 62$ The movement of gravity ends in singularity dismissing heat from eternity into infinity

Gravity is the result of the rotating movement of material condensing heat from the most expanded position of heat named as outer space $10/\ 7(4(\Pi^2+\Pi^2)) =112.795$ towards the centre of a star by a process we call centrifugal forcing of liquid which is gravity. This process works on the basis where iron $7/\ 10\ (4(\Pi^2+\Pi^2)) =55.3$ that is within the centre core of the star charges a movement of condensing heat by the dismissing of space-time which occurs through the intervention of copper $\Pi(\Pi^2+\Pi^2)$ directing the flow to singularity at a point where the proton $(\Pi^2+\Pi^2)$ ending space by meeting singularity uniting the governing singularity with the controlling singularity $\Pi(\Pi^2+\Pi^2)$

Every star (even **a midget such as the Sun**) is a gas giant going down to a Black Hole because the **Black Hole is where infinity ends the Universe.**

Hydrogen layer
Helium layer
Carbon layer
Iron core Iron Fe $_{56}$
Silicon core
Carbon/Helium development layer

The inner limit in the star centre space is Π^0

Outer space has a displacement of $10/\ 7 \times 4(\pi^2 + \pi^2) = 112.8$

Centre space has a displacement $7/10 \times 4(\pi^2 + \pi^2) = 55.27$

The outer limit in space is $10/\ 7 \times 4(\pi^2 + \pi^2) = 112.8$

The IRON INNER CORE of a STAR required producing GRAVITY:
The inner core of all stars producing gravity has to be Fe$_{56}$ to produce gravity that spins within a copper range. The relevancy indicating the value of $7/10 \times 4(\pi^2 + \pi^2) = 55.27$ proves the statement and the measure where displacement of space-time brings about depletion of heat is $\pi (\pi^2 + \pi^2) = 62$. This is what reduces space in conjunction with singularity where the atoms produce a dismissing value that the space-time can sustain with enabling the flow of heat through space. In the one limit of the six sided Universe no element can sustain duplicating above the value of $10/\ 7 \times 4(\pi^2 + \pi^2) = 112.8$ which is where the prosodic table ends going to **above** $7/10 \times 4(\pi^2 + \pi^2) = 55.27$ within the star inner core. Dismissing space beyond that capability will no longer contribute to duplicating space-time of the atoms involved. Only the iron atom producing and maintaining a displacement value of 55 – 56 can produce gravity by being on the edge of demising space time while maintaining duplicating which is gravity and in our Universe only stars with an iron inner core has the ability to bring about gravity. Gravity can only achieve a displacing relevancy by being concentrated at $7/10 \times 4(\pi^2 + \pi^2) = 55.27,$ and that produces a potential difference that brings about gravity within the inner star core where gravity accumulates. This then relates directly to the second value of the Titius Bode value of $10/\ 7 \times 4(\pi^2 + \pi^2) = 112.8$ that limits outer space in the three-dimensional and six sided boundaries of what forms our Universe as outer space forming the value of $7/10\ \Pi^6 / 6 = 112,162$. That is the outer relation to the inner relation set by the core in ratio to the outer space securing a position

for the star identity in the space limits and is an indicator of the balance in space-time displacing potential of the star. In every star there is this flow towards the centre firstly of every individual atom but also as a combined unit flowing towards the centre of the star and the dismissing of space in every atom centre brings about the forming of a relation as a group within one unit structure we call a star. This flow is there because we gave it the name of gravity and gravity is the result of all the atom protons dismissing space during a specific time duration that holds direct relevance to space and time. This is interlinked by singularity forming a relevance between what is the governing point serving singularity and what is the controlling point serving singularity and as such then has a linking that is invisible to the naked eye. In young stars the core ability is yet to develop and in such stars the gradual reducing comes about as layers support the effort little developed inner core. The space reduced becomes a unifying effort from all the atoms in all layers from the outer (hydrogen$_1$ and helium$_2$) through the carbon / oxygen centre and the silicon layer down to the iron core and even going down further into space-time obscurity where the atoms as a group combining their effort acting as one atom. An atom securing one proton will provide much more space by being volatile and with more movement secure a much better field to flow of heat coming into contact with the atom. That leaves space the opportunity to support the demising of space of all the individual protons by substituting the loss with a new supply of space that becomes converted to heat than would an atom supporting 56 protons within one containing centre. The stepping down or concentration of heat gives the spinning assistance of the young stars a weak heat envelope sustaining the spin effort of the yet underdeveloped inner core.

A question that immediately springs to mind is why the atom would be the broker of this point where the periodic table stars. It has to do with the relevancy forming the atom by the chain the electron / neutron / proton forms in displacing / dismissing space-time.

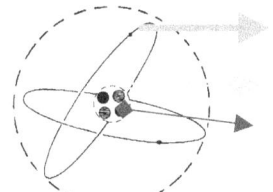

ELECTRON : Unoccupied space-time is 3

NEUTRON : Occupied space-time is ($\Pi^2\Pi$)

PROTON : Densified space-time is ($\Pi^2 + \Pi^2$)

The electron, the neutron and the proton are very commonly accepted by the role each sub atomic particle plays. But galactica and stars are just more cosmic atoms playing their part in the very same way, as do the sub atomic particles.

Unoccupied space: This forms the atomic relevance of **3,** which is where the ratio of space moves to the ratio of liquid in space. This is bringing motion in contact with **unoccupied space**.

Occupied space: Then forms the atomic relevance of $\Pi^2\Pi$**,** which is where the ratio of liquid space moves to the ratio of solid space. This is bringing motion in contact with **occupied space**.

Densified space: That forms the atomic relevance of $\Pi^2+\Pi^2$**,** which is where the ratio of solid space moves to the ratio of **densified space** or motionless space. This is removing motion by disallowing contact with space and then forming space less ness.

Space less ness: That forms the atomic relevance of $\Pi^2+\Pi^2$**,** which is where the ratio of solid space moves to the ratio of densified or motionless space. This is confining motion in a position being part of eternity.

This forms the atomic relevance of $(\Pi^2+\Pi^2)(\Pi^2\Pi)3$

One must not envisage this space duplication and space dismissing on the level where Einstein placed his vanishing and flat going Universe because it is on another plane. One must not see half a person following his other half into the future of a line of the same man continuously leading on in time. At the level we live and breathe and the level we can see light and even the space we have and live in has become solid. The liquid forming the atmosphere around the Sun is in relevance to the cosmic gas a solid. The liquid that was about during the Big Bang was as solid as the particles that formed the atoms with the only distinction applying was what moved and what did move in relevance. A particle entering the atmosphere meets with a solid space that is so unbroken it puts the material back in time by billions of years by turning it into photons. The suspending of space by material in motion takes place at the most intricate of places locating where singularity brakes into space- time. The difference between the density applying within the atom and the photon is the density applying the relevance. The one performs in relation to infinity and the other is controlled as eternity.

The atom like the star is a sphere and on the inside of the sphere there is the point where space dismisses into singularity because of the way that nature designed the sphere. The sphere secures a point where no space is possible therefore no motion can duplicate such a space. Then as space-time develops stronger locations flowing to the other border of the sphere, space-time becomes more defined and secured. On the outside the number of spots per volume of space-time developed is far reduced to the space claimed because the vacancy in space can only be represented by singularity that is the securing aspect of space-time. The duplication and the dismissing is such a small part of the space-time singularity secures that in the dynamics where a triangle must hold a different value from that of the half circle, which in turn has to hold a different value to the line, the difference there is in the forms secures the definite solidness of space in time. One must not see an object breaking composition because at a displacement factor of $(\Pi^2+\Pi^2) \times \Pi = 62$ the walls of space comes tumbling down on the flow of time, but only at a point where the governing singularity developed to a point that the displacement can secure the survival of the appointed located singularity the star developed throughout its life span. In the atom the density factor $(\Pi^2+\Pi^2)$ that serves the governing singularity Π° meets with the controlling singularity Π and by density meeting the governing singularity Π° and uniting with the controlling singularity Π space as we see it comes to a closure at $\Pi(\Pi^2+\Pi^2)$.

Gravity forms the limit to singularity at $10/7 \times 4(\pi^2 + \pi^2) = 112.8$

The **atom constructs a line** that will direct space-time displacement and that positions the atom structure formed by the **electron-neutron-proton** a flow capability of $(\Pi^2 + \Pi^2 + \Pi^2 + \Pi + 3) \times \Pi = 112.31$.

At this point the dismissing of space-time displacement entered the **Universe into a six dimensional** status holding the **sphere (3 dimensions)** within the **cube (3 dimensions)** to form $7/10\Pi^6/6 = 112.162$

The atom forms a connection between the governing singularity Π° by lining the **proton ($\Pi^2+ \Pi^2$** up with the **neutron $\Pi^2 + \Pi$** and then getting this lot in line with time formed by the **electron + 3**)
That then comes into a margin **Π that the controlling singularity forms** and all of that comes in place at the end of the periodic table. = 112.31.

In outer space and near to outer space regions as the Earth is the atom confirms the security of space duplicating through time. The atom sustains a space-time duplication of the atom as it is combining the total value of the factors forming singularity. The factors related to the atom as a unit that is securing space–time relating to singularity to the form of $(\Pi^2+ \Pi^2 + \Pi^2 + \Pi + 3) \times \Pi = 112.31$. These factors are holding the atom apart from singularity by confirming space-time which is the atom in another relation other than the normal $((\Pi^2+\Pi^2)(\Pi^2\Pi)3)) = 1836$ and that is the point the sphere secures space in the six dimensions or space walls formed by the Universe when the Universe came to $7/10\Pi^6/6 = 112.162$. It is only when the requirements of maintaining singularity surpass the confirming the maintaining singularity in the atoms inner core stability that the double proton confirms in dismissing space-time without the required duplication also thereof will singularity dismiss the structural security that space-time in the double spinning proton can provide. With me suspecting that there may be those that has super ambitions as space whirl constructers living out there and have plans to trick the Universe by building some double space whirl all you being the brave hear this: In the event where there is a person with a super elastic imagination that foresee a situation where such a person can create or establish two of himself in the same space in the same time with some mathematical formula he has dreamed about, then such a person must first demolish his body down to fit a position where any triangle in his body will measure the size any half a circle that will take forming a straight line. Therefore all those out there planning on their next combining of space whirls they anticipate creating to split time or bridge time or just annoy time in space, remember that once you reach the space you occupy no more than a straight line you may carry on with your ambitions and plans because as a true Newtonian you indulge in fairy tales and know little to nothing about science.

The breaking down of space by the inability to create motion is present in every atom and particle just as the duplicating of space is in every atom and molecule. Since all stars are group representatives of innumerable atoms within the space-time confinement the governing singularity lays claim on the dismissing of or duplicating of space-time that is present. It is the ratio in the atom or stars favouring the presence of one or the other that allows the atom or star the characteristic it shows. If the duplication of space –time is prevalent as it is with Xenon, then notwithstanding how massive the protons' grouping is, it still seems that space will duplicate much more frivolous than would the diminishing of space-time be a feature of the element. It is the ratio that gravity establishes placing relevancies against relevancies that produces space-time or even the lack thereof as in the case of a Black Hole. Where the Black Hole lost all ability to establish motion within that atom by providing spin it deflected all movement onto outer space to enable the cosmos to move in relation to singularity never moving. This is because of its securing of singularity maintaining does the proton then ejects all intended movement into outer space way outside the star that then becomes an overdeveloped atom and cast the conducting of movement to a position in outer space because the motion of the double proton is no longer valid within the well developed singularity governing the Black Hole.

It started with a spot Π^0 that then formed the dot Π, because that is the only form, size and dimension, that is holds mathematical logic and that will allow our brain to accept what would form as the first value flowing from singularity. From the one dot had to come a second phase of dots and a third lot of dots. The dynamics of such dots are smaller than we can understand because such a dot is in negative relation to what we see Π to be, and the deeper we delve in finding the smallest fragment where space started, and that is the spot where time is still eternal as much as we can accept eternity to be. This we find in the aligning of planets where the one dot from which the aligner stems becomes the reference to the distance applied between the aligner and the original dot, or governing singularity or structure in charge of holding position to all orbits following.

The reason why we should first locate the spot holding the dot is because we can only work from that point forward. By working forward we have to work backwards from the position we are in to locate where we are heading. The cosmos started at a point and where such a point is, we will find the Universe and where the Universe one day will end. Everyone knows where the Universe is, because we can see where the Universe is, but if we can see where the Universe is, then we should find the centre of the Universe in that spot. Einstein theoretically positioned the point of beginning at a place he indicated where singularity should be. With the cosmos the size it is and space so large compared to our smallness then we have no chance in finding the centre of the Universe. The Universe started where singularity is and singularity is the sure indicator of the centre of the Universe because it joins all there is in 1^0. With all spinning objects holding singularity we then have located singularity in as much as finding the centre of the Universe. The Universe started with a dot forming. That answer arrived from taking mathematics back to a point of being the smallest possible position, far smaller than we may be able to calculate form as we return for a moment to a time before Mathematics developed.

My approach might seem unconventional but through the abandoning of the accepted, it enabled me in locating the precise location of a Universal singularity forming a connecting basis of the Universe (this I say with some degree of confidence). The smallest figure there can be must be a spot that forms ultimately a dot. The only value a spot can have being without space is Π^0. The dot forms by connecting singularity as a natural development Π, which is the only form that leaves all the options open to extend in any and in all directions, should the opportunity arise. The only mathematically sensible option about extending a line from the dot will be non-bias progress in all directions equally in order to give a meaningful flow of mathematical equilibrium. This is what we find in the investigation of space in light of the Hubble expanding concept where the Universe seems to grow outwards from all angles like a bread rising.

The obtaining of singularity is part of my rejecting of the idea that outer space is filled with nothing and that I do by replacing the Newtonian "nothing" with something being the dot holding singularity. In outer space we find distance meaning there is something there and this was proved a very long time ago but science chose to ignore it for a very long time. A person that went by the name of Empedocles proved that something fills the clepsydra as the water spouted from the clepsydra. Empedocles reasoned that it can't be "nothing" that was filling the vacancy the water left in the clepsydra because as the water flow could be controlled by stopping what was coming in and if it was nothing that came in the water would still be flowing out. The water was being pushed out by air and the water was running out because no sooner did one stop the vacancy being filled then the water stopped rushing out. With the clepsydra or "water thief" Empedocles deducted that air was composed of innumerable fine particles, and then he broke the thought that what we now know is air, was also believed to contain nothing being altogether a space filled with nothing until proven to be wrong so many years ago. Never did science learn anything which they could take as a lesson learnt back then and was able to take this knowledge to the future and also out onto outer space. If there is space, there cannot be "nothing" as space is something because something is filling the space there is between cosmic objects. The claim of singularity filling the Universe as a mathematical truth becomes obvious when observing the connection between the half circle, the straight line and the triangle, which could also promote all the qualities lurking behind the pyramid. Consider the connection between 180^0 sharing and then one may realise much of the pyramid mystique becomes less spectacular in considering the very basic in mathematics being the Law of Pythagoras on which all mathematics arrived. Once the idea that the water thief brought to mind was eliminated by some human intelligence not processing the data then the matter was left at that. Newtonians took nothing and shifted that nothing out to an area we think of as outer space. That is why they say we now find nothing in outer space. They say there in outer space is nothing but an atom here and there and even the atom is covered in nothing. Now I ask you how bloody logic is that coming from those that are supposed to be the most brilliant minds humans can offer.

I wonder why the nothing landed there in outer space. Could it be that the reverse came about because there was no visible "water thief" in outer space stealing the water from the clepsydra and if they can't see the thief then there can be no thief! Then they had to fill the Universe with what they understood and because of that man then thought because there is no "water thief" there has to be nothing? The very limit of man's suspicions came into practice as man had nothing to fill the Universe with therefore man used nothing to fill the Universe. Man has always been extremely good in flying from one outer edge to another and if the water thief proved something was present in outer space then the mere absence of a water thief must therefore prove that nothing must be in outer space. When they couldn't put ether in outer space as a filling substance, they replaced ether with the nothing they then had to offer when they had to reject ether as a substance. It seems much easier to shift nothing from what they have to what they then use than to find what should replace ether after ether was removed from space and replaced with nothing. But what is space as such. What can space be, because with explosions we can clearly witness space created from heat. Our Newtonian inspired science culture prevents us from admitting what our vision tells, but the truth is in what we see and we see space being released as heat produces a "shock wave". That "shock wave" is nothing less than space created from heat released and then expanding. The space that the release of heat creates re-establishes the position and location of the entire space it refills. We have to brake free from culture of the past and a rigged mind set narrowing our vision. We have to learn to see the Universe with our minds and not our eyes as we can see in the presence of the Black Hole we cannot see. The Black Hole is only visible by presenting invisibility. We know about the Black Hole because we can't see the Black Hole. Why would that be?

Because of the manner in which the Universe initially started where singularity formed in the relation to be controlled that what was uncontrolled because of the control forming contraction, heat moved from the uncontrolled to the controlled material and after that the overheating brought about expanding and the controlled brought about cooling. As space expanded it seems that the progress in space favoured overheating by expansion resulting in much more heat releasing into the Universe. However as increase in space of heat relieving the initial concentration this brought about an increase in space because the expanding of heat brings about movement and the expanding coming from heat realises cooling as a by-product of the movement in expanding. In short the expanding of outer space is movement and the movement is about controlling heat and therefore the expanding brings movement that brings cooling. It may seem that the expanding of heat is space uncontrolled but the expanding has properties that deliver even more deliberate control and cooling than that which remained controlled by singularity secured inside a unit such as atoms or as a star or any form of containing material. The container had borders coming about from motion that the unit employed and such motion set the forming structures apart from the rest of space. Once again we see $a^3 = T^2k$ being correct from the start. There is more heat in space uncontained than space contained in some cosmic unit in heat that is volumetric contained and secured. Therefore to restore balance there must be a position where singularity reduces space faster than heat can fill space. This is brought on by the value of time forming a value of Π within material where time forms 3 outside material. The difference of $\Pi - 3$ is 0.141592653 and when that is duplicated by 7 we have singularity 0.991. This proves that the contraction will finally outlast the expansion as all becomes 1. At such a point singularity is taking longer to reduce the space than it takes the heat to fill the space and therefore the space reducing takes longer than filling the heat in space. The space flickering that announces contraction takes longer than the expansion causing the heat to turn to space. Before the heat expansion can begin in progress the contraction already completed the motion successful. This question proves fusion between materials to be the answer. By primarily having 112 protons being equal to what was needed to form the three dimensional Universe the proton neutron electron chain initiated forming atoms, the atom could from that point on start a list of atoms that would dismiss and would displace space within and outside material the material by number all had different functions down to single hydrogen particles. Changing the proton displacement value into a number of protons and placing this into an atom of single proportions will for instance stop light moving while having 56 protons in any atom that is housing 56 neutrons in one accumulative structure. All of this secures space-time to the value of $a^3 = T^2k$ which is $\Pi^3 = \Pi^2\Pi$.

($\Pi=\Pi^3/\Pi^2$) or in Kepler's terms $k = a^3 / T^2$ presents the duplication of space while
($\Pi^2=\Pi^3/\Pi$) or in Kepler's terms $T^2 = a^3 / k$ presents the destruction of space and
($\Pi^0 / \Pi = \Pi^2 / \Pi^3$) or in Kepler's terms $k^0 / k = T^2 / a^3$ presenting the final act or the demise of space. Infinity presenting $k^0 / k = T^2 / a^3$ becomes totally dominating when there is a displacing ratio of **6** (materials Number) in the square of space **10** which then presumes the value of **60**. Space will remain duplicating as long as there is space available to convert to heat and as long as there is heat that can be converted to material and material available to transform to singularity. Nowhere is their having a free ride coming to any part or dimension of Creation. There is built in only a lot of hard work and a dear price to pay for every inch gained or lost in growth.

In this mentioned ratio between the dismissing and the duplicating of space through motion, stars form by accumulating material in a giant sphere and keeping the atoms secluded from outer space. All the atoms within the star that are forming the space within the star that is forming the star are as much the star as the star is all the atoms combined. Since all the atoms in the star work towards a mutual goal as to provide the star with the required singularity security by providing the maintenance that brings about survival to components in the star, the star is as much one atom in cosmic space as it is representing all it's atoms in the unit being one as much as all the atoms that are individual cosmic structures.

When the flow of space is exceeded by a certain specific number of proton abilities to dismiss the space, the dimensional walls keeping the space in form, no longer can sustain the flow of space by substituting the demise with a flow of space. We have to remember that the initial motion was equal to the initial expanding that was in turn equal to the initial space a^3 that developed. The distance k that came about was the same value as space a^3 and the motion of the space T^2. The Universe divided innumerably as it remained one structure. Relevancies came about that excluded no possibilities and whatever one may think of being in the Universe came into place through relevancies between innumerable factors all acting in groups that remained one. There was no space but the space created in that cycle motion. There was the motion that provided the expanding in the straight line that was precisely the same value in the half circle and that brought about the triangle also to the identical value as the other two and was securing the space that was precisely the same as the other two factors. Reflecting on matters we still find this very same trend applying to light at the present time we live in. In the development space grew because the diminishing was only half the growth and size rather than direction became a major influence. The direction is the foundation and came in as a bases in the very first instant. The second repeating instant changed the scenario. With time progressing the distance k^1 will tend to lag behind as the rotation T^2 has to compromise for more space a^3 involved. With more space coming about the circle that had to produce the rotation was at the start equal to the expansion distance and therefore $T^2 = k = a^3$. Since then the ratio changed to $a^3 = T^2 k$. That is how space/time relates according to the original calculations Kepler introduced.

$1/k = T^2/a^3$. We know that there is a demise of space relating to the growth in space proving that when distance k reduces space a^3 will do the same and

$k = a^3/T^2$ when k expands it will produce space in relation to motion.

$T^2 = a^3/k$ When k demise the growth in distance or relevance, k expands pushing time T^2 which is where time will increase by the square but distance k will diminish by the single therefore time T^2 will grow faster than space a^3, which is the result of k will diminish.

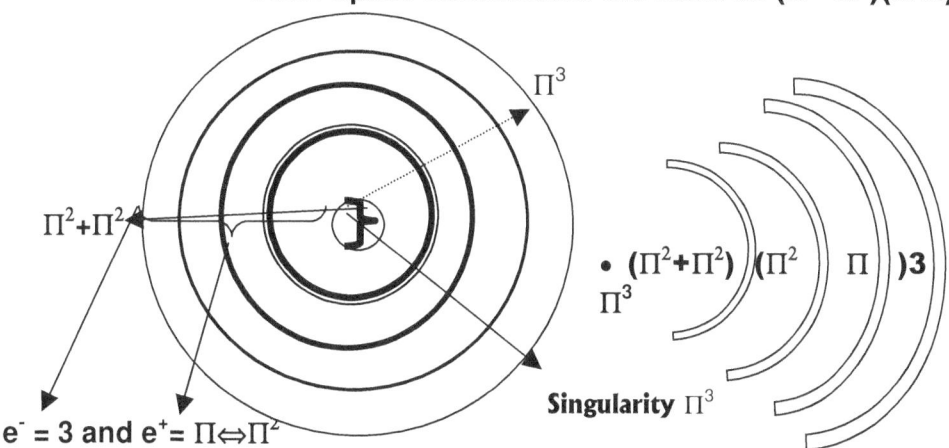

Time forming space = Π^3 = 31.0061
Outer space substantiate the atom as $(\Pi^2+\Pi^2)(\Pi^2\Pi)$ 3

Every layer in the star represents one factor in the atom since the star is just another cosmic atom securing strings of atoms that as a unit aims for one goal and that is to secure one singularity within the star.

$e^- = 3$ and $e^+ = \Pi \Leftrightarrow \Pi^2$

Singularity Π^3

The Universe is what the cosmos has in it and the smallest particles form groups by association that determine the larger congregations. The atoms in the star produce the galactica that cradle the developing star until such a time as the atoms can generate a release from the galactica and the movement of the atoms become responsible for the movement of the star. In that way it is not that surprising the atoms do not form a general mixture of all sorts and sizes but form in association. The star forms particular elements in groups of the same kind in layers where that particular element forming that particular layer carries the responsibility as to what function such a layer should provide. The star has the responsibility to install the medium through which the time in infinity can reconcile with the time in eternity. It is to that end that stars form alliances bonded by era of cosmic development where during certain specific eras of development the star contracts allowing what does not contract to expand. The star is shrinking in space as much as space is expanding in relation to the star. The brilliance in design of the cosmos will surpass the human intellect until eternity ends in infinity and still human minds will be incapable to fathom the entire process. The Mastermind behind Creation will forever leave all human intellect seemingly stupid. Since the star has atoms the atoms follow the general outlay of the star in terms of the outlay of the atom because once a design is part of the cosmos it remains part of the cosmos since it has no where to go but to remain and repeat within the cosmos.

At the beginning a trend was set that applied ever since throughout Creation. As the space increased the time ratio decreased since the distance in relevancy reduced in relation to the available space and that is the relevancy I simulate in the atom's ratio of space-time being $(\Pi^2+\Pi^2)(\Pi\Pi^2)(3) = 1836$. But as the star takes time in space back towards the earlier scenario conditions applying to the atom will reduce the space it holds and as such the time will bring the compromising aspect to the changing ratio.

Mass has no part in any process but space-time demise and the relevancy brought about as such plays the only part. $(\Pi^2+\Pi^2)(\Pi\Pi^2)(3) = 1836$. At first with the first motion producing the first space, it took every proton in the entire Universe to suppress the space created in that motion. One movement found the ability to dispose all the space created but also created all the space there was. It was an innumerable number of protons forming the first atom. The first atom was the Universe and the atom became the Universe. The atom still is the Universe and will remain the Universe till the end.

During this when material formed as cosmic solids parted from cosmic fluids at the time relevancies came about that we afterwards named sub atomic particles. At first through motion the double movement represented by the proton came in position when **$7^2 + 1^2 = 50 \times 2 = 100 = \sqrt{100} = 10$** and this was when seven points formed a relation with 10 by involving the Roche factor to form Π^2 on both sides of the divide. But the relation was in conjunction and not against therefore not in division and so the total became in addition $\Pi^2+\Pi^2$ and not a result of dividing such as 7÷10 or the other way around. Then the Universe was blessed with spinning space destroying double components allowing heat no escape. In the very beginning of creation some of the protons had to capitulate to allow the cosmos to survive. Some protons had to remove one square of $\Pi^2+\Pi^2$ to allow movement. From this came the neutron $\Pi^2\Pi^2$ at fist presumably duplicating and later dividing $\Pi^2\Pi$**x 3**. But every time it remained in relation to singularity extending, which is Π. In the Universe we enjoy the relevancy of the atom, which stands in direct relation to the atom holding space in the motion that duplicates the space.

If the electron has this electron's mass of $9,109 \times 10^{-31}$ and the proton has that incredible bigger mass of $1,673 \times 10^{-27}$ then that makes the proton a warping 1836 times more massive than the electron. Even in these differences in size and in gravity, how can science connect size and mass? With this information and knowing where the subatomic particles are one can clearly see that gravity is about space dismissing and what dismisses space more, the action of the proton? The proton is the place in the Universe where I now am referring to which is where mass is created because that is where gravity is generated. Size just cannot be the cause of mass that produces gravity but rather more the other way around.

This became the atom $(\Pi^2+\Pi^2)(\Pi^2\Pi)(\Pi^0+\Pi^0+\Pi^0) = 1836$ and the atom formed stars that still act in accordance with and to the atomic relevancy.

$(\Pi^2+\Pi^2)(\Pi^2\Pi)(\Pi^0+\Pi^0+\Pi^0) / (3\Pi^2) = 62.01$ is the end of all light within the atom.

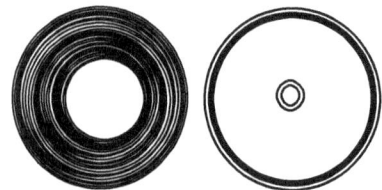

On the outside $(\Pi^2+\Pi^2)(\Pi^2\Pi)(3) = 1836$ **applies because**
$(\Pi^2+\Pi^2)(\Pi^2+\Pi)(\Pi^0+\Pi^0+\Pi^0) = 35.75 \times \Pi = 112.3$ **and heat is expanded**
On the inside $(\Pi^2+\Pi^2)(\Pi^2\Pi)(3) \div 10\Pi = 58.44$ **is where space ends**
...and $(\Pi^2+\Pi^2)(\Pi^2\Pi)(3) \div \Pi^3 = 59.21$ **space collapses**

In the very centre $(\Pi^2+\Pi^2)(\Pi^2\Pi)(\Pi^0+\Pi^0+\Pi^0) / (3\Pi^2) = 62.01$ heat collapses as infinity freezes space and stops all movement or gravity or electricity **($3\Pi^2$)** because at 62 is where all heat ends movement
Every line has a defined value of $\Pi^o\Pi$ that brings a relation about consisting of space-time or material Π^3 =$\Pi\Pi^2$. This brings about that in eve layer a different relevancy is in hand all being formed by the governing singularity Π^o forming the controlling singularity Π that applies an individual singularity Π^3 at a point that the mutual singularity $\Pi\Pi^2$ holds by controlling movement.

Following the process and seeing the influence of singularity should bring about a pattern that may lead one to a pattern of how the required heat formed and how the intended heat transformed to space. Density depends more on proton number arrangement producing specific form in relevancy as to merely and only having mass as factor that contributes to the forming and development of stars in the cosmos. The evidence is so clear that mass has nothing to do with gravity but density have everything to do with gravity. Density is the volume of space in numbers used to fill material in ratio with numbers of space per volume not filled with space. It is matter versus space in every sense there are. This came about before the Big Bang took place and before space was formerly space and time was formally motion. It was a time when singularity set relevancies moving from Π^O to Π

The Black hole reached the point of singularity. Singularity is a point where **k = Ω** placing **a³= Ω** and also **T²= Ω**. I feel sorry for the mathematicians but by using their number crunching this time they cannot take their brilliant formulas out the cupboard and play with the calculations for a while because their formulas are just too small to be fit for purpose. It is too small or too large for mathematics and what goes on the Black Hole only fits into thought. It is by using intelligence that we can detect singularity governing the Black Hole by controlling outer space. On Earth we may give it a purpose in value to what is being one molecular mass but this doesn't even apply in the strict sense everywhere on Earth because mass is different at different locations. Gravity is about the space surrounding the atom holding the heat relevancies diminishing as the space within the star compromises in favour of heat accumulating by measure of concentration. This is quite coming into place within the gas structures and is starting to apply in the solid stars.

Within the star the atom will follow the demising trend on space set by the star

All the while the atoms have to comply with the rules within the star as a demand on the atomic space claim sets new gravity related standards. This increase in gravity goes far beyond the speed of light. At a point the reducing of space becomes so demanding that the factor of light finding an ability to apply motion disappears as the massive structure draws even light towards the centre because at that stage the photon no longer has the ability to duplicate space as it displaces space. The photon must then surrender space due to a lack of adequate motion applied at a relevancy of 56.6. At the beginning when a star establish independence from the galactica outer space which is hotter than the star itself, it is not the heat but the motion as the totality of all the protons working as a group within the secure unit of the infant star that dismisses space-time and the total displacement finds a focus in the centre of the star.

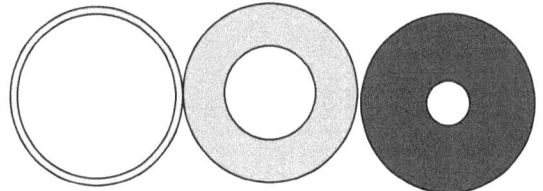 As the star captures space from outer space it diminishes all space including atomic space.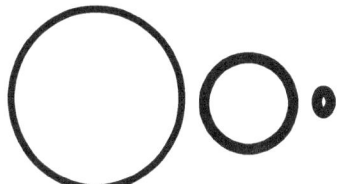

At first the star may only demand a reducing focus on the 3 or the Π to become independent and secure defined borders or atmospheres but as the star develops through the intense centre it forms, the protons will grow and through fusion bring about much more active displacing that eventually forms fusion. The fusion to which I refer is not exactly the Newtonian fusion where they wish to push something into something else and gain another form of atom. The fusion Newtonians try to use is miles off from what happens in nature because nature freezes together in fusion while Newtonian madness try to push as to heat and burn particles together. The more protons there are in the least space there are will bring about the strongest gravity there are. Well the concept about this is not new but the fact that the protons freeze space into reducing is new. The fusion to which I refer is the growth in space within the atom as the atom generates material by accumulating heat as a controlled substance. This is how stars progress in time.

As the star development progresses, the dominant gravity generating protons found in one location, begins to form within the centre of the star where the major heat is accumulated. The centre bubble holds young stars before birth in a frozen state because the atoms has not the ability yet to place a governing singularity that could hold the mutual singularity in a unit filled with individual atomic singularity. As the stars develop and form a mutual singularity by spin the value of the galactica governing singularity applies Π in a flatness that favours the movement of individual stars and not a round sphere such as we have in the centre.

The stars are at first frozen inside the mutual singularity that forms the controlling singularity responding on the governing singularity. In this stare the star remains until the star begins to form an individual singularity as a governing singularity because the mutual singularity, which is the atoms, formed an individual singularity granting the star independence by initiating a personal gravid independent of the movement of the galactica as a controlling unity

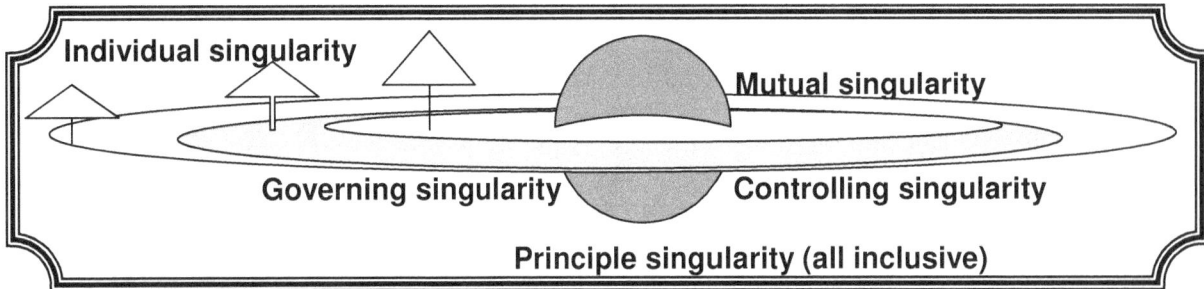

There is a shift that takes place moving cosmic space from where the slid at first was on the inside of the bubble in the centre of the galactica t the outside where a rim shape spiral forms the galactica and where such spreading is moving outwards. The stars start as a frozen liquid concentration still in an under development and in a frozen state and as the stars start to develop it moves into a rim of sorts. Then as the star is developing the inflow of space-time shifts the movement or focus towards the star since the star is generating an individual flow of space-time still in relation to the principle singularity of the galactica but progressively developing the individual singularity of the star and then to the middle sectors and eventually to the centre of the centre. In this the focus of the displacement gradually moves from a massive number of single proton atoms to a massive number of atomic protons. The quantity of protons efficiency move over to form a focus on to the quality in proton numbers in one unit of a centre and then the dominant atom displacement will not be Π but it will become Π^2, later $3\Pi^2$, $3^3+3\Pi^2$ and so the centre progressively develops.

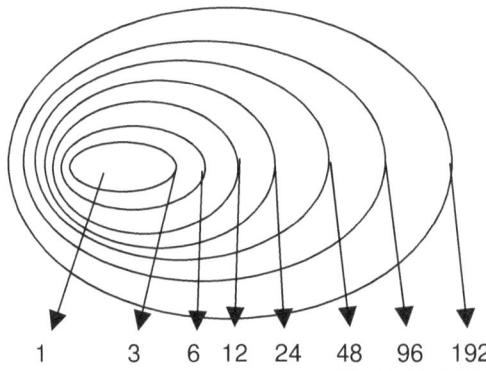

Planet	Mercury	Venus	Earth	Mars	Ceres	Jupiter	Saturn	Uranus
Bode's Law distance	4	7	10	16	28	52	100	196
Actual distance	3.9	7.2	10	15.2	28	52	95	192

The Bode's Law is a numerical sequence announced by J.E. Bode in 1772, which matches the distances from the Sun of the six planets then known. It is also known as the Titus-Bode law, as it was first pointed out by the German mathematician Johann Daniel Titius (1729-96) in 1766. It is formed from the sequence 0,3,6,12,24,48,96, and 192 by adding 4 to each number. The planets were seen to fit this sequence quite well – as did Uranus, discovered in 1781. However, Neptune and Pluto do not conform to the 'law'. Bode's Law stimulated the search for a planet orbiting between Mars and Jupiter that led to the discovery of the first asteroids. It is often said that the law has no theoretical basis, but it does show how orbital resonance can lead to commensurability. The importance that becomes known is the sequence the Ties – Bode law saw in the number arrangement of 3; 6; 12; 24; 48; 96 etc. The incorrect application of the Titus Bode law lies in subtracting the figure of 3 from 10 leaving 7. The other way of reasoning is to add four each time to the firs value of three starting with 3 and so on. The true significance of the Titus-Bode law is that it points directly to a circular growth of 7 stages. The 7 relating to 10 is a precise derogative of the Roche limit or the Roche limit is a precise derogative of the Titus Bode principle because he two systems interlink.

Forget about the fancy pictures showing smart spheres circling a Sun where the circling is nicely and evenly spread being the perfect matching distances apart because that picture is as much rubbish as any other Newtonian misconception trying to bring misunderstanding as to promote Newton's mass pulling gravity nonsense. Every planet forming a new orbit puts the planet at a distance that doubles the distance there is between the inner planet and the distance he planet forms from the next planet.

The relevancy changes while the size stays exactly the same.

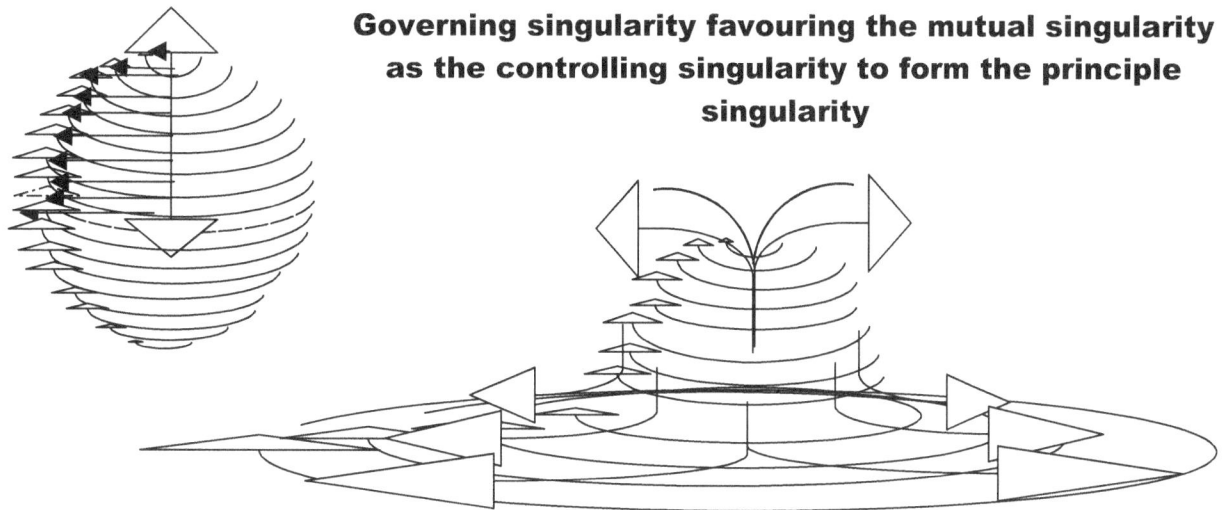

Governing singularity favouring the mutual singularity as the controlling singularity to form the principle singularity

Governing singularity favouring the individual singularity as the controlling singularity to form the principle singularity

When a star transfers the governing singularity relevance from the galactica centre to the controlling singularity of the star by connecting the centre of the star to the spinning of the star as in $\Pi°\Pi$ and moves this from the galactic governing singularity to connect $\Pi°\Pi$ in relation to the star that starts to spin, this action enables the star to gain and to maintain an independent worth in the governing singularity that the star then develops within the centre of the star by which the star charges gravity $\Pi\Pi^2$ by independent spin in order to form a border limiting the space Π^3 of the star from the rest of the galactica and this gravity forms the star. It is the atoms that start to spin releases the star from the frozen cocoon in the centre of the galactica and this freedom is that which the movement of the star then forms. When the mutual singularity becomes strong enough to form an atomic individual singularity the stars seeks independence from the galactica centre as it drifts outwards.

The Titius Bode law holds the values it represents in accordance to $\Pi°\Pi$ manifesting singularity from which develops gravity $\Pi\Pi^2$ and space.

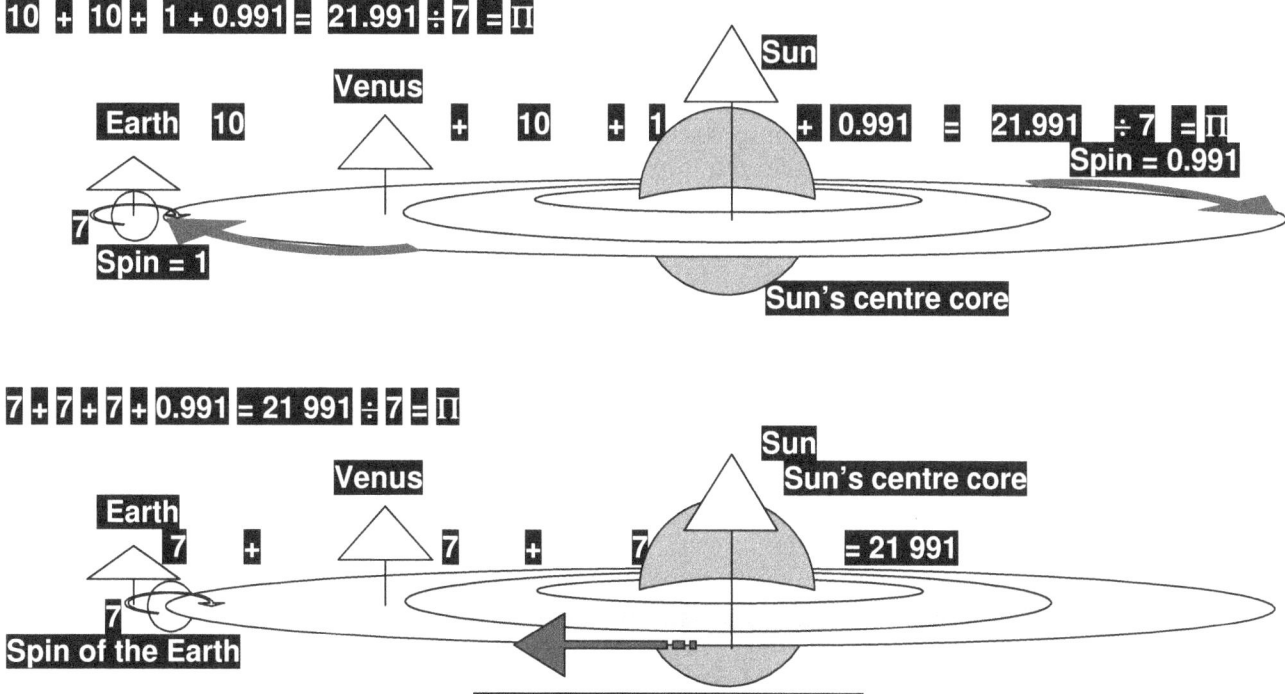

The Absolute Relevancy of Singularity used to explain THE COSMIC CODE

In the Titius Bode law we find the value of 7 in duplication (distances from the Sun doubling in relation to positions held by planets and the positional location the Earth has as 10 in relation to the Sun.

Planet	Mercury	Venus	Earth	Mars	Ceres	Jupiter	Saturn	Uranus
Bode's Law Earth distance	4	7	10	16	28	52	100	196

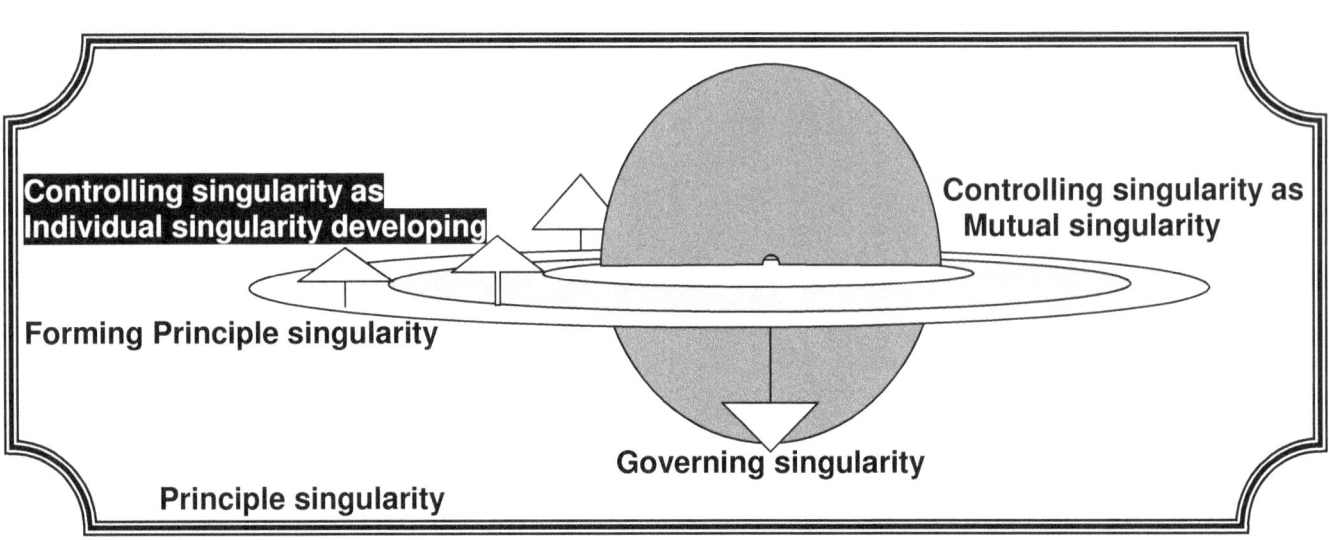

The value of 7 doubling and 10 forming space only applies as it does when the viewer is on Earth and connected to the Earth's governing singularity. When the observer is on any other planet the 7 and 10 values will apply in relation to that position and the Earth would have a different settings altogether. The fact of the Titius Bode law is more accurate and better proven by the Universe as an applying law than the contradictory Newtonian misconception about mass forming any validation in the Universe is beyond dispute, regardless of Newtonian misplace culture sentiments. The Titius Bode law is a fact and the way that the Titius Bode law proves that $\Pi^{\circ}\Pi$ forms the building block of space-time is a fact being far above any reprehension about validity. This is the manner applying in which space-time accumulation grows. It grows by the measure of Π and therefore every galactica grows by the measure of Π.

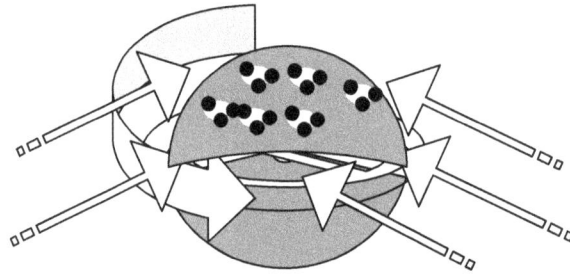

At the beginning of a galactica that is yet to develop and is still under construction the stars within the galactica centre are locked inside the galactica in a cocoon in the same manner as atoms are within the star spinning while the stars wrapped up within the galactica are in a frozen state of forming unit with the spinning galactica. Betelguese is one such an example of not being a star but a galactica not yet developed.

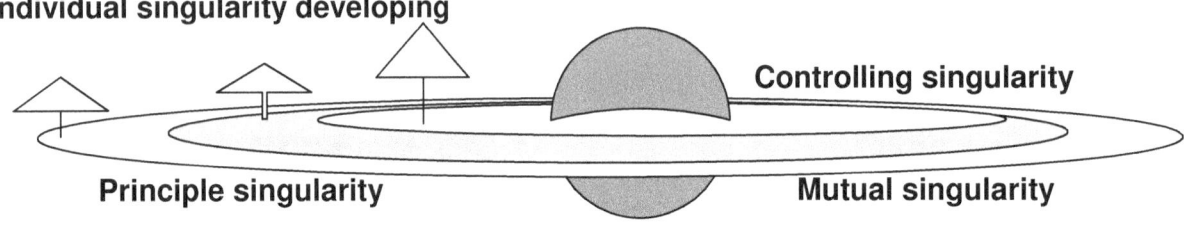

The galactica at that stage acts in the same manner as a star does with the sole difference that the stars do not independently spin inside the round galactica while they are cocooned in a frozen state while in the case of real stars the atoms will still spin even being cocooned within the star. The reason for this is that atoms are Universes and the rest are subsidiaries acting only to house the atoms through the atom's development of the Universe.

Initially the sun was part of the centre. There was a mutual singularity holding the sun as particle of the space-time of the mutual singularity. To the outside were other stars with larger inner-Core-Values and had more matter surrounding their individual singularity

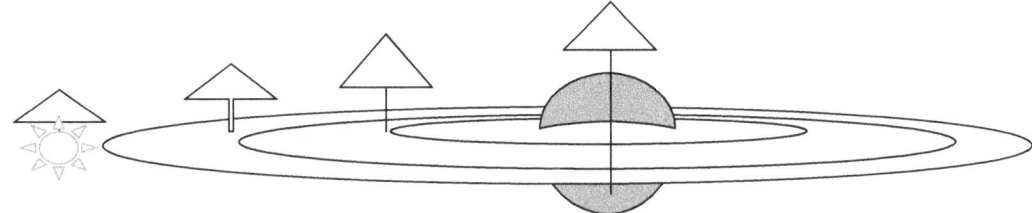

The relevancy changes while the size stays exactly the same but for the changing relevance .The star gains independence as the atoms charge a more prominent governing singularity within the centre of the star the controlling singularity find a way to form $\Pi^2 \div 2$. This puts space forming between stars in the galactica in a position of the controlling singularity being larger that $4\Pi^2$. The stars then start to rotate as the mutual singularity allow the star to find independence by establishing individual singularity as the atoms drive a generated governing singularity. Before the Roche limit validates the position of solids in relation to liquids within the star everything within the galactica centre is spinning as a liquid unit and even the solids are performing as a liquid. Remember that even sand can form a liquid state although sand is a solid, with the difference being in relativity.

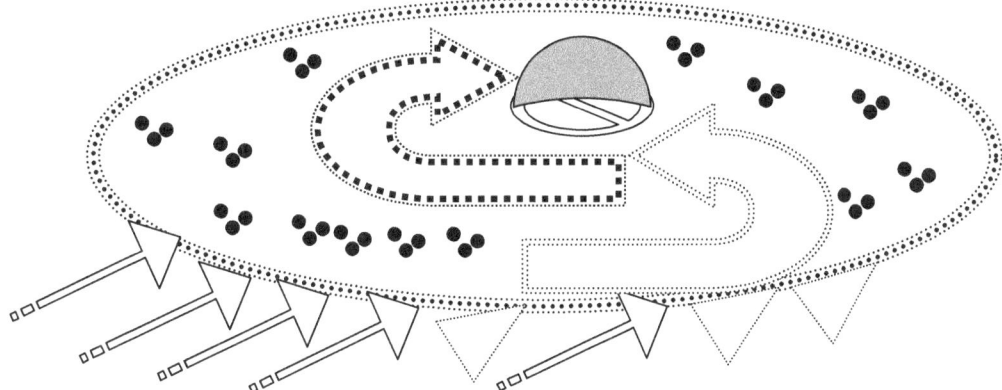

When the Roche limit establishes a differentiation between solids and liquids within the newly developed stars spinning, the shift takes place from the focus at first forming a bowl where a controlling singularity is harbouring all under developed stars within the centre and this moves out in a ring to where the stars then form a round rim on the outside as the rim of the galactica then becomes flat. This is because the individual stars b spinning then gains an independent state through an individual singularity where such focus is on every star spinning and therefore contracting heat as gravity. In the first stage of galactica development and at the start of development the roundness of the galactica harbours the stars while the stars form in the centre. At that point the stars in the centre are still under development and in a frozen state and the entire galactica acts as a unit concentrating heat by a mutual singularity. Then as the star is developing the inflow of space-time shifts the movement or focus from a plasma unit which it is at first towards the individuality of spin of stars since the star is then generating an individual flow of space-time towards an independent governing singularity of independent stars forming a ring while this lot is still in relation to the principle singularity of the galactica but progressively developing the individual singularity of the star and then to the middle sectors and eventually to the centre of the centre. The process can be traced back to the way that satellites encircle the gas planets such as Jupiter Neptune and the others.

The validity of this applying is proven by the way the Roche limit applies as a cosmic entity in terms of the so called "sound Barrier" Fledgling stars apply the same starting procedure when igniting an inner core as the aircraft do that goes through the "sound barrier', which is one process desperately in need of a new name.

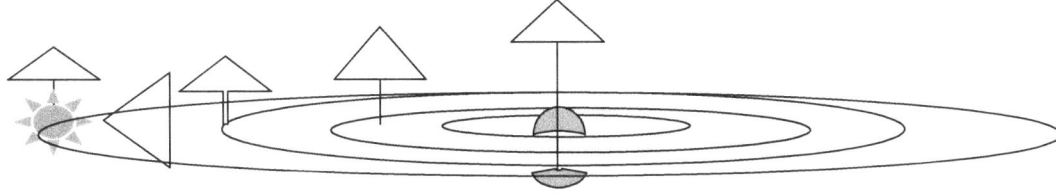

This reality is that stars that do not yet displace space-time by virtue of forming an independent governing singularity forms part of the galactic mutual singularity and confirms a controlling singularity in terms of the

galactic governing singularity in terms of forming a structural sphere. This places the star in a frozen surroundings freezing it in time within the centre galactica and I am of the opinion that some big stars such as Betelguese is a galactica waiting to transform the $\Pi°\Pi$ aspect to individual star formation $\Pi\Pi^2$ still to launch its independence Π^3. The sphere holds the fledgling stars safely protected in a wrapper and out of harms way until such a time that the space Π^3 releases by gravity Π^2 forming Π. This has nothing to do with mass because the stars don't have any mass because the stars are floating in a cosmic frozen liquid.

The Hawkins Black Hole is just another Newtonian forms far outside of reality. Every galactica grows governing singularity that the

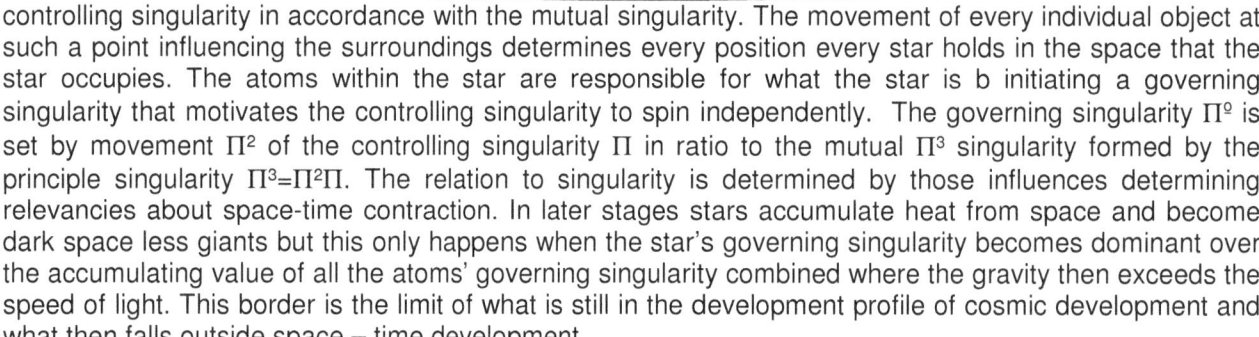

misconception spawned by ideas that individually in accordance with the determines the limits of

controlling singularity in accordance with the mutual singularity. The movement of every individual object at such a point influencing the surroundings determines every position every star holds in the space that the star occupies. The atoms within the star are responsible for what the star is b initiating a governing singularity that motivates the controlling singularity to spin independently. The governing singularity $\Pi°$ is set by movement Π^2 of the controlling singularity Π in ratio to the mutual Π^3 singularity formed by the principle singularity $\Pi^3=\Pi^2\Pi$. The relation to singularity is determined by those influences determining relevancies about space-time contraction. In later stages stars accumulate heat from space and become dark space less giants but this only happens when the star's governing singularity becomes dominant over the accumulating value of all the atoms' governing singularity combined where the gravity then exceeds the speed of light. This border is the limit of what is still in the development profile of cosmic development and what then falls outside space – time development.

By duplicating the space of any particle sharing space within a larger cosmos structure such as an atom inside a star or a human inside the Earth there are two relations applying. At this point I must indicate that I entirely disagree with Stephen Hawking on the matter of a Black Hole being in the centre of a Galactica. Yes…a Black Hole there is but it is there because the principle singularity controls the galactica governing singularity through the mutual singularity giving the controlling singularity such a controlling prominence. The principle singularity forms a mutual singularity that is charged by the application of the individual singularity that is forming the mutual singularity bondage which can spread through the controlling singularity to such a point that all the movement of all the material within the galactica charges the galactica governing singularity into forming a Black Hole. The Black Hole is not an independent entity centred on its own in the centre of the galactica but is the result of all the combined movement of all the material and atoms spinning within the entire galactica and is the result of all movement that forms a controlling singularity by charging the governing singularity. If you want to know how strong must the applying singularity be that forms by gravity a Black Hole, then all the known and unknown stars in such a galactica combine their movement in order to form a controlling singularity that then results in a Black Hole forming as a governing singularity within the centre of that galactica.

The galactica and the star that ended as a Black Hole started off equally in dynamics but much different in mass distribution. The star that eventually became the first Black Hole united the mass into a single point serving where all the atoms united in one inclusive governing singularity it had forming the controlling singularity, while the galactica distributed the mass it had in separate cluster packets I call proto stars.

The Black Hole and most galactica started off on the cosmic developing journey at the same time period but where the release of the star was as a unit the galactica release of stars was in stages being time driven. Both had equal singularity generated, except for the way the material compliment produced the governing singularity by motion of duplication and proton cluster density. The atom is the star and the atoms unite to form a star. While the atoms are controlling the governing singularity that is maintaining the star, in such a developing star of which the Sun is one, the gravity that the star will produce will be slower than the value of the electron and thus lower that the speed of light. In the proto phase before individual singularity comes about, the galactica froze all atoms preserving the atoms in units that later on became stars. In stars falling outside galactica development the stars developed through the atoms compressing space up to the speed o light and then even much later preserved singularity to become Black Holes as the gravity exceeds the speed of light up to a point where infinity reunites with eternity.

Any Black Hole forming as an independent structure in the centre of a galactica cannot be possible since the galactica presents more motion than what is in the entire galactica present with all anonymous hidden matter included. This I say because the Black Hole is a star that has gone the full circle from the time that it was forming as a developing star and then went through all the stages to a point where it has developed back to singularity while all galactica is a unit holding lots of stars all being somewhere in the process. In the centre of the galactica there is the very opposite of what the Black Hole is and which is something that still has no name given yet. In that centre the governing singularity forming is so much in charge by all the spinning of material forming a controlling singularity, it has not yet released space-time but is feeding on space-time formed by material within the principle singularity inside its realm of influence. To maintain form the governing singularity is removing space-time from the Universe and charging the yet to be developed space-time within the structure that is still in a state of pre Big Bang conditions. Since that is not the same as a developed Black Hole gone through the cycle, material will escape outwards in seemingly a very small amount since there is an outflow of space-time towards the outer limit of galactica, as the displacement will require space duplication that causes ventilation. This comes as growth accumulates heat around structures spinning and by concentrating heat a directional flow of heat is established where such a flow of heat alters the shape of the galactica in accordance to the applying controlling singularity at such a point. Black Star relevancy in equilibrium to governing singularity

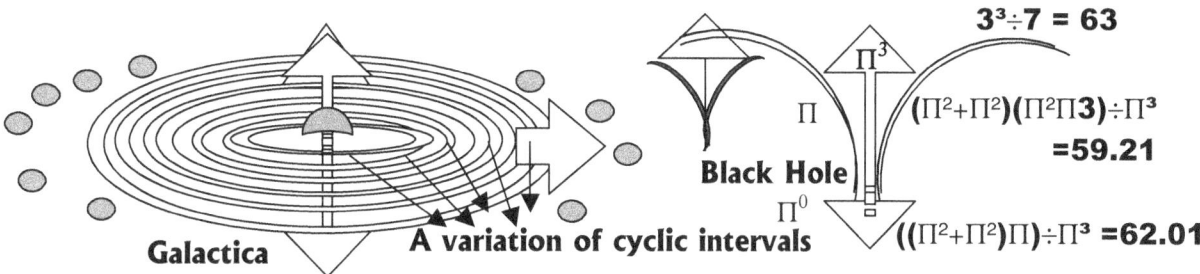

In the galactica many stars form units and the units form partitioned identities in clustering around a unifying governing singularity by spin that holds a controlling singularity, which forms the spin as units or stars. The units are at first clusters of material being stars-to-form that by spin become stars. The clusters or pre-stars or proto stars preserve atoms that provide atoms a wide range of units to form as the atoms can group together in layers. Every layer forms a star within a star serving the greater function of the star by providing a lesser and limited function in the totality of the start. The volatile outer layers form a role of displacing or moving through heat and the inner layers of the stars form a role of dismissing or contracting heat. Every layer is a time-capsule there to serve the star in a specific time developing faze and is to be discarded as the ability to dismiss becomes greater than the ability to displace within the star. Having layers that help the development of stars through fazes in that this method is giving every star a personal significance and time period of development within the galactica whereas in the Black Hole, by eventually becoming a Black Hole all the material that once were then finally becomes the same in total as forming one united point and being one point that is holding singularity at a governing singularity Π^0 while having the controlling singularity formed without space but in presenting a space less controlling singularity Π.

At the end when the star is developing towards a Black Hole, the variation of space occupation is much less using only atoms with much more protons in numbers per cluster unit or elements as they are more widely called and eventually only the atoms that can dismiss and displace to the proton value of between 55, to 62 remains in use. By such a time the electrons has vanished a long time ago and the neutrons went away and before collapsing into the eventual Black Hole only the protons in the atomic - clusters from 55 t 62 remains viable. In the galactica, the time factor gives a variety by which the different material forms by which stars can develop and the atoms form many equal points all preserving the governing singularity within the galactica that becomes the controlling singularity represented by the star's governing singularity, where each element is serving the star by being grouped by ability to perform space-time dismissing and displacement in a layer designed and designated to fore fill just such a purpose. In the star at first there is a range of cyclic periods which is actually time cells in which to develop the star through stages and give a range of singularity developing periods and the name by which we think of those are layers.

This comes from the density and the duplication in every star where every layer is in a position to function as a collector of liquid as to provide that star with the required cooling. Through the spin the atoms provides space-time by contraction that forms the unit called a star and they group to form a star through connecting singularity. The quality of spin of all the atoms would achieve a governing singularity that connects to an atomic governing singularity acting as the star's controlling singularity that in turn would connect to a governing singularity of the galactica serving as a controlling singularity of the galactica and so on…the connection is never-ending. In the Black Hole all the atomic material that connects to the centre governing singularity of the star eventually becomes space less and the same, but all atoms then produces one unit. In the galactica this process of development is quite the opposite and before the process starts the cluster forms a unit but then later on it becomes a multitude of stars in cluster as it then forms spirals around the galactica. With the massive proton numbers that forms proton clusters in stars a very high demand is placed on dismissing and little is on duplication whereas at the time the Black Hole is in developing, all movement is within space having space collapse on the Black Hole where then we see only dismissing with no emphasis at all on duplication of material. In the galactica this is the very opposite where little is promoting the dismissing factor as almost only duplicating is present since all material is spinning around the galactica centre. By stars being developed it seems as if every aspect in the galactica is waiting for development by having applying the duplication and having the dismissing factor developed. Every layer is a phase ad when that fazes done the singularity forming the atoms in the layers goes singular as the layer disappears. When these fazes becomes redundant and the layer becomes obsolete, the star goes into a flashing faze where these fazes are named as variable stars and pulsars and Cepheid's and holding (I suppose) more names that the alphabet can spawn.

The cluster of stars in the galactica reaches maturity in singularity development as they start to spin independently. This fact we can see from moving objects on Earth where the same process is used by life moving objects. We can detect this being the case by an aircraft showing the precise attitude when reaching the "sound barrier" or the Roche limit factor $\Pi^2 \div 2$. One must keep in mind that life can only manipulate and apply cosmic law because life can create nothing at all and therefore if it is there and then only can life use it.

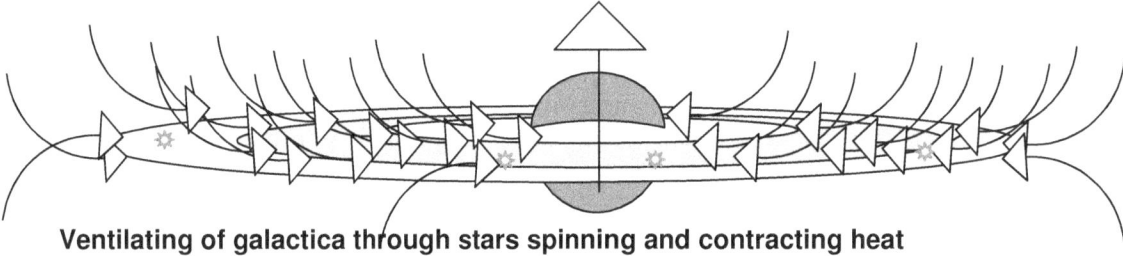

Ventilating of galactica through stars spinning and contracting heat

It will be the task of science to see where the cosmos use phenomena instead of telling the cosmos it should have mass and that it is mass that creates gravity and all the other Newtonian bullshit. When the movement of the released star finds a strong enough gravity to start the spin and break free from the frozen state in which it is, such maturity finds release from the centre governing singularity as being part of the controlling singularity, much the same way as stars forms the controlling singularity in relation to the galactica governing singularity, however although finding independence by enabling individual singularity, it never strays out of cyclic developing order and remains part of a galactica governing singularity, however much the relevancy changes. In such a release the stars by spin moves away from forming a mutual singularity in relating to the controlling singularity still forming as part of the principle singularity to form independent singularity granted by gravity in spin while forming part of the principle singularity while still being also part of the mutual singularity within the galactica. In the case of the Black Hole there is one unity that form one governing singularity where all atomic singularity becomes void of space and remains as Π attached to such a centre holding the governing singularity Π°.

The governing singularity within the star centre is strong enough from the start to allow individual development of all the atoms to dismiss space completely in the unit and all material works as a unit as the star develops towards achieving one united singularity Π as Π°. The mass however connecting to the main singularity in the galactica is the same as that which connects to the singularity and if there forms a governing singularity within a galactica large enough to represent a Black Hole in the centre it is mind blowing to think what number of atoms once were in the star to turn such a star into the Black Hole. The Black Hole that Hawkins saw forming in the centre of the galactica he was investigating represents the governing singularity of every atom that was spinning within the star and that then is the equivalent of all the atoms acting as the controlling singularity within the confinement of the principle singularity being the containing mutual singularity on behalf of the governing singularity forming the galactica. In the Black Hole singularity makes the atom structures redundant as it unites all proton singularity into one unit forming something we still need to name. The galactica has the proton mass more widely spread and by development brings space that spins which then by spin parts the formed units to form individual clusters with many different centres attaching to a centre singularity but with much less confining value.

The cluster we call stars as the stars form individuality by moving away from the centre singularity that keeps the galactica united. As the singularity governing the star finds atoms developing, it would move away from the centre attachment of the galactica and progress to achieve independence in the same way as the spinning top tries to achieve independence. It is this developing law that we mimic in the top we spin. The Black Hole eventually destroyed all means to duplicate material and was set on dismissing space-time. In a book of mine called **"STARSTUFFIN" ISBN 0-9584410-3-0** I describe and explain how the two developing stars progress. (A galactica is just another big star with spawning abilities as it develops stars while staying on route to self-destruct.

By releasing stars when the time comes for such release of a star to occur, it keeps cosmic development on course and the release of stars are matched by the development of the Universe in its entirety. The galactica releases stars in line with a matching of the governing singularity in recognition of the Universal expanding of heat forming space. The process is interlinked by the most precise matching there could ever be. All released stars will eventually, in far off time to come, end their development as Black Holes but that is in time to come). I show there is a time coming in the eventual future of comic development so many eternities from now where every atom will finally form a star and that star will collapse into forming a Black Hole. I am of the opinion that most if not all so called giants stars are galactica that has not indicated any form of growth because of the little intensity in dismissing there is and the high value of duplicating the unit shows. The difference between the two options that the stars had when the galactica and the Black Hole-to-be formed was one in the form of the galactica concentrated on duplicating and by duplicating developed pockets of forming dismissing eventually and the other star expired duplication long ago and only concentrated on dismissing. But it is important to realise that every atom is a black hole because every atom absorbs light in a very slight degree. That puts every star including the Sun and the Earth on the road to become a black hole in the future.

There are two definitions we can use when looking at such a growth. We can look at the space not holding material that grew in size in which the stars froze their development by remaining behind all because of a lesser developing singularity or we can focus on the stars growing and with that push the space much more into expanding. By maintaining a controlling singularity in terms of the governing singularity the star freezes cosmic development as it comes about in its search for independence and matching that to cosmic progress it holds the atoms governing singularity to match the value of what the cosmos had before and during the time the star became independent by spin.

As the cosmos grows in space, the cosmos expanding progresses just as much as the star was what the star is reducing in space and the space that the star left behind is the same space as that with which the cosmos expands. This ratio is the ultimate relevancy. In that there are young stars still to develop which can include a grain of dust and then there is not that young stars that at present might be neutron stars and pulsars but it has no implication on the particles in volumetric size holding mass and as such in terms of cosmic development the object produces gravity. Say if an object had the intensity of 1 kg in outer space (and this is just plain speculation in order to drive home a point), the object will be 1000 times more dense on the outer edge of the Sun or it will be Π^2 more intense on the very inside. That object will be 1836 times as intense in a neutron star or 6188965056 million tons in a Black Hole. This comes from the manner that the star manages to destroy space and redirect the space to fluid heat or the solidity of frozen space as matter really is. In the Black Hole it reduces much further as it claims the singularity, which the object had, and destroys all space and all time there ever was.

The normal developing of galactica suggests not the Newton formula but much rather space dividing and stars on the rim developing.

There is and there can be no such a thing as "dark matter". What would make matter "dark"? If the material is "light" it then has a higher concentration of light than where we are at present. That puts that object in a colder environment and in higher moving surrounding than where we are. If the material is "dark" our light is moving towards that position and that makes that area move slower that what we do. That area is therefore hotter and more expanded than where we are. It is a question of relevancies applying by movement in relation to "standing still" or moving faster and moving slower. From the Sun we would be so dark on Earth we would be invisible but by only using the naked eye Pluto is so dark it is "invisible" to the normal human perception. It is because "space" there is much more expanded out there than what it is where we are and if it is more expanded it is moving slower making that area hotter. Then we have Mercury of the approximate same size but are very visible because it is more compressed in that area and therefore more visible than where we are. The space through which Mercury moves is denser and the space through which Pluto moves is less dense. Everything forming the entirety we call the Universe and everything in it works on movement in relevancy…movement faster than and movement slower than. It is space occupied moving through space unoccupied and moving though space unoccupied leaves a ratio or a relevancy. Movement faster than would represent heat formed in colder vicinity where the heat seems hotter but because of concentration is colder and with the faster movement thereby is producing a concentration of heat whishing to expand by moving outward and going to a larger area. The darker area moves slower, which forms a wider circle with less concentration of heat and thereby gives a hotter environment that moves slower.

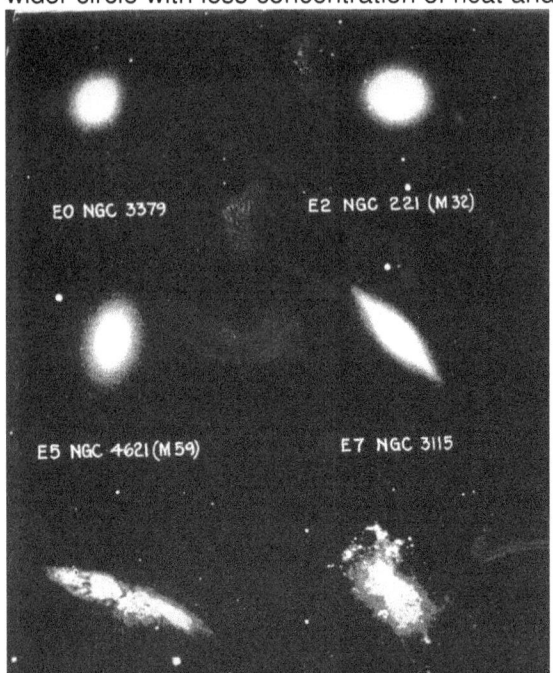

It is clear to view that EO NGC 3379 as well as the one next to it E2 NGC 221 (M32) has the perfect "bubble" formed as round as any star. The next two E5 NGC 4621 (M59) as well as E7 NGC 3115 is forming an oval shape and for that there too has to be a very good reason.

The bulging forming of a sphere shape bubble is due to a mutual singularity applying throughout that spins as a unit sealing off what is inside from that, which is outside and the cooling of the unit, is formed by one circling movement of the entire sphere. The effect of going oval from a sphere shape has to be as a result of individual stars performing as independent s8ingularity forming the controlling singularity that charge an independent governing singularity to connect to the galactica centre which is vested within the centre of each star in the circle.

At this point the top springs to mind where by spin the top establishes governing singularity that charges a connection to the controlling singularity that orders spin from independent gravity

Then finally the centre fragments and the shape the galactica had before goes to pieces as the individual stars take charge and promote independent galactica regions.

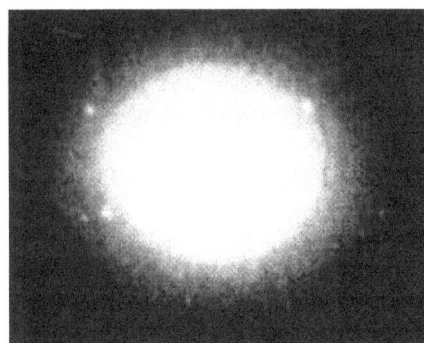

At first the galactica holds the form suggesting the ever-popular sphere. This implicates gravity being present as gravity is NOT about pulling and pushing but about diminishing space and therefore applying the strongest gravity in the centre where the least space is located. It shows galactica once started out as proto-stars but since then became a breeding zone for developing stars.

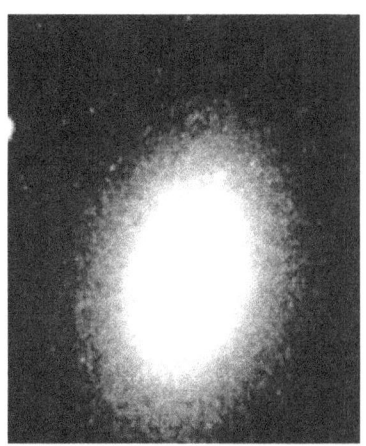

It can be seen that individual star already form independent circling on the rim of the round galactica as the stars are driving through an independent governing singularity forming a controlling singularity valuing an independent Π

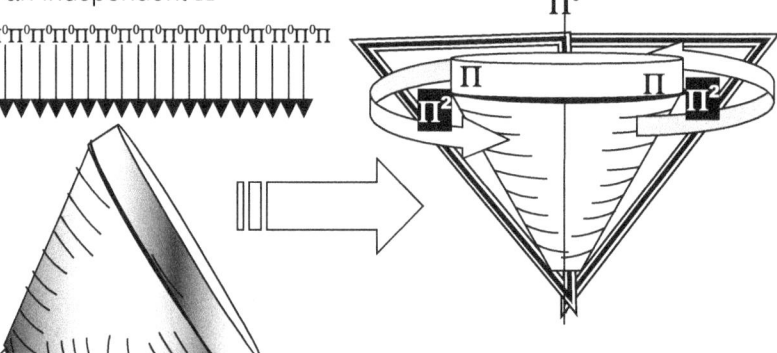

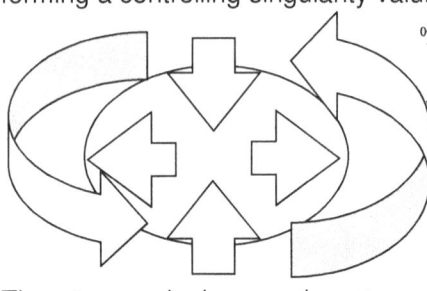

The top spinning and not spinning holds the key to any question that might be asked in proving the explaining I growth by accumulation and the form that the galactica see how different individual provide. The spin of the galactica promotes this brings about a change in the shape and takes on. From the photo one can clearly stars start to shift towards the outside and start to promote individual singularity as the stars break free from the original mutual singularity that held all the stars in one container. As the spin moves about an axis that charges a governing singularity, it is at this point that we can learn from certain behaviour of cosmic principles such as the top that spins and the so called "Sound Barrier" how the that mutual singularity will react on movement of individual structures and how in that manner the core of the star will then grow by gathering heat much the same way as the spinning top will react on movement set within boundaries on Earth. There are only four principles applying that are responsible for gravity and life has to manipulate these principles to make do with applying movement. By "breathing" the stars develop the galactic through investing in the governing singularity Π° in terms of a controlling singularity Π. This effort pulls the controlling singularity into forming a ring instead of the sphere, which is what the mutual singularity advanced. The individual singularity reforms the status and form the galactica takes on. Then there forms dark and light areas but there are no dark and light in the Universe. There are hotter parts that seem dark because there is more expanding thus less movement and then there are lighter areas where it is cooler and that is because more movement is active.

The developing process shows directly how the stars develop on the inside and this shapes the form of the

galactica by stars forming which is progressing outwards from the galactic centre and as such shapes the galactica as the development progress through the stars growing. Stars holding position on the edge of the galactica consume more space than the inner centre can construct space by developing. If Newtonian principles of matter grabbing matter did apply, such development was not possible because in the galactica all that forms would collapse. The shape however tells a different story altogether than the Newtonian picture. In accordance with the Titius Bode principle the growth of space between stars doubles in value each time and therefore where the galactica is at the widest at the centre of the sphere the star will double the growth as the space growth progresses outwards from the middle of the galactica. This will bring about an oval shape developing.

The effect shows the more dynamic stars is growing by movement accumulating space to the outside of the centre that forms an circular oval ness and as they grow the spinning stars accumulate and concentrate heat at a higher rate and the space then in the rings grows faster than what the inside stars can produce by occupying more space. In that the development shows that in the oblong shape galactica take on.

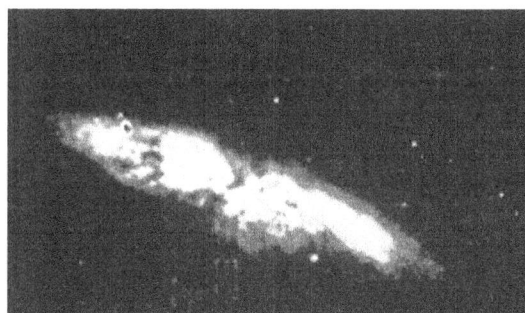

With insufficient prominence serving the original galactic governing singularity that will maintain the galactic form through the controlling singularity; this allows the mutual singularity to break up as the activated individual singularity forms a responding governing singularity that fractures the galactica into many petitions. Every star core forms a governing singularity that charges an accumulation all possessing different points sharing one controlling singularity. By the spin developing the stars, there is a lack to commit the form in securing the inner stars position to heat and space as bonding the space to the mutual singularity, the outer stars

show much more dynamic propulsion leading to more dismissing of space which then allows the galactica to disintegrate and become malformed. This is the curvature of space-time happening in the reverse where the governing singularity on the inside puts different prominence to the outside altering Π while leaving the galactica to disappear into obscurity. It then becomes clear that the galactica fragments into many pieces and each form a governing singularity charging not one, but many points holding principle singularity in charge of different galactica structures.

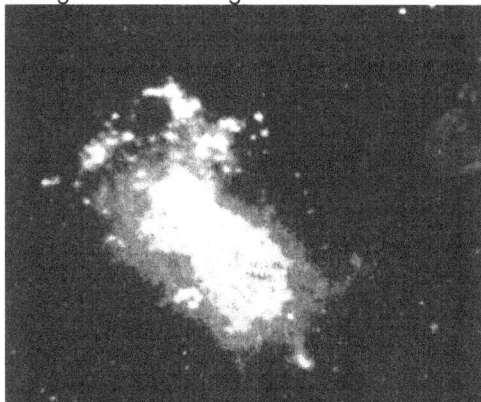

The photo presents a very typical development pattern of how galactica development takes place through the developing of stars within the galactica. One can clearly see the heat cloud on the outside become a misty shade of what it once was as the stars on the edge dissolve the density that first protected the stars before they became active by establishing individual singularity relevancies. However, also by establishing different points serving a governing singularity that charges different positions holding a controlling singularity, the emphasis that this shift in governing singularity brings about, is that the mutual singularity splits to favour many sectors where each mutual singularity sector holds individual singularity in prominence by charging one position to each of the sectors developing with a governing singularity presenting a controlling singularity difference.

In this manner the galactica is a cradle in which stars are hatched before birth but the birth of stars forming the way they are has nothing to do with magic of mass pulling dust into rock or whatever Newtonian fancy fairy tail stories would be believed, but they are as they came about in the instant the Big Bang took place. What will be was when the Big Bang brought form to a dimensionless singularity based Universe. By the spin with which the Big Bang gave conception to what then came in place, relevancies concatenated and material connected form to positions and this formed a Universe to come.

The following diagram that I am about to show is part of the cosmic code in reference to star development concerning the working principle of every layer and it is showing the working relation between the displacing of heat and the dismissing of heat in regard to the <u>controlling singularity</u> acting as a reference point to the <u>governing singularity</u> in regard to the <u>individual singularity</u> forming the <u>mutual singularity</u> applying within each specific layer.

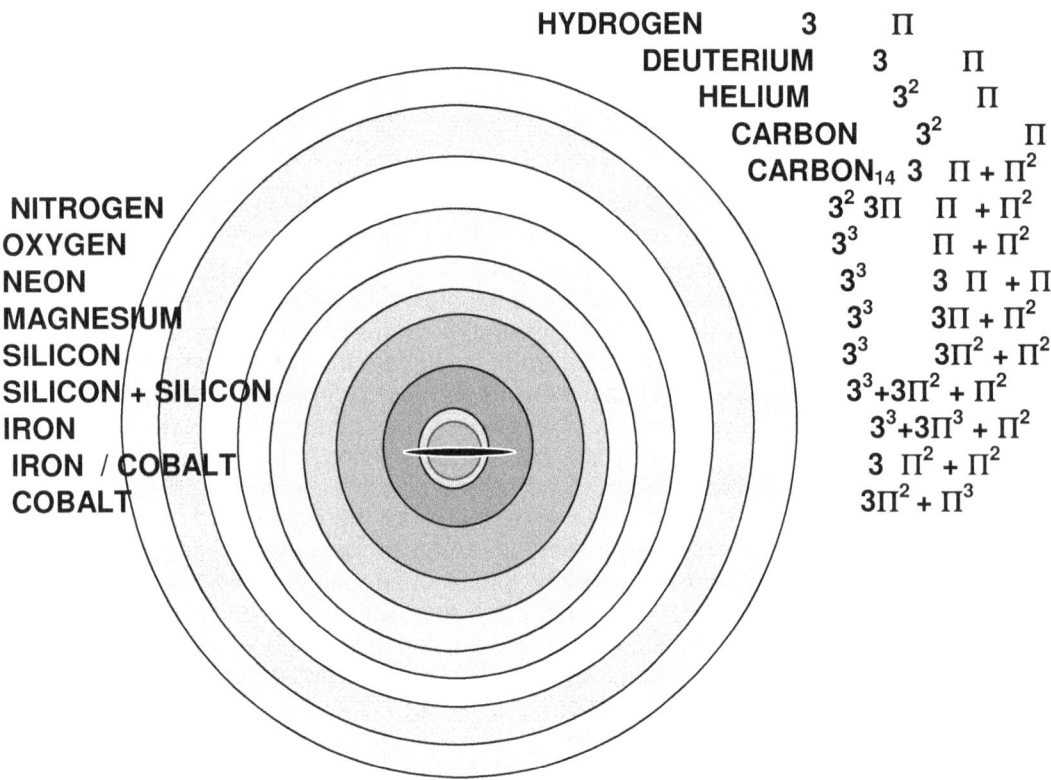

	HYDROGEN	3	Π	
	DEUTERIUM	3	Π	
	HELIUM	3^2	Π	
	CARBON	3^2	Π	
	CARBON$_{14}$	3	$\Pi + \Pi^2$	
NITROGEN		3^2 3Π	$\Pi + \Pi^2$	
OXYGEN		3^3	$\Pi + \Pi^2$	
NEON		3^3	$3\ \Pi + \Pi^2$	
MAGNESIUM		3^3	$3\Pi + \Pi^2$	
SILICON		3^3	$3\Pi^2 + \Pi^2$	
SILICON + SILICON		$3^3 + 3\Pi^2 + \Pi^2$		
IRON		$3^3 + 3\Pi^3 + \Pi^2$		
IRON / COBALT		$3\ \Pi^2 + \Pi^2$		
COBALT		$3\Pi^2 + \Pi^3$		

THE INTAKE OF HEAT IS **AND** THE SPACE-TIME CONSUMPTION IS

HYDROGEN	$3^3 \times \Pi^2$		3	Π	
DEUTERIUM	$3^3 \times \Pi^2$		3	Π	
HELIUM	$3^3 \times \Pi^2$		3^2	Π	
CARBON	$3^3 \times \Pi^2$		3^2	Π	
CARBON$_{14}$	$3^3 \times \Pi^2$		3	$\Pi + \Pi^2$	
NITROGEN	$3^3 \times \Pi^2$		3^2 3Π	$\Pi + \Pi^2$	
OXYGEN	$3^3 \times \Pi^2$		3^3	$\Pi + \Pi^2$	
NEON	$3^3 \times \Pi^2$		3^3	$3\Pi + \Pi^2$	
MAGNESIUM	$3^3 \times \Pi^2$		3^3	$3\Pi + \Pi^2$	
SILICON	$3^3 \times \Pi^2$	3^3	$3\Pi^2 + \Pi^2$		
SILICON + SILICON	$3^3 \times \Pi^2$	$3^3 + 3\Pi + \Pi^2$		⎫	
IRON	$3^3 \times \Pi^2$	$3^3 + 3\Pi^2 + \Pi^2$		⎬ **EXCEEDING**	
IRON / NICKEL	$3^3 \times \Pi^2$	$3\Pi^2 + \Pi^2$		⎭ THE SPEED OF LIGHT	
COBALT	3^3	$3\Pi^2 + \Pi^3$		Ω	
COPPRER		Ω			

I include this to demonstrate that the Cosmic Code has far reaching implications but I will not dare to try and explain this as part of this introduction book. In order to make this understandable I need more or less five hundred pages just to explain this what is portrayed in the sketch and there are so much more interpretations that is not part of this format. That I leave for the day I have funding to print **"Matter's Time In Space: The Theses"**

There are several large misconceptions that Newtonian science carries and cherishes and while we are on the topic there is one in specific at this point I have to mention. It is the way Newtonian science view time and time in relation to what space actually is. I am not pondering on the others disagreements I have with Newtonian views because then this book too will be ignored as all the others were ignored. There is this one however at this venue that I wish to address in the context of the subject we are dealing with. Again I come back to what I said before about time and space and that space is being what time forms as the history of what time was and I wish to address the subject because of the misinterpretation science has about time and space. Space is what my body fills and what the body we call the Earth fills. Space is what time cannot penetrate but through singularity connecting. Space is what material fills and the rest is the image left by time to form penetrable space that in truth is not for real. Occupied space-time is true space while unoccupied space-time is images of memories of time in the past laid down by the governing singularity as to be represented by the controlling singularity of a relation with the governing singularity that is forming the mutual singularity of what the principle singularity was of everything filling the cosmos at the instant the image was laid down again in relation to the governing singularity. Every point that forms while being the centre of whatever spins and by spin then forms a centralised governing singularity stands connected to whatever forms part of the cosmos that also fills the need to be moving, where all of that mentioned is then any and every instant of time moving space forward becoming the controlling singularity in terms of any specific governing singularity.

Time is in the instant of any centred singularity placing relevancies at that precise instant and the rest that seems to be space is images representing singularity in terms of relevancies as relevancies that then is applying singularity as the positions once was at the instance the relevancy was placed. Time is not focussed in what we see but it is cemented in what was when we were there where we could be allocated a point as a reference in terms of every point in the Universe being the reference of where we were. We are in time when we hold a position and the object is in reference (not the space) at the time the relevancy in singularity applied. If I try to shoot at a fast moving object and I aim at the point where the object is at the time I am pulling the trigger, I have no chance in hitting the object because I am aiming not at the object any more but at the image that the object left at the point in relevance to where I was when I took aim. Time shifts the object while I focus on the memory of where the object was in the past, but in the meanwhile time moved the object to a point I cannot yet see. What I see is an image of what time had already moved. The time part is in singularity putting the object in relevance without any space applying seeing that singularity then charges relevancy in the instant of application while the space part is the image portraying the part at the present when I was pulling the trigger aiming to shoot at an object that was no longer filling the space I see as the object. If I wish to hit the object I must disregard the image I have of the object and aim to where I realise the object must be at the point when hitting the object. I have to ignore space and think via singularity. This places the object I see that is in my imagination, which is to be ignored in favour to where the object is in my intellectual reasoning, and that proves the concept of time and space being separate entities.

The time is in the singularity while singularity places the space according to the positional instant when the relevancy applied. Everyone thinks of space-time as the same but it is the very opposite of what could be the same. The coin back and front is the same thing but the "head" part is the very opposite of the "tail" part. The one is the action that is taking place and the other is the reflection of the action that took place in the past as time went on. Space forms as singularity placing time in the governing part of singularity puts space in the controlling part of singularity as a reference to where time allocated the governing singularity and everything having an independent governing singularity will become part of the controlling singularity of any point at any instant that formed the governing singularity at that specific position.

The entire misconception involving the ideas Newtonian science carries, which includes atheism no less, is that science promotes the idea of when not finding proof of any concept put forward as the truth, then according to the philosophy of Newtonian science not finding proof does not mean that finding no proof proves the idea as being wrong but merely states there is no proof yet while the idea remains still very much viable. Then this could also mean that madness promoting fairies and a Big Foot and other forms of aliens is there to be accepted and until it is not proven that it is there it then is there all the while. Finding no proof doesn't mean disproving it. This is the grossest form of stupidity I have come across because not finding proof means whatever should be proven is not there and while it remains unproven whatever is not proven is not there. I have no idea where science has gone mad but it went off the rail completely. This is somehow coupled the affectation about life coming from other sources or places than the Earth and it is in place in order to prove atheism correct by finding half ape half man species to confirm and prove Darwinism while Darwin remains correct until Darwin is proven correct. That is stupidity beyond the brainless. Darwin is as correct as Little Red Riding Hood because Darwin is as well proven as Little Red Riding Hood and we just have to locate talking wolves that form some missing link but while we search for talking wolves Little Red Riding Hood is a statement of much to believe in.

Life found on Earth could not have come from any other place than here on Earth because gravity will not permit life to come from some far away galactica because nothing can travel via intergalactic travel. Here is the reason why there could never be intergalactic travel. Since my focus that I have though the body I use as my controlling singularity which then is in reference to me in life filling the governing singularity, the space I hold is what I am because my body fills space which is what connects my life or movement to the cosmos. Filling my body from forming a point holding the governing singularity (and the governing singularity holds time in the instant but no space at all) and that connects my body which is not me but only part of my controlling singularity, that puts all the controlling singularity in relation to a position that every atom has, not only in my body but in terms of every atoms being part of the entirety of the solar system which is connected to the gravity of the Sun and while everything in the solar system is connected in the same way to the Sun by forming the controlling singularity of the Sun holding the governing singularity, I will be unable to leave the solar system at any point in time because I will be unable to break free from the governing singularity by which the Sun controls singularity within the entire solar system as wide as it goes and that includes up to and even beyond Oort's cloud. My atoms forming my body as well as that which includes whatever atom finding itself in the solar system and has a governing singularity forming part of the controlling singularity that is connected to the very centre of the Sun which then presents the governing singularity in the Sun and that keeps me a part of the solar system is preventing me to leave the solar system.

If I wish to leave the solar system the forward propulsion I aim to charge by moving my body must be stronger than the controlling singularity vested not only in the body forming the Sun, but in every atom within the Solar system because every atom in the solar system then forms part of the governing singularity of the Sun and that gives the Sun the gravity that connects me to the solar system. I can never leave the solar system but so can no object ever leave the space which such an object forms apart of. This says that we can't embark on inter galactic space travel and aliens can't visit us from far away places because it is physically impossible to do so. The space that my atoms fill is in movement connected to time by the electron going around at the speed of light and since the relevance applying between my rotating movement of all the atoms representing me as my controlling singularity and the speed by which I displace space in a linear mode, my locomotion possibility is such that I can't move faster than what the gravity of the Sun would permit because by going forward I am circling around the governing singularity just because I form the controlling singularity. The Sun as a star is the solar system and that is what all the stars are the Universe. In the centre within every star, and that even includes the Sun, there is a point that already has gone to form a Black Hole no matter how inconsequential that point at this stage might be. To leave the solar system the speed any object must reach that tries to leave the solar system, must be stronger than the speed of light since it must be stronger than the gravity forming the future Black Hole in the centre of the Sun. This is because f gravity applying and gravity applying has nothing to do with mass pulling mass but is the compressing of totally expanded over boiling heat into such a density the heat freezes into the abyss.

In outer space the relevancy between the atom's electron forming the controlling singularity and atom's proton, which is what connects to the governing singularity, is a ratio of 1836. This has to do with the resistance heat shows to depress and to cool or freeze by movement. In this we find the ratio of relevancy in accordance with singularity to be $((\Pi^2+\Pi^2) \times (\Pi^2 \Pi)3) = 1836$. It is at that point that the relevancies start to apply because any point above and beyond $((\Pi^2+\Pi^2) \times (\Pi^2 \Pi)3) = 1836$ the cosmos is flat because the cosmos is $((\Pi^0)=(\Pi) \div ((\Pi^3 \Pi^2)$ which is a flat Universe. It is by establishing movement $(\Pi^2)=((\Pi^3 \div \Pi))$ that space $(\Pi^3)=((\Pi^2\Pi)$ comes about. There the spin 7/10 and 10/7 establishes gravity because the relevancy between the governing singularity and the controlling singularity which also is formed as outer space from where gravity starts $10/7 (4((\Pi^2+\Pi^2) =112$ and to inner space $7/10 (4((\Pi^2+\Pi^2) =55$ where gravity flows towards. The point where the flat Universe ends is where space goes dimensional at $7/10((\Pi^6) \div 6))= 112$. As the star moves through outer space and is in contact with outer space in more than one way. The star is the atom that forms the star. By moving through outer space the star is disturbing outer space as it is by the same token removing outer space. Outer space is pushing against the star by measure of the star moving through outer space and in that movement the centrifugal movement compresses outer space as outer space moves towards the star. The maximum displacement in duplication and contraction that outer space can accommodate is the total sum of the atom in relation to singularity which is the value of the proton added to the value of the Neutron added to the value of the electron where this lot forms the governing singularity as $((\Pi^2+\Pi^2)+((\Pi^2 + \Pi)+3)=35.75$ and then placing this in connection with the controlling singularity forming at $\Pi=112$ where this total starts the cosmos at $((\Pi^2+\Pi^2)+((\Pi^2 + \Pi)+3)=35.75 \times \Pi= 112$ and with that placing a relation between the governing singularity and the controlling singularity it forms the atomic displacement limit equal to that of what 112 protons to one cluster would have. With the motion that this relevancy generates that is the maximum expanding there can be when duplicating. However motion stands in relation also to contraction. The contraction is freezing of heat into a state of liquid coming from a gas. At 112 the state of singularity is expanded at a maximum and cannot cope with more heat than 112 protons will manage to control in one atom cluster. But relative to that must be a cold where such a cold will not hold space under a specific level of freezing. Beyond a specific limit the cold of space freezes time into singularity.

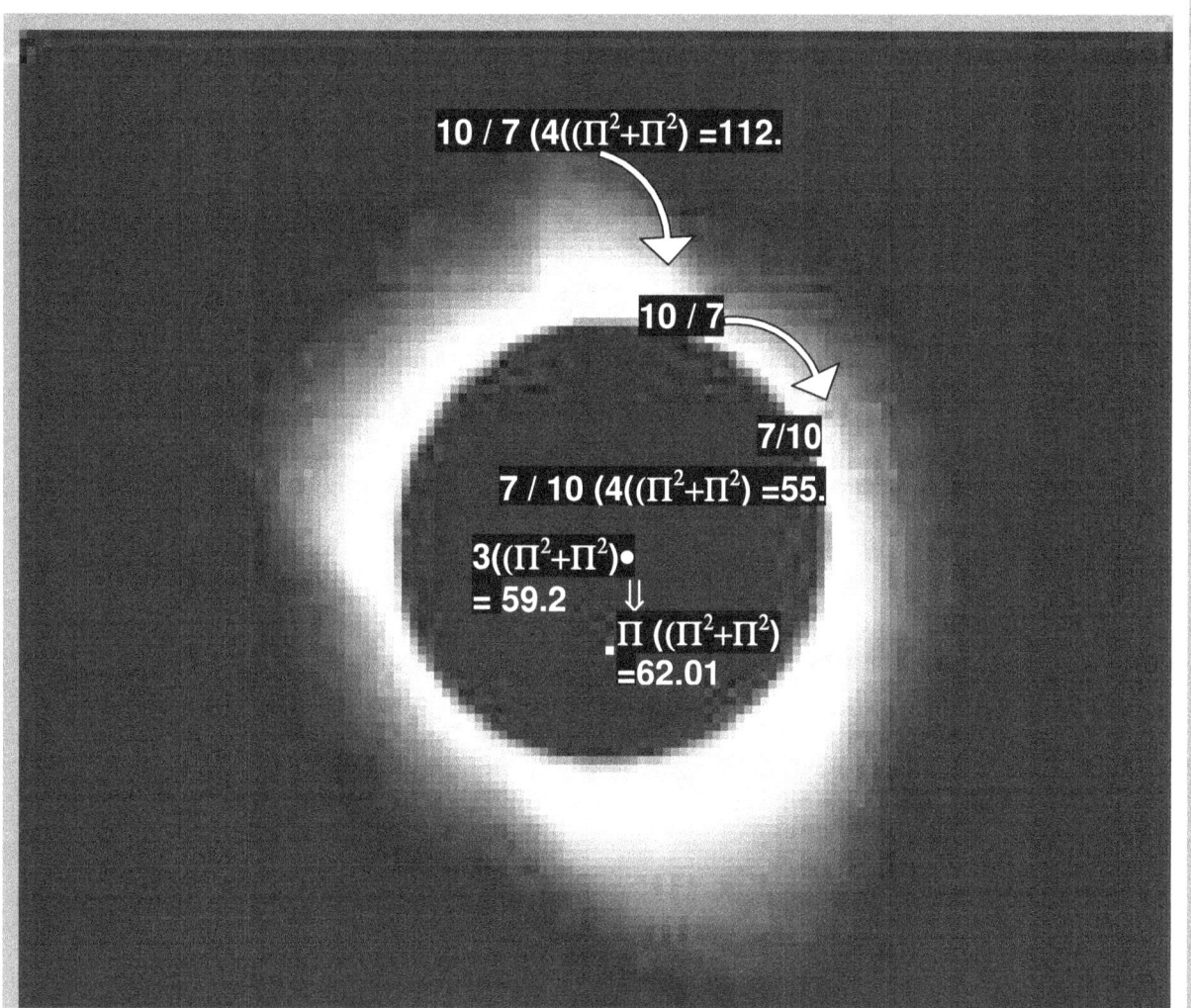

In a young star movement by displacement is all-important since it needs the contact there is in spin between outer space and the star. In such a scenario the demand on space flowing will be much more beneficial to the flow where all the atoms comprise of hydrogen and helium, which is much volatile. This is also the main substance in the outer regions that we find large super giants have. However a star just cannot die because a star is not a coal fire that can run out of fuel or flammable material. In the event where fifty percent of the star holds iron$_{56}$ and the rest is composed of silicon $_{26}$, the demand on space flow will be at a prime and the heat envelope that will support such gravity flow coming from such demand will not likely allow any fluids, which is what the photon is to escape from the star. In the relation between solids $\Pi=\Pi^3 \div \Pi^2$ which is the material spinning and thus thrusting outwards into space and what is then in relation cosmic liquid $\Pi^{-1}=\Pi^2 \div \Pi^3$ which is what the spinning is contracting, the contraction will become so fast the liquid will become so dense that what we see as the liquid will cool so rapidly (exceeding the speed of light) that the liquid will freeze solid by movement. Then the concentration of protons overshadows the flow of liquid substance and while the neutron is also a liquid the flow eventually overshadows even the liquid of the neutron by far and the star will become darker. The star absorbs the liquid quicker than the speed by which any liquid could ever escape. Such a star going darker does not die because it is not life or a coal stove that can go out. There is no time period where it consumed all the available fuel as is the case with coal stoves that was used during and just past the Elizabethan age. The fuel stars use is available in unlimited quantities forming volumes that only time extending to eternity can consume. At that point the governing singularity within the star placed a higher demand on movement on the governing singularity forming as result of atoms spinning and the governing singularity of the star generated a higher gravitational flow than what the atom's electrons could establish.

The star did not diminish the space and the space did not outgrow the star. It is a relevancy where the one factor represents a compromise to substitute for the other factors changing the relevancies applying. It is the space that grew as much as it is the star development that fell away and started in initiating independence from further development by securing sufficient heat within the inner core to provide the motion that will produce such independence from its surroundings. The diminishing of space takes place in every atom, deep within such an atom where the proton forms a conjunction with the inner core per ratio of the number of protons acting as one unit.

The result of the product is the accumulated space dismissal accumulated by the centre and totalling the complete effort of all atoms involved in the unit to form the total value dismissing or duplicating effort ability of the star. In every atom there is the dismissing of space that is fed from the top or outside of the star that spirals the diminishing value downwards towards the circle and into a centre. The dismissing has little effect in the immediate vicinity of the atom as the removing of space is compensated by the supply of space from positions where more space is available. The flow of space from outer space will substitute the dismissing of space. But in the centre of the star where all the heat accumulates and gravity is at the very prime, where there is no space left such diminishing will have an accumulating effect gathered from all the atoms' accumulated efforts that cannot be substituted by the flow since the flow comes from every proton in the atom within the star housing all atoms and protons as one unity.

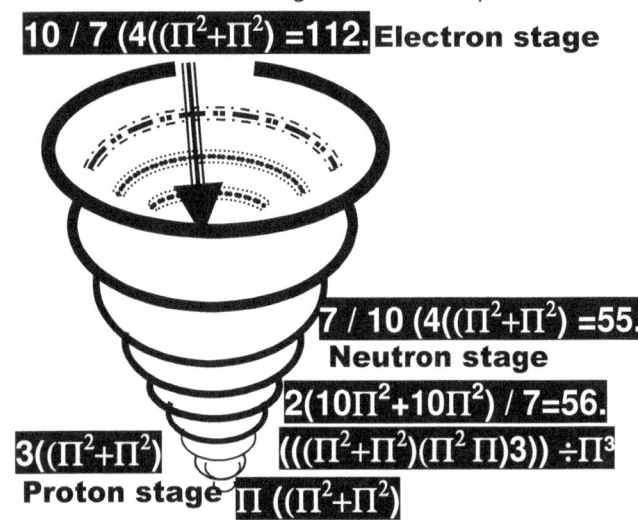

In the star such as the Sun, the Hydrogen and Helium layers brings about a volatility by the rapid movement it could sustain as to collect as much almost un- compromised heat or weak gravity producing ineffective concentrated heat because the density factor of the star's gravity is still very much under developed. As the gravity increases its dynamics the rotational dynamics reduces in favour of the accelerated downward spiralling gravity because in that the dynamics go on to favour at first the neutron stage and later only the proton stage. From that point the proton stage disappears as the atom become singular $(((\Pi^2+\Pi^2)(\Pi^2 \Pi)3)) \div \Pi^3$

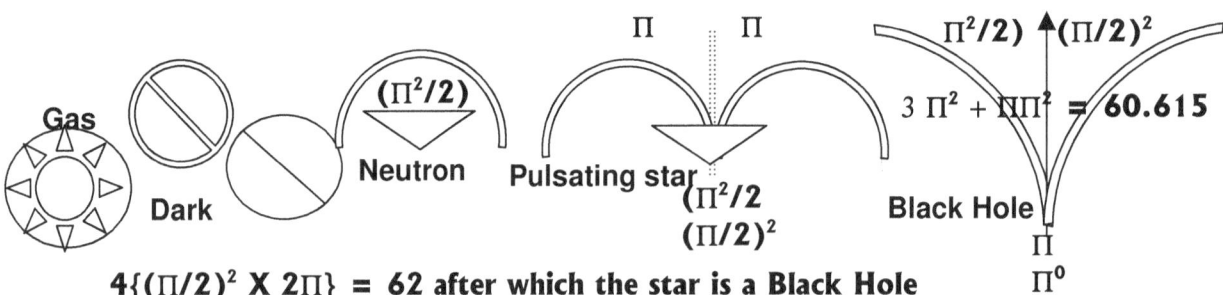

The star is on a predestined route that only diverts if the movement of the material within the layers don't bring a coherency level to the governing singularity and through the governing singularity the spin brings the correct adhesiveness to every layer.

Every star is on the inside many different stars because every layer holds a different (**k**) or relevance making the space in the star totally different from every other layer in the star. This is because every layer has different motion in relation to the governing singularity and therefore has a different gravity confining space. The layer is the result of the gravity effort of all the atoms in such a layer and therefore the space in that layer will bring about the time factor that produces the proton cluster relevancy.

Time that runs forms no space but forms singularity and that runs where space does not even begin to exist. Where time is there are no correlated positions in terms of placing a point at a certain angle and that is what Einstein saw as the Universe going flat. It is not space going flat but time is flat and that misconception makes physics become true or remain ridiculous. Time holds points on a line whether the points on the lines are coming as a result of the past (which is space) or is the result of time duplicating to form the present the line holds points on a line. That Universe is beyond us. Time has no space that forms but places points already existing or newly positioned dots in terms of a line and lines may run parallel but still forms lines in terms of all other lines. There is no way I can explain this because whatever we try to interpret forms space and by this never holding space I can't explain it as much as I can't visually explain the half circle being equal to the triangle being equal to the straight line. This is how gravity or electricity runs. The line forms a connection between **10 / 7 ($4((\Pi^2+\Pi^2)$ =112 and 7 / 10 ($4((\Pi^2+\Pi^2)$ =55.** It is the relevancy between 7 and 10 that alternates giving a condensation of heat or pace-time or whatever name goes best and that is the Titius Bode law, which is the very same law that Newtonians frown upon and say it doesn't exist. The relevancy applies as much to the star as it applies to the atoms forming the star. In this we have the integrity of the star's forming and that is where the dismissing or duplicating is represented. Where the Universe goes dimensional there will be 112 dots and where the again goes singular is at a point where space forms by the governing singularity connecting to the controlling singularity Π at Π (($\Pi^2+\Pi^2$).

The atoms displacement value places further premiums on the space-time integrity that the individual atom forms as a coherent bonding within the layer giving the layer a specific duty to function and to displace the required space-time the inner core demands to dismiss. A line must form that is unbroken by which singularity cam connect **10 / 7 ($4((\Pi^2+\Pi^2)$ =112 and 7 / 10 ($4((\Pi^2+\Pi^2)$ =55.** Not only does the atom require a dismissing proportionate to what the outer layer could cope with which then is the required displacement running through the entire star $2(10\Pi^2+10\Pi^2) / 7 = 56$, $3((\Pi^2+\Pi^2) = 59.2$ and $(((\Pi^2+\Pi^2)(\Pi^2\Pi)3)) \div \Pi^3 = 59.2$ but also it stands in relation to what demand the inner core may require in order to dismiss heat towards singularity in order to maintain the cooling of singularity and in that the cooling the entire star demands. The dismissing becomes finally resolved as the space demand within the star annihilates all atomic motion of individual atoms and the neutron, which is the representing of motion in the atom, abandons the unit of the atom to reproduce space in the manner the photon did in normal stars. In the end there are by then no hint of any photons left because a lack of motion brought on a total demise of photons. The star is not dead! The relevancy $(((\Pi^2+\Pi^2)(\Pi^2\Pi)3)) \div \Pi^3 = 59.2$ indicates that the atom $(((\Pi^2+\Pi^2)(\Pi^2\Pi)3))$ has lost the third-dimensional integrity $\div \Pi^3$ at forming **59** dots on the connecting line. Eventually only the collapsing of space can sustain the proton activity still present in the star as singularity sets in and diminish all motion activity within the star.

As explained previously the mass indicating the relevancy of 1836 comes from the fact that the proton lags in time forming space by motion to singularity, which is motionless. Explaining this concept or the following concept will also take too much time but what I mention in this page I have in a book of more than five hundred pages that covers the whole aspect and has the name of **AN OPEN LETTER ON "STARSTUFFIN'_" ISBN 0-9584410-3-0, which forms a chapter of "Matter's Time In Space: The Theses".** I will just quickly introduce you to the thought. What we see from the outside is just the opposite of what is applying on the inside of the Universe where the "inside" is singularity being in relation to singularity which forms the "outside" by eternal movement. Considering the idea that all information we receive is written in that light and that is why all the information we receive from the Universe comes to us by means of light. Because of light not representing the cosmos but forms an image of the cosmos the light by which we see is a mirror reflection of what is taking place. I shall quickly mention the most basic idea of this concept: The motion produces time and the time brings about space. Light is heat expanding while singularity compresses by condensing heat. Gravity is the compressing and anti – gravity, the mythical substance Newtonians dream of is light, which expands because it heats.

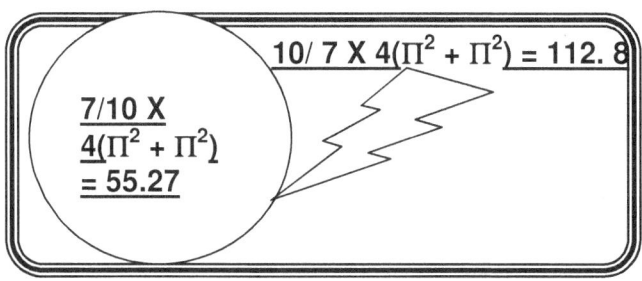

Gravity is the concentration of heat running from outer space with a displacement value of 112 to the inner star that has to have a displacement value of at least 55 to ensure the flow of gravity generated by the motion of the object within the boundaries of these two limits. The contraction of heat forms a process we think of in terms of as gravity. It stars where it is most expanded.

With an inner core displacement of less than the required 55 the star would not yet have arrived at the point of securing an individual singularity in the presence of outer space at 112. The potential difference needed to generate gravity is 112 coming down to 55 and that proves that all stars have to have an iron inner core spinning in copper.

In outer space the limit on the atom is at maximum $10/7 \times 4(\Pi^2 + \Pi^2) = 112.8$ and at minimum within the centre of the star it is $7/10 \times 4(\Pi^2 + \Pi^2) = 55.27$. Space or heat is condensed from $10/7 \times 4(\Pi^2 + \Pi^2) = 112.8$ to $7/10 \times 4(\Pi^2+\Pi^2) = 55.27$ by the rotational movement of the star and moreover of the inner core of the star. This value where gravity stars is initiated by the singularity linking value of the atom forming a relative chain of **$((\Pi^2+\Pi^2)+(\Pi^2+\Pi)+3) = 35.75$** giving this relevancy to the controlling singularity performing as Π.

$((\Pi^2 + \Pi^2) = 19.74$
+
$((\Pi^2 \Pi) = 13.01$
+   3
= a total 35.75
x Π =   112.31

19.74 + 13.01 + 3 = 35.75
35.75 X Π = 112.31
The atomic total times singularity is the maximum displacement achievable

One property I give indicates the relevance of the atom $((\Pi^2 + \Pi^2) = 19.74$ while the other concerns the flow density of outer space in relation to the centre of gravity within the star **10/ 7 X 4(Π^2 + Π^2) = 112.**. The two are very much linked but indicating how, would be a task that exceeds the purpose of this letter

The atomic total of 35.75 x Π (singularity) = 112.31. That is the limit placed on the atom within the boundaries of what we consider to be the Universe. That will remain a unit as singularity forms space-time and stars the Universe off.

The same formula applies to setting the boundaries that limits the possible duplication as it limits the boundaries after which space-time goes flat. The relevancy is **4(Π^2+Π^2).**
From outer space the atomic relevancy is as follows

(Π^2 + Π^2) Represents the proton in relevancy to singularity Π through out the universe

4 Is the time aspect of spin creating motion that is creating space.
x Π **(singularity) = 112.31**. That is the limit placed on the atom within the boundaries of what we consider to be the Universe. That will remain a unit.
From outer space the atomic relevancy is as follows:

(Π^2 + Π^2) Represents the proton in relevancy to singularity Π through out the Universe

4 Is the time aspect of spin creating motion that is creating space

10/ 7 Is the **space (10)** in which the **material (7) spin** according to the Titius Bode principle.

7/10 Is the **material (7),** which **spin through the space (10)** according to the Titius Bode principle.
Outer space has heat secured at **10/ 7 X 4(Π^2 + Π^2) = 112. 8** while the star through motion generates a requirement to heat that establishes a flow of **7/10 X 4(Π^2 + Π^2) = 55.27.**

The outer wall of outer space is **10/ 7 X (Π^6) / 6 = 112. 8** while the position that the atom demands space is the value iron have as a potential difference. It is in the **7/10** and the **10/7** that the limits are placed. It is seven spinning about in ten crossing singularity by turning about the inner core of the star.

The factor of **10Π** being in relation to Π^3 is a direct translation from Kepler's formula $a^3 = T^2k$. By substituting the symbols used with the actual value of Π the symbolic message transforms to specific values applying.

10Π Is space square 2(5) (T^2) in relation to singularity Π (k) being equal to space Π^3 (a^3)

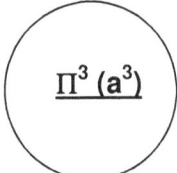

In outer space the motion **2(5)(T^2)** of the material **Π^3(a^3)** keeps space in dimension **Π(k)**. But this motion produces a relation that apply to material groups such as stars relating to space holding groups such as outer space which I refer to as geodesic space in more advanced books.

$a^3 = T^2k$ is Π^3 (a^3) = **2(5)(T^2) Π(k)**
When using the atomic relevancy I refer to the proton relevance in space in example ((Π^2 + Π^2) and then how space will relate to accommodate the atom as the atom as a group facilitate the star and accommodate the star's unifying requirements.

In the expression **10Π** relating to Π^3 it is space flowing towards the star centre in approximately an equal manner as volts flow from space to the Earth or Neutral or the governing singularity or whatever name there is to choose. One must see the 10Π not for what we read into the numbers as such but what it represents. The number 10 is the square of space that stands for space outside gravity in the place of Π^2 and is the square of space relating to the ten positions in relation to singularity dot as Π. That is the space in which the motion is providing the establishing of gravity Π^2 in relation to the creation of space by motion Π^3.

The star on the inside cannot support space up to equal or beyond **2(10Π)** before ultimately collapsing the space dimensional support of 6 sides in the square of space **(10).**

On the other hand the space in the geodesic secures the presence of the atom holding space up to the ability of 112 protons displacement secure **10/ 7 X 4(Π² + Π²) = 112. 8**. This is theory because we well and truly know that it is actually 5(Π²+Π²) (Π/2)² (3/5) = 244 which is the number of neutrons and protons that will allow Plutonium the ability to remain a constructed atom on Earth being within our Universe. But as one can clearly see it is as volatile as no other element and is on the very edge and it might be valid in some large star but is completely superficial and man-made on Earth.

When a star favours **3** and a multitude of three, the star is still in a process where it will favour more the duplication of space. As the star develops, it will ever increase as it moves through the ranks of being liquid **(Π²Π),** the favour to the proton **(Π² + Π²)** where more space is dismissed than space is duplicated, because the motion of the star also diminishes progressively towards the end phase of the star, where the star will once more be as motionless as the governing singularity is that forms the entire star. The cosmos comes together at a displacing limit of 112 protons per atomic unit. It values time to space at **10 /7(4((Π² + Π²)) = 112.795** on the space limits while motion breaks down within the centre of the star at half that being at **7/10 (4((Π² + Π²)) = 55**

Space receives six sides at **7 /10 (Π⁶) / 6 = 112.16** which is where gravity starts **10 /7(4((Π² + Π²)) = 112** while also that is where singularity forms the spinning atom **((Π² + Π²) the proton +(Π² + Π) the neutron + (+ 3) space = 35.75 X Π singularity = 112.313**

The cosmic cube we live in is singularity that is six sided **7 /10 (Π⁶) / 6 = 112.16**

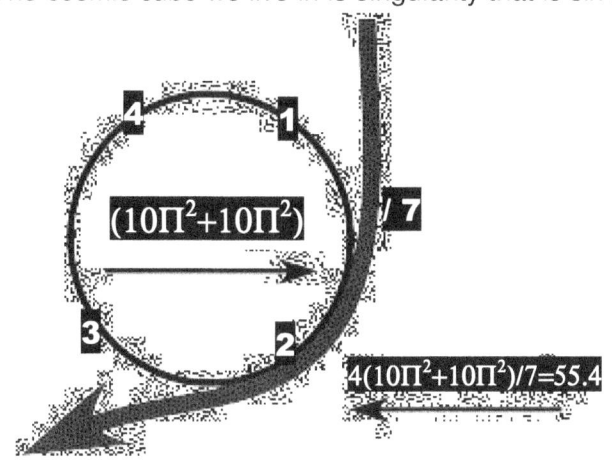

To decipher the code one goes about as follows. The value that generated gravity or if you wish to call it electricity will be as follows: 4(10Π²+10Π²)/7=55.4
4 The 4 indicate the time presence influencing the value. Four always indicates time
(10) Represents the square of space in outer space in the square already
Π² Indicated the square forming gravity
(10Π²+10Π²) Is the double proton in multiplication with the square of space on both sides of the Universe.
/ 7= Is representing the value of the sphere.
55.4 Are the maximum protons or connecting points lining up that is displacing space-time working in one space less unit.

The four quarters of time (4) holding the square of space (10) in the double to gravity Π² also in the double on both sides of the Universe (+) in a proton relation (Π²+Π²) still maintaining the shape of the sphere (/7) is representing the flow of space-time by duplication in relevance to space-time dismissing that comes to a limit will be the maximum that the form of the sphere can sustain. After increasing the flow, the sphere as form will collapse although space-time will still remain and be concentrated but valid. It is where gravity as we know gravity ends because that is where any star's liquid atmosphere meets the density of the electron and the density equals the speed of light. After this value the speed of light will no longer sustain an electron in the atom, which by that time has been compressed to its limit as it was during the Big Bang. That is the end of the electron where the star is all electricity with no space-time being less than the value of the electron.

The fact that the displacement equals the iron proton "mass" explains why iron Fe_{56} is the only element that has the ability to generate electricity because by bringing focus on an iron Fe_{56} core through motion contributes to employ the Coanda effect where electricity is generated as long as copper $_{62}$ is used to demolish space-time. Every electric generator is a star core in the micro.

In this we have the three space preconditions there are at the present time within which motion of the atom can occur using motion that is lesser that the speed of light. It is the present Big Bang arrangement conditions set in accordance with singularity applying during the event of this era. This forms the borders of atomic development and singularity recouping of lost heat through distorting and retarding time miss forming space and bringing about space in the process.

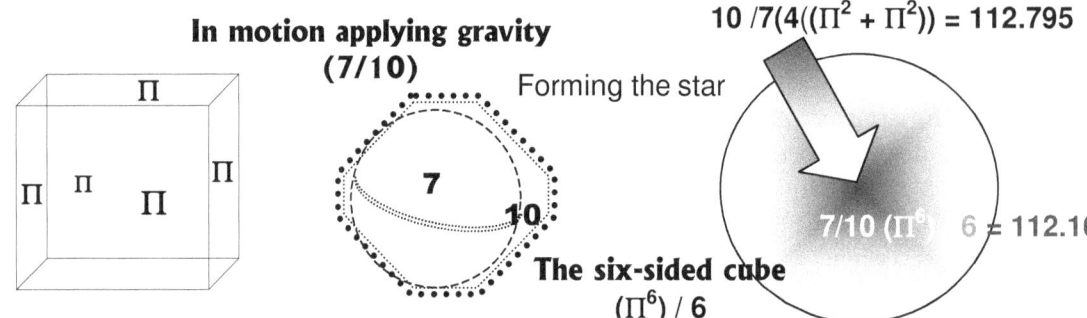

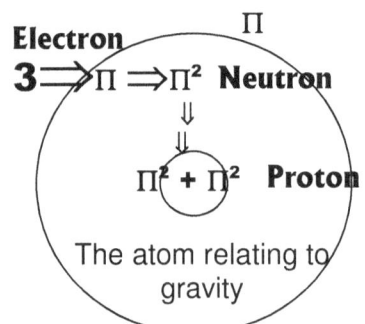

In motion applying gravity (7/10)

Forming the star

$10/7(4((\Pi^2 + \Pi^2)) = 112.795$

The six-sided cube $(\Pi^6)/6$

$7/10 (\Pi^6)/6 = 112.16$

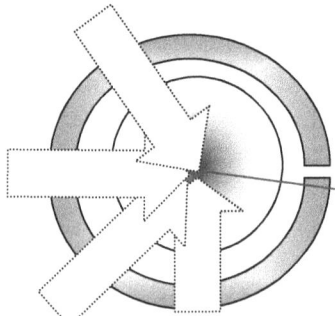

The atom relating to gravity

From the value that outer space can support being the sum total of the particles forming the atom $((\Pi^2 + \Pi^2)$ the proton $+(\Pi^2 + \Pi)$ the neutron + (3) space = $35.75 \times \Pi$ singularity = 112.313 the star deliberately reduces the atomic space or the subatomic constructed space as the star intensifies motion and that reconstructing of space-time changes the qualities of the atom from what we presume the atom to be, to suspending the atom beyond the boundaries of $7/10 (\Pi^6)/6 = 112.16$. The converting of space-time from outer space through gravity to the star centre is the same route electricity follows.

Electricity and lightning is the absolute epitome of the Coanda effect where the Coanda effect is precisely the manifestation of light following the exact principles of the Coanda effect and the **Total Internal Reflection** is also miming the same principle as the Coanda effect which is vivid proof of space-time $a^3 = T^2k$ (the Coanda effect in acting principally by using the flow of photons instead of atmospheric heat). **Total Internal reflection** is only about applying motion by the flow of space-time (in this case water running) through the atmosphere but in the case of the phenomenon we call the **Total Internal Reflection,** singularity captures light holding the flow of light honest to a specific centre as does the Coanda effect and by setting borders the boundaries light is restricted to as singularity sets limiting boundaries to the flow of photons. But that is what electricity is; it is only creating space-time accelerated motion with much intensity added and it links a line than is concentrating space-time as it accelerates space time through the displacing differentiation which one finds in stars between copper dismissing space and iron accelerating heat directly to singularity. It is only with much more intensity. All it is, is the Coanda effect forming electricity and lightning as the Coanda effect.

Electricity and lightning is gravity reduced by the intervention of the phenomenon we know as the Coanda effect where the Coanda effect is the establishing of a more intense dynamic point representing a new point of a controlling singularity dynamic.

Iron forming a centre or an iron core precisely as the Earth core forms in iron form a centre core.

The copper field coil breaks down space-time by dismissing space-time as the 63 factor $(3^3) \times 7$ excels the space-time as fast as the electron will, as the space-time has to flow through the copper in the event where the iron being in motion causes the flow and charging the flow to the equal time set as the photon has. That is electricity. It is taking space-time directly to the centre of the Earth because the motion $T^2 k$ excited the space a^3 to a level that gravity is within the Earth core. With the displacement of iron being 55 + the iron atom has the capacity to dismiss space and by doing that it has the ability to generate such proton motion as to remove space all together from a selected area on conditions of motion producing a connection with singularity. Such connection we call electricity, which is very concentrated gravity. The diminishing of the space is a product of gravity, which results in electricity and for that reason all stars in the era we are in must have an $Iron_{56}$ core in order to establish the generating of gravity. Without the iron core no gravity will flow between is $\underline{10/7 \times 4(\Pi^2 + \Pi^2) = 112}$ and is $\underline{7/10 \times 4(\Pi^2 + \Pi^2) = 55.27}$ which puts the core at a value of sustaining gravity or electricity. It is the iron core Fe_{56} that generates gravity and gravity cannot be conducted nor can it be generated without the connection provided by movement. The centre liquid of the Earth is the same as what we call an electromagnetic field. The electromagnetic field is the reducing of space between the element by name of copper, **(63)** where we find copper exceeding the border of $2\Pi^3$ and that of iron Fe_{56} that is reducing space-time to establish gravity or electricity being whatever name you wish to use. There is a host of other elements also playing a role but I am not venturing into that at this point. At $2\Pi^3$ the space-time as far as the sphere goes will become motionless and being motionless it exceeds what the element iron can establish by way of gravity. The elements that then takes charge is producing the motion to generate the gravity exceeding that in the micro as it then exceeds that it also does in the macro, and the concentration of space-time carries the human name as electricity.

Gravity is electricity because electricity is the flow of heat from a gas source to singularity by charging iron **{7/10 (4(($\Pi^2 + \Pi^2$)) = 55)}** forming the artificial core exactly as the Coanda effect will charge singularity by applying motion through the influencing of total space reducing which copper can manage having the specific space-time displacing value. The influencing of copper $(\Pi^2 + \Pi^2)$ **X Π = 62.0** breaks down space-time as stars do in the core centre. That is the reason why only iron can excite to charge electricity and only copper can dismiss space to a time equal to the flow of photons. All phenomena used in the Cosmos are the precise same thing using the precise same principles in a more intense or lesser intense gradient. Still it is all about singularity charging the control and the flow of space-time through motion where a liquid flows through space to a solid iron core that is influenced by copper.

Plutonium holds at 94 and as an atom almost falls outside space-time reality $3\Pi^2$ as it is on the very border with a possible increase in displacement of $5(\Pi^2+\Pi^2) (\Pi/2)^2 (3/5) = 244$. In the Sun however the dimensional change is $10 \Rightarrow 10\Pi^2$ in comparison with our change of $10 \Rightarrow 3 \Rightarrow \Pi$. With the Universe being $7/10\Pi^6 /(6)=112$ and the Sun at $\$ =(\Pi^0) 10\Pi^2 = 98.696$

In the star the balance bringing about space-time flow is in the iron displacing limit of an atom not holding more that 56 protons because the atomic relevancy is **7/10 X 4($\Pi^2 + \Pi^2$) = 55.27** whereas the neutron reaches the value of **2 × $3\pi^2$ = 59,22** the double photon value **($3\pi^2$)**, it will respond by returning to a space-time value. This is as far as the atom will go down eternity and no further. The space holding light will collapse at $3^3 + 3\pi^2 = 56.6$ where space (3^3) forms the cube of time which is what we know as the Universe. However with $3\pi^2$ still in affect the rush of light is then going towards the centre of the star and the star will become dark. The star does not die because the star is not a coal stove coming out of Newton's steam age where fires either "burn" or "die".

Where we are, is not the only place that is possible in the Universe to be. This concept is as wide as the Universe is and I am by no means getting into that argument in this book because that argument covers 650 pages pf the book **Matter's Space In Time: The Hypothesis ISBN 0-9584410-3-0**. What I will refer to in the following few paragraphs is what applies to our Universe forming our space in our time concept. There are as many possibilities of different Universes all contained by one Universe as there are names for people on Earth and that even is underestimating the possible quantity by the indefinite, but I refer to the one I share with all my fellow Earthlings circling on route around a star we named the sun. In the Universe I am able to witness the proton holds $\Pi^2+\Pi^2$ giving a displacing of space in the duration of time as **19.74** of what ever you wish to name the measure.

Space-time displacement which also is motion that reconverts space from heat taking the heat back to singularity where this starts to achieve a specific duration in time putting such duration above what the Universe reserves as having the ability to duplicate. This stars with the layer hydrogen working through helium and the rest ending the first layer at Boron$_5$ and from they're the part stars where the concentrating of the heat becomes not expanding in movement but dismissing through oxygen and then nitrogen. This layer of C.N.O. is dismissing but moving the nitrogen expanding towards the centre via the silicon that contains heat. At that layer the gravity of the iron core has to establish a flow stronger than the expanding qualities of the oxygen/ nitrogen layers are to prevent a super nova forming. If the oxygen/nitrogen has more expanding capabilities than the contracting gravity the iron layers have, a super nova event will occur by blowing the layers into gas and liquid. The entire process works on spinning and thereby cooling or not spinning sufficiently and thus not cooling sufficiently leaving the layers to overheat and expand back to gas. No Newtonian gravity goes mad because gravity has no mind to go mental. Newtonians best effort in explaining a super nova is the state that the gravity has gone mad! Above **62** the flow of space-time reaches the epitome of time of space in motion. At the point the space can no longer sustain the flow in time to sustain the demand set out by singularity with a dismissing potential of **62** protons. After a displacement value of **62** protons in motion is reached the motion providing the space of space-time displacing returns to only being form without allowing space as we know to form space in the third dimension. The space held by the protons break down the dimensional wall created by motion and the atom of which only the proton remains collapses into singularity as the atom reaches $2\Pi^3$. At $2\Pi^3$ not even the proton remains as the atom forfeit all form and only singularity within the atom core is in place at that point. After this the atom as a unit completely destructs. After $2\Pi^3$ duplication stops to be a process since motion then exceeds the speed of light considerably and time shifts to what it was before the event of the Big Bang. Only singularity remains casting all other space-time out into outer space. The star then becomes a star holding no atom but only contains singularity on the inside. It becomes the all so famous Black Hole where all falls down a pit of space less ness into singularity without space or motion.

In the star in the Universe which we are in the proton number of those atoms forming the composition of gravity or the dismissing of space to become eternal in time is 7/10 (4(($\Pi^2 + \Pi^2$) =**55**. Coincidently this displacing value belongs to iron and therefore iron can produce electricity because when applying motion iron with the ability to displace space-time by using the combining motion of **26** protons and **26** neutrons has the ability to confirm space-time to singularity. For this reason stars must have an iron core and if the Earth did not have an Iron core our gravity was not able to generate electricity. What this means in short is that the star then can convert space to light by diminishing space through the contraction of space. Light is the failed product of diminishing, which brings about conversion of space-time to singularity.

At double the value of outer space **2(10Π) = 62** space within the star collapses since the compactness within the star starts to destroy the space that atom holds as **55.27**. The motion applying within the density of the star at that stage of developing is creating gravity with much more intensity, which then leaves any atom with no space or time to occupy. What then is used in form is not using space-time as we think of that which all atoms must do.

When a star finds the inner-core-value of applying atomic spin or motion to create (($\Pi^2 + \Pi^2$) X Π = 62.0 which has more collapsing than double than that of outer space 112.1÷2 = 56 we find all space and not only liquid in motion converts to a singular form at **2(10Π) = 62** and the result is that space depletes within the inner core of the star and the star will start to withdraw more heat from outer space than what outer space could supply where the star then establishes all light freezes from outer space directly to singularity. At $7/10\Pi^6$ / (6) = 112 the Universe stretches space-time to the limit we find ourselves in. For this reason atoms that are exceeding the mass of 112, (($\Pi^2 + \Pi^2$) **the proton** +($\Pi^2 + \Pi$) **the neutron** + (+ 3) **space = 35.75 X Π singularity = 112.313** and cannot fit in our Universe we have any longer. But it has nothing to do with mass coming from pulling, pushing or shoving. It is about motion exceeding eternity by compressing all heat into infinity. This is what the atomic number is that can apply motion within the atom centre by the maximum number of protons gathered as a group, which as one group can apply space-time displacement although the practical number of protons in one cluster is $3\Pi^3$. More protons that bring about a group motion will produce a collapse of the atom space. At a displacing value surpassing $2\Pi^3$ the dimensional walls of space leading on the motion forming time $7/10(\Pi^6)$ / 6 will collapse into the centre of the atom. Beyond $3\Pi^3$ no atoms form a unit because in the practical sense atomic motion cannot surpass the displacement value of $3\Pi^3$ where at such a point singularity starts maintaining space-time without the support of atomic structures and substructures.

The number of protons applying motion produces space dismissing and cultivating heat from space and in that process is the space in time that is returning to heat by duplicating space. The protons apply motion where there is just about no space and by the motion at that level the proton motion turns space to absolute heat where singularity then dissolves the heat. By reducing the space it intensifies the heat and that returns singularity to what it was when space came about long before the Big Bang presented space-time. And before the atom (($\Pi^2+\Pi^2$)(($\Pi^2\Pi$) 3) became the atom number used to form the atom in the development which brought about the atom. The reducing of space is 1836 times more at the proton (($\Pi^2 + \Pi^2$) than it is at the electron but the combining effort of displacing is the sum total of all the atomic parts.

When the proton (($\Pi^2 + \Pi^2$) and the neutron ($\Pi^2 \Pi$) is added to the **3** the electron produces the total dimensional sum of singularity and not as it should produce in duplicating space by the third dimension in the square being **6^2** as it should but instead it reaches a maximum of **35.75.** With the sum being **35.75** one can see where space will collapse or return to the form singularity provides if it exceeds the singularity connection there is between the atom in total and such an atom connecting to singularity Π. Then we arrive at the universal six dimensions of three sides in space and three sides in motion bringing about a totalling of (($\Pi^2 + \Pi^2$)=19.74+(($\Pi^2 \Pi$)=13.01 + 3 = 35.75 x Π (singularity) = 112.31 and that is the maximum atom displacing value outer space can tolerate before destroying the space holding the atom all together. Inside the star the proton maintaining a connection with singularity directly will produce the proton value of (($\Pi^2 + \Pi^2$) X Π = 62.01. This means that at this point within the star the protons and above this velocity time cannot duplicate space any longer and the wall of space erected by time collapses back into singularity. Space disappears because time cannot any longer sustain space. Any star having a space with a displacement exceeding the generating ability to displace space to the value or above the value of 62 protons in one secluded given space that is repeated as a unit by motion duplicating space will no longer have the ability to sustain the walls time provides space.

The limit is 62, which is **10Π** holding the one proton Π^2 on the one side in place and the **10Π** holding space in duplicated motion Π^2 and it is also double the space value of $2\Pi^3$. Therefore the gravity a star produces at a maximum point is (**10Π.+10Π.**) converting space to $(\Pi^2 + \Pi^2)$ in relation with singularity Π then collapses back to Π^0.

At a value of 6 X 10 time can duplicate space because 3 X 10 = 30 and that is more than singularity Π^0 extended to the square of space at **10Π** will tolerate. But when the displacing exceeds (6 X 10 +6 X 10) making the duplication of space 60, the walls start tumbling in or space is overhauling time as the one side catches the other side and $a^3 \neq T^2 k$

This **62.01** is the total and the maximum number of protons dismissing a value of Π^0 space-time after which the atom as a single unit or as a group in space and time and ultimately the star as a unit of the combining effort of all the atoms forming the star will abort space or the Universe will dispense of the star, which is all the same effort. The star then has grown back to the connecting with singularity and then forms a Black Hole. The principle may sound somewhat simple but it is quite involved in the total explaining.

The Black Hole that forms is within every star centre because that keeps the star in a unit above and beyond what is going on in outer space and as soon as it can bring about fusion by performing motion it will sustain a proton growth setting the star on its final journey to eventually form a Black Hole, although there are two stars even beyond the Black Hole. However I would prefer not to elaborate on that at this venture. The black Hole hides in every atom because through that tiny space not even being part of the Universe on our side, space-time dismissing is applying by killing space because of motionlessness. All atoms in our Universe are showing equal growth notwithstanding position or motion because all atoms in our Universe are in complying with motion as a result of the Titius Bode law of 10 /7 and 7 / 10 bringing the time split equal to all involved in space-time.

When the atom finds the motion within the star centre or star-core-centre starts to reach $(\Pi^2 + \Pi^2) \times 3 =$ **59.22** the neutron moves outside the atom and also outside the star. At a displacement value of **59.22** the atom in the star can no longer accommodate the neutron within the atom and the neutron motion slows down to a point where the Neutron motion is too slow to find accommodation in the atom and in the star. At $((\Pi^2 + \Pi^2) \times \Pi = 62.01$ the proton collapses, which is double the value of **10Π** as well as $2\Pi^3$, therefore Kepler's is doubled or duplicated to serve the motion increase. That is what space-time represents because space-time also represents **k** in variations of distances. It is the doubling of $\Pi^3 = 10\Pi$ to $2(\Pi^3 = 10\Pi)$ which then exceeds the duplicating ability introduced by Kepler. But we must not forget that it is also half the cosmic value of $7/10\Pi^6 / (6) = 112$. At a higher motion the proton moves to outside the atom and forms a proton motion by introducing the proton motion to the space surrounding singularity. It is then where it started with, the dot that claimed a spot in the cosmos. But the dot went further and grew a lot to become innumerable dots. Every dot is related in time spawning each and every spot becomes a dot relating to the past. Every spot and every dot is directly related as every spot is $(\Pi - 3) \times 7$ to grow into Π° that becomes Π.

The entirety rests on relevancy. As time moves on forming a line by implementing more dots in relation to the dots already there forming the history of time, which is what we call space, the area we call outer space receives many dots that time leaves as a footprint while the dots time leaves within material are less. Because there are more space in outer space than the space available in the concentrated material, time leaves more dots in outer space than time leaves in material. Therefore the space outer space gains supersedes the space that material gains and that makes material more compact or more and more dense in relation to outer space that is losing density by gaining more space. The space gained by the space occupied by moving material becomes denser in relation to the space in outer space losing relative density The space forming outer space is gaining more space than the space forming that which we see as material and the space material holds advances more in density through the loss of density in the space called outer space. Material is growing in space less than outer space is growing in space but material is becoming more and denser than the space in outer space because of relevancies applying. This leaves material with occupying more space while the space occupied are more compact in relevancy to outer space that loses density and this is moreover because of the relative loss of density in outer space brings a gain in density in material. The density in outer space is thereby lost and in that the density in material is gained by the loss of the density in space in outer space being more. When looking at a solid we have the governing singularity Π° forming the controlling singularity Π has no radius because the governing singularity Π° of every atom is the controlling singularity of the star's governing singularity and since all the atoms forming the star is the controlling singularity Π of the star, the atom applies as Πr° when forming the controlling singularity. In that the atom's space or radius brings no consequence to singularity because every atom's radius is in effect singularity applying. The dot also leaves one point every time on the dot forming the governing singularity and that confirms the point holding governing singularity in terms of many dots received by the spin of the controlling singularity in terms of the gain of endless space in outer space. In that material always grows as outer space declines in density and that forms the "Hubble Constant" that is no constant ever. **In a nutshell that is gravity**.

That is why the distance between the Earth and the Moon becomes more. That is why the circumference of the Earth becomes bigger. That is why there are Earthquakes and hurricanes. That is why a human grows and heals and that is why hair and nails grow. That also is why there is aging and eventual unavoidable death. The body never stops growing which brings about the inevitable decline of life's body structure

As time goes by everything on the Earth including the Earth and everything in the Universe around the Earth is gaining in space because that is what time leaves. Time leaves on dot holding no space in relation to the dots holding space because of constant movement applying. That is why everything in fossils seems to be bigger the further back the fossil goes in the history of the Earth. I saw a program where Newtonians were telling about millipedes somewhere in Scotland (I think) some many billion trillion years ago that once roamed the Earth and they were one metre wide…their prints they left on the rock spanned one mete across and Newtonians not only believe that but advocates this information as the truth!

If you are a scientist and you wish to gain fame then you must go for the outrageously insane theoretical jargon and ignore all possible prevailing sanity by promoting the absurd as the truth. If I was the author of something as mentally deprived hogwash as Harry Potter or the Da Vinci Code then I would have been a multi billionaire because I deliver the biggest insane hogwash the human mind could possible create. Such deplorable delusions are well received by the well read, but when I show Newton caped science and made a mockery thereof then I am seen as crossing into the corridors of totally madness. I am giving away a book for free also available from Lulu.com where I show just how ridiculous Newton's entire science are and I challenge every person that reads it to prove me wrong on just one point! Go to www.newtonsfraud.com and down load your copy of www.newtonsfraud.com for free. Then see who the jokers are and who are realistic and honest about science. There is not much difference in the science as to how they portray dinosaurs that lived sixty five million years ago and the science Harry Potter practises.

Everything in the past was many times bigger than what the same thing is in the present. According to the most accurate proven information Newtonian science can offer everything that lived in the past was back then when they lived many times bigger than the same specie is currently. The further back in the past the animal lived that they discover, the bigger the animal was in relation to what the same specie would be today. No one ever offered a reason why species shrank as time went by but the real giants that walked the Earth was living on Earth a very long time ago and as time went on they shrank in posture and by now they are small. Some of these statements exceed the ridiculous and goes completely the Hollywood-way of fantasizing in the magnitude of losing their wits totally as they go on a venture of publicity seeking. Every one goes for notoriety and advocating the bizarre without once trying to compose their insane publicity seeking by just sitting back and allow reality to bring some sane conclusion. According to those with the know how, T-Rex was as long as a rugby field (tail included), as wide as half a Rugby field, could destroy several house broken garden dwelling elephants, could run faster than a cheetah as was more agile than a meerkat. How do they fit all that into one idea and still remain sane?

I wish to repeat a statement I made in the beginning of this book. The theory I introduce here and now would never be accepted in my lifetime because science in the Newtonian way is bent on believing in the marvellous, the facts bordering the supernatural, the outrageously inconceivable and the magic of what can never be explained, although they claim to use facts. It is the marvellous that would prompt any person think that mass can create gravity, while how and why was never uncovered. It is bordering the supernatural to think that with nothing between stars, yet by the magic of mass, mass has an unexplainable ability to attract another star many astronomical units away. It is the outrageously inconceivable to argue that life started on Mars, then overcame the quite impossible to escape the gravity that Mars holds on all things being on Mars, and after overcoming the unthinkable by escaping the gravity of Mars, then made a dive for the Earth just to come and evolve over here. Science think they my have the ability to create a Black Hole in a Manmade atom-accelerator because science thinks of the Black Hole as the magic of what can never be explained and therefore that proves that science has no idea of what a Black Hole is and I can prove what a Black Hole is. That fact that I can explain what a Black hole is, becomes the same reason why the Wizards of Oz will never allow the explaining I present to be done in as simple manner as I have presented such explanation in the Cosmic Code. I explain every detail of information I present and I dissect every formula I produce. Yet I am ignored as if I was never alive…why would that be? I show the depth of the insanity we find in what they call Newtonian science and just how outrageous that is and all of that proof still it gets me nowhere. Go and see what lies behind mass bringing about gravity and one must conclude that only magic has to allow mass to produce a power of pulling.

Everything holding material grows by time leaving space as the history of time that went by. That is why we can see galactica so far way because time carries the light from there to over here by establishing $\Pi^\circ\Pi$. That growth we find is why every generation seems to be bigger than what the previous generation was. That is why the dinosaurs such as T-Rex were seems to us as being so big. T-Rex was no larger than about the size of a tiny kangaroo and walked much like the kangaroo but much slower that the kangaroo because the gravity prevailing at the time was immense in relation to what applies currently. I say much more about this as a topic and many other topics in the book available from www.gravitysveracity.com, which I named as **THE VERACITY OF GRAVITY.**
To find out more about other work on offer via my private publishing, press www.gravitysveracity.com and go there.
If you are interested in more as well as better-substantiated information you are welcome to order The following books that are ready to sell by private printing:
1) The Veracity Of Gravity
2) An open letter On Gravity Part 1 Volume 1 + 2
3) An open letter On Gravity Part 2 Volume 1 + 2
4) Newton's Mythology
5) Newton's Fraud
6) Sir Isaac Newton : A Conspiracy to Defraud Science

KOSMOLOGIESE EN ASTRONOMIESE TEGNIKA
P O Box 1093
Ellisras 0555
Limpopo Province
Rep. South Africa.   e-mail gravity@bosveld.co.za
After reading these books you are ready to

The Absolute Relevancy of Singularity

Outer space is $10/7(4((\Pi^2+\Pi^2)))$

Time collapsing space = $2\Pi^3$ = **62.01255**

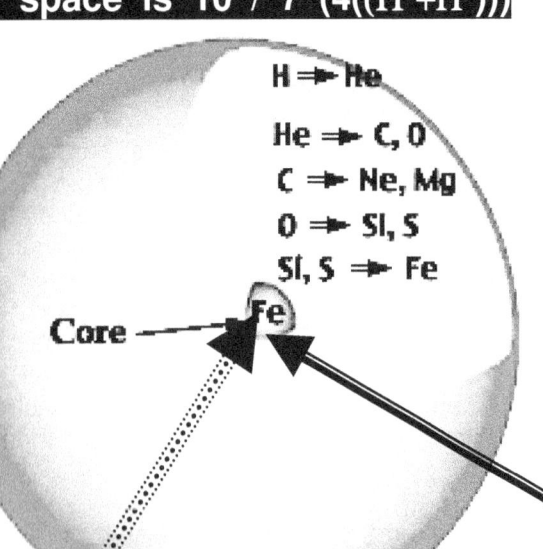

The condensing of space-time or the freezing of heat or the destroying of unoccupied space or the demolishing of time or whatever term is the favourite to connect to the movement of space in a motion called gravity condensing the motion down to contraction is in the following margins

$10/7(4((\Pi^2+\Pi^2))$ **to the lower level**
$7/10(4((\Pi^2+\Pi^2))$ **of space-time**

=112.79547 gravity expanding or motion
Inner space is $7/10(4((\Pi^2+\Pi^2))$=55.2697 gravity contraction

Light meeting singularity is
$3^3+3\Pi^2 = 56.6$

Elimination of space-time is $(\Pi^2(3\Pi^2)/5)$
ENDING AT $3(\Pi^2+\Pi^2) = 59.21762$

Elimination of time and space differentiation is $\Pi(\Pi^2+\Pi^2)=62.01$

Space reuniting with time is $= 2\Pi^3 = 62.01$

Outer space is $10/7(4((\Pi^2+\Pi^2)=112.79547$

Inner space is $7/10(4((\Pi^2+\Pi^2))=55.2697$

Gravity generated from outer space to inner space, which is done as a result of the collapsing or freezing of heat.

Final collapse of the photon space is $3^3+3\Pi^2 = 56.6$

Final collapse of the Neutron space $(\Pi^2(3\Pi^2)/5) = 58.44$

End of space of neutron space is $3(\Pi^2+\Pi^2) = 59.21762$

Final collapse of proton space is $\Pi(\Pi^2+\Pi^2) = 62.01255$

www.ingramcontent.com/pod-product-compliance
Lightning Source LLC
Chambersburg PA
CBHW080653190526
45169CB00006B/2102